Stem Cell Transplantation

Edited by
Anthony D. Ho, Ronald Hoffman,
and Esmail D. Zanjani

Related Titles

Novartis Foundation Symposium

Stem Cells

Nuclear Reprogramming and Therapeutic Applications

2005
ISBN 0-470-09143-6

Minuth, W. W., Strehl, R., Schumacher, K.

Tissue Engineering

Essentials for Daily Laboratory Work

2005
ISBN 3-527-31186-6

Deutsche Forschungsgemeinschaft (DFG) (ed.)

Research with Human Embryonic Stem Cells

Positions

2003
ISBN 3-527-27219-4

The World Life Sciences Forum (ed.)

Health for All – Agriculture and Nutrition – Bioindustry and Environment

Analyses and Recommendations

2005
ISBN 3-527-31489-X

Freshney, R. I.

Culture of Animal Cells

A Manual of Basic Technique

2005
ISBN 0-471-45329-3

Stem Cell Transplantation

Biology, Processing, and Therapy

Edited by
Anthony D. Ho, Ronald Hoffman, and Esmail D. Zanjani

WILEY-VCH Verlag GmbH & Co. KGaA

The Editors

Prof. Dr. Anthony D. Ho
Department of Medicine V
University of Heidelberg
Im Neuenheimer Feld 410
69120 Heidelberg
Germany

Prof. Dr. Ronald Hoffman
College of Medicine
University of Illinois at Chicago
Section of Hematology/Oncology
900 S. Ashland Ave.
Chicago, IL 60607-4004
USA

Prof. Dr. med. Esmail D. Zanjani
Department of Animal Biotechnology
University of Nevada, Reno
Reno, NV 89557-0104
USA

Library of Congress Card No.: Applied for

British Library Cataloguing-in-Publication Data:
A catalogue record for this book is available from the British Library.

Bibliographic information published by Die Deutsche Bibliothek
Die Deutsche Bibliothek lists this publication in the Deutsche Nationalbibliografie; detailed bibliographic data is available in the Internet at <http://dnb.ddb.de>.

© 2006 WILEY-VCH Verlag GmbH & Co. KGaA, Weinheim

Printed in the Federal Republic of Germany.
Printed on acid-free paper.

Typesetting Hagedorn Kommunikation, Viernheim
Printing Betz Druck GmbH, Darmstadt
Bookbinding Litges & Dopf Buchbinderei GmbH, Heppenheim

ISBN-13: 978-3-527-31018-0
ISBN-10: 3-527-31018-5

Contents

Stem Cell Transplantation. Biology, Processing, and Therapy.
Edited by Anthony D. Ho, Ronald Hoffman, and Esmail D. Zanjani
Copyright © 2006 WILEY-VCH Verlag GmbH & Co. KGaA, Weinheim
ISBN: 3-527-31018-5

Preface

The continuing enthusiasm for and controversy around stem cell research has been spurred by the establishment of human embryonic stem cell lines in 1998. This technology has opened up novel avenues for tissue engineering in organ transplantation. Never in the history of biomedical research have scientific discoveries stirred up such tremendous repercussions on a global scale. Stem cells have been compared to the "fountain of youth", that mankind has searched for since time immemorial. It has been speculated that out of stem cells, we might be able to produce all sorts of replacement parts for regenerative medicine.

Despite this world-wide enthusiasm and efforts, major fundamental issues have remained unresolved. For embryonic stem cells, the challenges are tumorogenesis, rejection by the host immune system, transmission of pathogeneic agents during cultivation, in addition to the continuing ethical debate. For adult stem cells, initial results intended to demonstrate the plasticity potentials have been severely challenged. Some of the initial experiments were not reproducible and others have demonstrated that nuclear or cell fusions might account for most of results interpreted to be due to transdifferentiation. In addition adult stem cells, if identifiable, are of such miniscule amount to be of no clinical relevance.

Nevertheless, stem cells derived from the adult bone marrow, i.e. hempatopoietic stem cells, have been used in the clinic already for almost 40 years for patients with leukaemia and hereditary immuno-deficient diseases. Within this time, blood stem cell transplant has evolved from an experimental therapy into standard of care for specific types of myelo- and lymphoproliferative disorders. Progress was, however, gradual and incremental and many groups have contributed. This development has shown that stem cell research requires resources, commitment and team work.

To bring stem cell technology into clinical practice for regenerative medicine, a thorough understanding of the basic principles underlying stem cell regeneration and regulation of self-renewal versus differentiation is absolutely essential. Research efforts in the next years should focus on the cellular and molecular mechanisms regulating "stemness" and the decision process involved in differentiation. Only through a fundamental understanding of these principles can we be able to acquire the power to manipulate a stem cell's destiny. This volume, Frontiers in Stem Cell Transplantation, deals with all the above mentioned challenges.

Stem Cell Transplantation. Biology, Processing, and Therapy.
Edited by Anthony D. Ho, Ronald Hoffman, and Esmail D. Zanjani
Copyright © 2006 WILEY-VCH Verlag GmbH & Co. KGaA, Weinheim
ISBN: 3-527-31018-5

Part 1 focuses on basic stem cell biology with an introductory chapter on clinical potentials of stem cells. This is followed by a chapter each on the epigenetic control of hematopoietic stem cell fate and the impact of micro-RNAs on stem cell biology and medicine.

Part 2 focuses on standardization and quality assurance of stem cell preparations with chapters on novel mobilization based on a precise understanding of the SDF-1α/CXCR4 pathway in stem cell lines derived from umbilical cord blood and bone marrow and the challenges associated with genetic manipulation of hematopoietic stem cells.

Part 3 focuses on the strategies which are on the threshold to clinical applications: large animal models testing the plasticity of human marrow-derived stem cells, a unique murine blastocyst model for studying transdifferentiation, animal models testing the potentials of MAPC, and mesenchymal stem cells as vehicles for genetic targeting. The last and fourth is on novel strategies using adult stem cells within clinical trials. Mesenchymal stem cells might serve as a unique immunomodulator, and this is dealt with in chapter 15. The clinical practice and the evidence for adoptive immunotherapy in hematologic malignancies are summarized in chapters 14 and 16.

Heidelberg, Chicago, Reno, April 2006
A. D. Ho
R. Hoffman
E. D. Zanjani

List of Contributors

Graça Almeida-Porada
Department of Animal Biotechnology
and Department of Medicine
University of Nevada
Mail Stop 202
Reno, NV 89557-0104
USA

Michael Andreeff
Section of Molecular Hematology
and Therapy
Department of Blood and
Marrow Transplantation
The University of Texas
M. D. Anderson Cancer Center
1515 Holcombe Blvd.
Houston, TX 77030
USA

Hiroto Araki
Section of Hematology/Oncology
University of Illinois at Chicago
900 S. Ashland Ave.
Chicago, IL 60607
USA

Iris Bigalke
Clinical Cooperative Group
Hematopoeitic Cell Transplantation
Department of Medicine III
University of Munich and GSF-National
Research Centre for Environment
and Health
Marchioninistr. 15
81377 Munich
Germany

Philippe Bourin
GECSOM and Département
de Thérapie Cellulaire
EFS Pyrénées-Méditerranée
75, rue de Lisieux
31300 Toulouse
France

Raymund Buhmann
Clinical Cooperative Group
Hematopoeitic Cell Transplantation
Department of Medicine III
University of Munich and GSF-National
Research Centre for Environment
and Health
Marchioninistr. 15
81377 Munich
Germany

Stem Cell Transplantation. Biology, Processing, and Therapy.
Edited by Anthony D. Ho, Ronald Hoffman, and Esmail D. Zanjani
Copyright © 2006 WILEY-VCH Verlag GmbH & Co. KGaA, Weinheim
ISBN: 3-527-31018-5

Annette Deichmann
National Center for Tumor Diseases
(NCT)
Im Neuenheimer Feld 581
69120 Heidelberg
Germany

Jennifer Dembinski
Section of Molecular Hematology
and Therapy
Department of Blood and
Marrow Transplantation
The University of Texas
M. D. Anderson Cancer Center
1515 Holcombe Blvd.
Houston, TX 77030
USA

Stefanie Dimmeler
Molecular Cardiology
Department of Internal Medicine IV
University of Frankfurt
Theodor-Stern-Kai 7
60590 Frankfurt
Germany

Luc Douay
Service d'Hématologie Bioloqique
Hôpital Armand Trousseau
26, avenue Du Dr Netter and
Université Pierre et Marie Curie
27, rue de Chaligny
75571 Paris Cedex 12
France

Peter Dreger
Department of Medicine V
University of Heidelberg
Im Neuenheimer Feld 410
69120 Heidelberg
Germany

Michael Dürr
Institute for Medical Radiation
and Cell Research
University of Würzburg
Versbacher Str. 5
97078 Würzburg
Germany

Christine Falk
Institute for Molecular Immunology
GSF-National Research Centre
for Environment and Health
Marchioninistr. 25
81377 Munich
Germany

Ulrich Fischer-Rasokat
Molecular Cardiology
Department of Internal Medicine IV
University of Frankfurt
Theodor-Stern-Kai 7
60590 Frankfurt
Germany

Stefan Fruehauf
Department of Internal Medicine V
University of Heidelberg
Im Neuenheimer Feld 410
69120 Heidelberg
Germany

Hanno Glimm
National Center for Tumor Diseases
Im Neuenheimer Feld 350
69120 Heidelberg
Germany

Brett Hall
Department of Pediatrics
The Ohio State University and
Center for Childhood Cancer
Columbus Children's Research Institute
Columbus, OH 43205
USA

Friedrich Harder
DeveloGen AG
Rudolf-Wissell-Str. 28
37079 Göttingen
Germany

Anthony D. Ho
Department of Medicine V
University of Heidelberg
Im Neuenheimer Feld 410
69120 Heidelberg
Germany

Ronald Hoffman
College of Medicine
Section of Hematology/Oncology
University of Illinois at Chicago
900 S. Ashland Ave.
Chicago, IL 60607-4004
USA

Jingqiong Hu
National Heart, Lung and
Blood Institute
National Institute of Health
10 Center Drive
Bethesda, MD 20892-1202
USA

Christof von Kalle
National Center for Tumor Diseases
(NCT)
Im Neuenheimer Feld 350
69120 Heidelberg
Germany

Gesine Kögler
Institute for Transplantation
Diagnostics and Cell Therapeutics
University of Düsseldorf
Medical Center
Moorenstr. 5
40225 Düsseldorf
Germany

Hans-Jochem Kolb
Clinical Cooperative Group
Hematopoeitic Cell Transplantation
Department of Medicine III
University of Munich and GSF-National
Research Centre for Environment
and Health
Marchioninistr. 15
81377 Munich
Germany

Katarina Le Blanc
Division of Clinical Immunology
Centre for Allogeneic Stem Cell
Transplantation
Karolinska Institutet
Huddinge University Hospital
141-86 Stockholm
Sweden

Georg Ledderose
Clinical Cooperative Group
Hematopoeitic Cell Transplantation
Department of Medicine III
University of Munich and GSF-National
Research Centre for Environment
and Health
Marchioninistr. 15
81377 Munich
Germany

Horst Lindhofer
TRION Research
Am Klopferspitz 19
82152 Martinsried
Germany

Nadim Mahmud
Section of Hematology/Oncology
University of Illinois at Chicago
900 S. Ashland Ave.
Chicago, IL 60607
USA

Frank Marini
Section of Molecular Hematology
and Therapy
Department of Blood and Marrow
Transplantation
The University of Texas
M. D. Anderson Cancer Center
1515 Holcombe Blvd.
Houston, TX 77030
USA

Mohammed Milhem
Section of Hematology/Oncology
University of Illinois at Chicago
900 S. Ashland Ave.
Chicago, IL 60607
USA

Albrecht M. Müller
Institute for Medical Radiation
and Cell Research
University of Würzburg
Versbacher Str. 5
97078 Würzburg
Germany

Christopher D. Porada
Department of Animal Biotechnology
and Department of Medicine
University of Nevada
Mail Stop 202
Reno, NV 89557-0104
USA

Felipe Prósper
Hematology and Cell Therapy Area
Clínica Universitaria
Universidad de Navarra
Mail Stop 202
Av. Pio XII 36
Pamplona 31009
Spain

Olle Ringdén
Division of Clinical Immunology
Centre for Allogeneic Stem Cell
Transplantation
Karolinska Institutet
Huddinge University Hospital
141-86 Stockholm
Sweden

Matthias Ritgen
Department of Medicine II
University of Schleswig-Holstein
Chemnitzstr. 33
24116 Kiel
Germany

A. Kate Sasser
Department of Pediatrics
The Ohio State University and
Center for Childhood Cancer
Columbus Children's Research Institute
Columbus, OH 43205
USA

Christoph Schmid
Clinical Cooperative Group
Hematopoeitic Cell Transplantation
Department of Medicine III
University of Munich and GSF-National
Research Centre for Environment
and Health
Marchioninistr. 15
81377 Munich
Germany

Manfred Schmidt
National Center for Tumor Diseases
Im Neuenheimer Feld 581
69120 Heidelberg
Germany

Timon Seeger
Department of Internal Medicine V
University of Heidelberg
Im Neuenheimer Feld 410
69120 Heidelberg
Germany

Luc Sensebé
Service Recherche EFS
Centre-Atlantique
2 Blvd. Tonnellé
BP 52009
37020 Tours
France

Belinda Simoes
Department of Haematology
University of Sao Paulo
Ribeirao-Preto
Brazil

Christof Stamm
German Heart Institute
Augustenberger Platz 1
13353 Berlin
Germany

Michael Stanglmaier
TRION Research
Am Klopferspitz 19
82152 Martinsried
Germany

Gustav Steinhoff
Department of Cardiac Surgery
University of Rostock
Schillingallee 35
18057 Rostock
Germany

Dirk Strunk
Division of Hematology
and Stem Cell Transplantation
Department of Internal Medicine
Medical University
Auenbrugger Pl. 38
8036 Graz
Austria

Matus Studeny
Section of Molecular Hematology
and Therapy
Department of Blood and Marrow
Transplantation
The University of Texas
M. D. Anderson Cancer Center
1515 Holcombe Blvd.
Houston, TX 77030
USA

Johanna Tischer
Clinical Cooperative Group
Hematopoeitic Cell Transplantation
Department of Medicine III
University of Munich and GSF-National
Research Centre for Environment
and Health
Marchioninistr. 15
81377 Munich
Germany

Julian Topaly
Department of Internal Medicine V
University of Heidelberg
Im Neuenheimer Feld 410
69120 Heidelberg
Germany

Catherine M. Verfaillie
Stem Cell Institute
University of Minnesota
420 Delaware Street SE
Minneapolis, MN 55455
USA

Wolfgang Wagner
Department of Medicine V
University of Heidelberg
Im Neuenheimer Feld 410
69120 Heidelberg
Germany

Peter Wernet
Institute for Transplantation
Diagnostics and Cell Therapeutics
Heinrich-Heine-University Medical
Center
Moorenstr. 5
40225 Düsseldorf
Germany

Ting Yang
Clinical Cooperative Group
Hematopoeitic Cell Transplantation
Department of Medicine III
University of Munich and GSF-National
Research Centre for Environment
and Health
Marchioninistr. 15
81377 Munich
Germany

Esmail D. Zanjani
Department of Animal Biotechnology
and Department of Medicine
University of Nevada
Mail Stop 202
Reno, NV 89557-0104
USA

Part I
Stem Cell Biology

Stem Cell Transplantation. Biology, Processing, and Therapy.
Edited by Anthony D. Ho, Ronald Hoffman, and Esmail D. Zanjani
Copyright © 2006 WILEY-VCH Verlag GmbH & Co. KGaA, Weinheim
ISBN: 3-527-31018-5

1
Clinical Potentials of Stem Cells: Hype or Hope?

Anthony D. Ho and Wolfgang Wagner

1.1
Introduction

The present enthusiasm for and controversy around stem cell research began with two breakthroughs: (i) the successful cloning of "Dolly" by Ian Wilmut, Keith Campbell and coworkers in 1997 [1]; and (ii) the establishment of human embryonic stem cell (ESC) lines by the laboratory of James Thomson in 1998 [2]. Without any doubt, these technologies have opened up novel avenues for tissue engineering and organ transplantation [3]. Never in the history of biomedical research have scientific discoveries spawned such tremendous repercussions on a global scale. The ability to rejuvenate or even replace defective organs and the tissues of the human body has been a centuries-old dream. Stem cells have demonstrated their potential to develop into practically all types of specialized cells and tissues in the body, and have therefore been compared to the "fountains of youth" that mankind have searched for since time immemorial. Recent discoveries using both adult and embryonic stem cells as starting cell populations have led to speculations that out of such "raw material" we might be able to produce all sorts of replacement parts for regenerative medicine. Hopes are high that many age-related degenerative disorders such as heart disease, Parkinson's disease, diabetes, and stroke could some day be cured by stem cell therapy.

1.2
What are Stem Cells?

All life forms begin with a stem cell, which is defined as a cell that has the dual ability to self-renew and to produce progenitors and different types of specialized cells in the organism. For example, in the beginning of human life, one fertilized egg cell – the zygote – becomes two, and two becomes four [4]. In these early stages, each cell might still be totipotent – that is, a whole organism can be derived out of each of these cells. Within 5 to 7 days, some 40 cells are formed which

Stem Cell Transplantation. Biology, Processing, and Therapy.
Edited by Anthony D. Ho, Ronald Hoffman, and Esmail D. Zanjani
Copyright © 2006 WILEY-VCH Verlag GmbH & Co. KGaA, Weinheim
ISBN: 3-527-31018-5

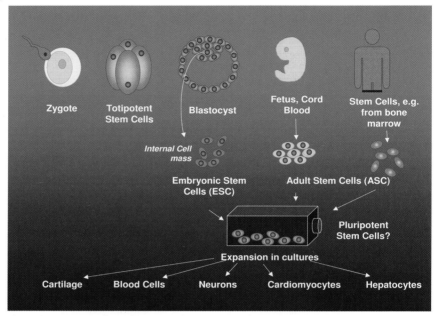

Figure 1.1   Sources for embryonic and adult stem cells.

build up the inner cell mass, surrounded by an outer cell layer forming subsequently the placenta. At this stage, each of these cells in the inner cell mass has the potential to give rise to all tissue types and organs including germ cells – that is, these cells are pluripotent (Fig. 1.1). Ultimately, the cells forming the inner cell mass will give rise to the some 10^{13} cells that constitute a human body, organized in 200 differentiated cell types [5]. Many somatic, tissue-specific or adult stem cells are produced during fetal development. Such stem cells have more restricted ability than the pluripotent ESC and they are multipotent – that is, they have the ability to give rise to multiple lineages of cells. These adult stem cells persist in the corresponding organs to varying degrees during a person's whole lifetime.

1.3
Stem Cells and Regeneration

Lower life forms have amazing prowess of regeneration which mammals and especially humans woefully lack [6]. Upon decapitation, planaria (e.g., a flatworm) will regenerate a new head within 5 days. Hydra, a small tubular freshwater animal that spends its life clinging to rock, is able to produce two new organisms

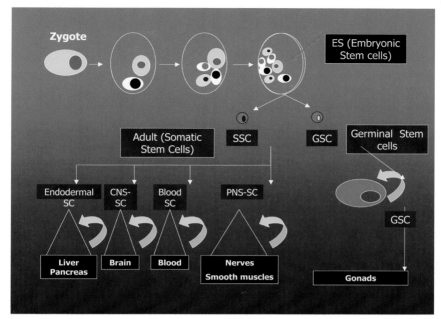

Figure 1.2 Embryonic stem cells (ES) are derived from 5- to 7-day-old embryos and are pluripotent. Pluripotent stem cells can also be derived from germinal stem cells (GSC) and possibly from some somatic (adult) stem cells (SSC). During embryonic development, tissue-specific stem cells (SC) give rise to the mature, differentiated cell types that constitute the specific organs with special functions.

within 7–10 days when its body is halved. After losing a leg or the tail to a predator, a salamander will recover with a new limb or tail within a matter of days.

Mammals pay a high price for climbing up the evolutionary ladder, and have lost comparable regenerative power. Those animals with staggering regenerative potentials are either in possession of an abundance of stem cells, or they can convert specialized cells into stem cells on demand. For example, it has been estimated that some 20 % of the planaria consists of stem cells, while hydra is a "kind of permanent embryo" [6]. Salamanders use a completely different mechanism; when they need a new limb or tail, they convert an adult differentiated cell back to an embryonic undifferentiated one. These cells then gather at the site of a severed organ and form a blastema, which regenerates the missing part. An understanding of the cues and molecules that enable the stem cells to initiate self-renewal, divide, proliferate, and then differentiate to rejuvenate damaged tissue might be the key to regenerative medicine.

To a limited extent, humans can rejuvenate some types of tissue, such as the skin and the bone marrow, but are nowhere near as proficient. The regenerative power is associated with an adequate presence of stem cells in these organs – that is, epidermal stem cells in the skin and hematopoietic stem cells (HSCs) in the bone marrow (Fig. 1.2). Moreover, regenerative potential of the skin and marrow declines with age [7, 8]. An understanding of how ESCs differentiate into various tissues and how adult stem cells can be coaxed to replace damaged tissue could therefore hold promise for cell replacement of tissue repair in many age-related degenerative disorders.

1.4
Adult and Embryonic Stem Cells

In 1998, the group of James Thomson reported on the establishment of human ESC lines. Human ESCs used for research have been extracted form embryos created by *in-vitro* fertilization. Some 40 cells forming the inner cell mass at day 5–7 after fertilization are transferred to a culture dish lined with feeder cells. After culturing and replating for several months, these cells might maintain their self-renewing ability without differentiating into specialized cells, and give rise to ESC lines that could, in theory, replicate for ever [9–11]. Thus, ESCs have the potential to form most – if not all – cell types of the adult body over almost unlimited periods.

As mentioned above, the adult body has a small number of adult or somatic stem cells in some tissues and organs [12–14]. Such adult stem cells (ASCs) have been known to possess the ability to regenerate the corresponding tissue from which they are derived. Hematopoietic stem cells (HSCs), for example, continuously regenerate the circulating blood cells and cells of the immune system during the life span of the organism. Based on animal models, many studies have recently claimed that ASCs might exhibit developmental potentials comparable to those exhibited by ESCs [14]. More recent reports, however, have severely challenged the interpretation of the initial results, suggesting the "plasticity potential" or "trans-differentiation" of ASCs [15–18]. Hence, ASCs have the ability to regenerate the tissue from which they are derived over the lifespan of the individual, while ESCs have the potential to form most, if not all, cell types of the adult body over very long periods of *in-vitro* cultivation. ESCs seem to demonstrate unlimited potential for growth and differentiation. The use of ES-derived cells for transplantation, however, is associated with hazards and ethical controversies. In animal studies, undifferentiated ESCs can induce teratocarcinomas after transplantation, and they have been shown to be epigenetically instable. Pre-culturing of immature ESCs in conditions that induce differentiation along a specific pathway might reduce the risk of tumor genesis. Animal studies have also shown that only donor ESCs after a specific differentiation stage would be accepted by a fully grown animal. ESCs must be primed towards a predefined differentiation pathway before transplantation. Such cultures are likely to contain a variety of cells at different stages of development, as well as undifferentiated ESCs. Purification of the cell preparation is necessary before clinical use could be considered.

1.5
In the Beginning was the Hematopoietic Stem Cell

The concept of stem cells was introduced by Alexander Maximow in 1909 as the common ancestors of different cellular elements of blood [19]. It took, however, almost another 60 years – that is, in 1963 – before McCullough and his coworkers provided unequivocal evidence for the existence of stem cells in the bone marrow [20, 21]. In a murine model, their series of experiments demonstrated that, first of all, cells from the bone marrow could reconstitute hematopoiesis and hence rescue lethally irradiated recipient animals. Second, by serial transplantations, they have established the self-renewal ability of these cells. When cells from the spleen colonies in the recipients were harvested and re-transplanted into other animals that received a lethal dose of irradiation, colonies of white and red blood corpuscles were again found in the secondary recipients. Based on these experiments, HSCs were defined as cells with the abilities of self-renewal as well as multilineage differentiation. This discovery marked the beginning of modern-day stem cell research. Only in recent years have other somatic stem cells been identified in tissues with a more limited regenerative capacity, such as the liver and the brain [22, 23].

The first *successful* attempts of using bone marrow transplantation as a treatment strategy for patients with hereditary immunodeficiency or acute leukemias were performed during the late 1960s [24–27]. The original idea was to replace the diseased bone marrow with a healthy one after myeloablation. Without the benefits of present-day knowledge of immunology and supportive care, morbidity and mortality rates associated with the treatment procedure were then high [27]. Nevertheless, the results were considered encouraging as compared to those obtained with conventional treatment options. Bone marrow transplantation has in

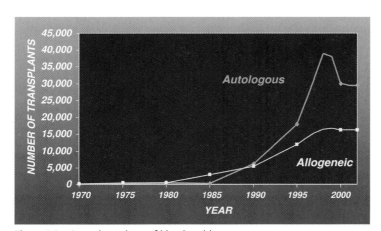

Figure 1.3 Annual numbers of blood and bone marrow transplants worldwide (1970 to 2002), as registered by the International Bone Marrow Transplant Registry.

the meantime been proven to be the only chance of cure for some patients with leukemia and some hereditary diseases [28]. Its success was due to the presence of HSCs in the marrow graft, which were able to reconstitute the blood and immune systems after myeloablation. Although initially identified in the marrow, HSCs could also be found in the peripheral blood upon stimulation, such as during the recovery phase after myelosuppressive therapy [29] or after the administration of cytokines [30, 31]. Such HSCs obtained from the peripheral blood or isolated CD34+ cell populations have been used successfully in lieu of bone marrow to reconstitute hematopoietic and immune functions in the recipients [32, 33]. According to the International Bone Marrow Transplantation Registry, blood stem cell transplantation now offers chances of durable cure for at least some 27 000 patients each year as a treatment strategy for various cancers, marrow failure, or hereditary diseases (Fig. 1.3) [34].

1.6
Trans-Differentiation of ASCs

Parallel to encouraging developments in ESC research, numerous studies have reported that ASCs might exhibit developmental potentials comparable to those exhibited by ESCs. In one of the first studies in murine models, Ferraris et al. reported that unmanipulated bone marrow cells were found to participate in the muscle regeneration process when injected into skeletal muscle that was chemically induced to undergo regeneration [35]. Furthermore, bone marrow cells that have engrafted in the muscle were also involved in the repair process if muscle injury was experimentally induced again at a later time. Since then, many authors have reported that stem cells within the marrow of mice possessed the ability to form differentiated skeletal muscle fibers, and that even cardiac muscle cells were able to regenerate by recruiting circulating marrow-derived stem cells. Eglitis and Mezey [36] showed that bone marrow cells were able to differentiate into microglia, astroglia and neurons within the central nervous system. Stem cells from the rat bone marrow have been shown to give rise to hepatocytes in recipients with artificially induced hepatic injury [37]. Other authors have confirmed that bone marrow-derived stem cells probably participated in hepatocyte restoration [38]. Multi-organ, multi-lineage engraftment by a single bone marrow-derived stem cell with HSC phenotype has been reported by Krause et al. [39]; indeed, their data have provided one of the few indications that multiple tissues could develop from a single hematopoietic tissue-derived stem cell. The magnitude of engraftment was, however, minuscule such that the biological relevance has been questioned.

ASCs from several nonhematopoietic tissues have also been reported to produce cell types other than those from the tissue in which they reside. Bjornson et al. showed that neural stem cells could produce a variety of blood cell types after transplantation into irradiated hosts [40]. The observation that adult neural stem cells might have a broader developmental potential has also been reported by Clarke et al. [41]. The latter group showed that neural stem cells from the adult mouse brain could contribute to the formation of chimeric chick and mouse embryos. These adult neural stem cells gave rise to cells of all germ layers.

1.7
The Plasticity of ASCs: All Hype and no Hope?

More recent reports, however, have severely challenged the interpretation of the initial results suggesting the "trans-differentiation" of ASCs [15–18]. For example, in the experiments described by Bjornson et al., the cells from neurospheres that were dissociated and transplanted were passaged 12 to 35 times in the presence of growth factors prior to transplantation [40]. One possible explanation for the loss of specificity of neural stem cells is that they were transformed during their *in-vitro* passaging. It has long been established that cells growing in culture, even of defined, permanent cell lines, can spontaneously change their gene expression pattern and state of differentiation, giving rise to clonally stable "trans-differentiated" sub-lines [42]. ASCs in culture, when exposed to extreme pressures to trans-differentiate, might generate cells with genetic instability and with features of unrelated cell types. Efforts to repeat this experiment has been reported by Morshead and coworkers. The latter group confirmed that transformation of primary neural stem cells did occur during in-vitro passaging, but they could under no circumstances observe any contribution of neural cells to the blood cell lineage [18]. These authors concluded that trans-differentiation could not be proven. Studies conducted by Ying et al. and Terada et al. then provided evidence that cell fusion between somatic stem cells (SSCs) and ESCs occurred spontaneously upon coculturing *in vitro* [16, 17]. Both groups cautioned that such hybrid cells with tetraploid nuclei and characteristics of both SSCs and ESCs could account for the proclaimed plasticity potentials of ASCs. To verify the trans-differentiation potential of hematopoietic stem cells (HSC), Wagers et al. have generated chimeric animals by transplantation of a single green fluorescent protein (GFP)-marked HSC into lethally irradiated nontransgenic recipients. Single HSCs robustly reconstituted peripheral blood leukocytes in these animals, but did not contribute to any nonhematopoietic tissues, including brain, kidney, gut, liver, and muscle. These data indicated that "trans-differentiation" of circulating HSCs and/or their progeny is an extremely rare event, if it occurred at all [43]. Schmittwolf et al. demonstrated that only through modifications of DNA and chromatin could they establish long-term, stable and trans-differentiated hematopoietic cells from neurosphere cells [44]. Almeida-Porada et al., however, have provided new evidence that trans-differentiation did occur without cell fusion, especially under physiological conditions of the developing fetus, albeit at much lower frequencies then previously claimed [45] (see also Chapter 8).

Most of the experiments performed thus far have focused on the dramatic changes in the destiny that is, differentiation program of ASCs. Trans-differentiation, or in some rare examples plasticity, seemed indeed possible under highly selective pressure from the microenvironment. There is, however, an absolute paucity of data on the cellular and molecular processes involved in the complex cascade of (trans-) differentiation. The first step, which is migration of the ASCs towards their niche and communication with the surrounding cells in the microenvironment, has not been elucidated adequately. Evidences at cellular and molecular levels show that re-programming along a different differentiation pathway are lacking. It is also not known how the newly acquired differentiation program can be maintained. Indeed, until these processes are known, it is premature to translate the observations in animal models into clinical trials.

1.8
The Battle of Two Cultures: ESCs versus ASCs

In both self-renewing as well as differentiation potentials, ASCs have been proven to be far inferior to ESCs. In injury models, ASCs from an allogeneic donor (e.g., from bone marrow), might be responsible for some of the reconstituted cells in the recipient's organs of another ontogenetic derivative. Cell and nuclear fusions might be largely responsible for this phenomenon. When trans-differentiated cells within evidence of fusion could be identified, they were of such minuscule amount as to be of no clinical relevance. Thus, many have come to the conclusion that only ESCs could hold promise for the future (Fig. 1.4).

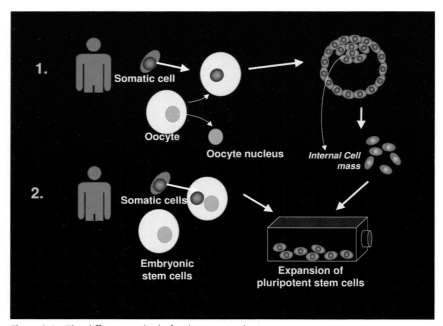

Figure 1.4   The different methods for therapeutic cloning.

Although a number of countries have since permitted the use of public funding for ESC research, opponents of this approach regard cells derived from sacrificed embryos as being close to cannibalism. On the other hand, advocates of ESC research pointed out that unwanted embryos derived from *in-vitro* fertilization clinics are continuously destined for disposal worldwide. If parents agree to donate embryos, it would not be ethical to deny their use for research purposes that target at identifying novel strategies to treat incurable diseases. Another critical challenge for the clinical use of ESCs or cell preparations derived thereof is the development of tumors, especially teratocarcinoma. The therapeutic potential of ESCs is also hampered by the threat of contamination from serum products and live feeder cells of animal origin. Serum-free and feeder layer-free systems have been used successfully by some groups, but the results have yet to be reproduced and confirmed. Thus, the debate pro and contra ESC research goes on from state to state, from country to country. Within the European Union, no consensus could thus far be reached. Whereas ESC research is strictly regulated in Germany and Austria, the U.K., the Netherlands, Belgium, Spain and Italy – and recently also Switzerland – took a much more liberal stance and have permitted ESC research under specific criteria. The U.K. has also been one of the first countries to have permitted therapeutic cloning.

1.9
The Challenges for Stem Cell Technology

One of the major challenges for the application of ESC and ASC technology is the establishment of standards and definition of stem cell preparations. The heterogeneity of the starting population renders comparison of results between different groups difficult, and this might account for the lack of reproducibility of some of the initially reports using ASCs. The significance of establishing standards and guidelines for clinical applications can best be demonstrated by the evolution of bone marrow or blood stem cell transplantation from a highly experimental procedure to the standard strategy that it is today [14, 28]. With the significance of hematopoietic tissue transplantation as curative treatment for hematologic malignancies and marrow failure, the need for *in-vitro* assays to identify human hematopoietic progenitors has increased. However, in order to infer that any *in-vitro* assay measures stem cells, the properties of the cells analyzed *in vitro* must be compared with those of repopulating units tested *in vivo* [14, 46]. Repopulating units were estimated in transplantation models and could be performed only in animals (for a review, see [13, 46]). Colony assays, including those for long-term initiating cell (LTC-IC) and myeloid-lymphoid initiating cell (ML-IC), have been developed that might serve as surrogate markers for the repopulating potentials of the stem cells present in a given population. Surface markers, such as CD34, CD133, Thy-1, HLA-DR have been shown to be associated with the "stemness" of cell preparations, while CD38 plus a whole range of surface markers have been associated with lineage commitment. Hence, many groups have used the CD34+/CD38– and lineage-negative population as representative for primitive progenitor cells for hematopoiesis.

Despite all of the efforts made throughout the past 40 years, no *in-vitro* assay has ever been considered adequate for the identification of HSCs [13, 14, 46]. Hence, there is no appropriate substitute for the repopulation assay in murine transplantation model after a lethal dose of irradiation [20, 21]. Clearly, this experimental approach cannot be used to estimate human HSCs. However, the immunocompromised mouse model (i.e., SCID mouse model and variations thereof), or the *in-utero* sheep transplantation model at a time when the animal is tolerant to human HSCs, have been proposed for estimating the repopulating potentials of human HSCs [47]. Preparative protocols for acquisition, *in-vitro* cultivation, expansion and differentiation along specific pathways of ESCs or ASCs have thus far been extremely heterogeneous. A precise characterization and standardization of ESCs as well as ASC preparations and the progeny cells derived thereof represents a *conditio sine qua non* for future development and for comparing the results from different research groups. Hence, there is an urgent need for establishing robust standards and developing a catalogue of marker profiles for the definition of stem cells and of their differentiation products. Such efforts will be described in Chapters 3 and 7.

1.10
Regulation of Self-Renewal versus Differentiation, Asymmetric Divisions

A hallmark of stem cell activity is the dual capacity to self-renew and to differentiate into cells of multiple lineages. Thus, the ability to divide asymmetrically might be regarded as a unique feature of stem cells. A central question in developmental biology is how a single cell can divide to produce two progeny cells that adopt different fates. Different daughter cells with different functions can, in theory, arise by uneven distribution of determinants upon cell division (i.e., due to intrinsic factors) or become different upon subsequent exposure to environmental signals (i.e., due to extrinsic factors) [47]. Recent advances in the understanding of asymmetric division of stem cells in *Drosophila* and murine models have provided some insight into human stem cell development.

Evidence from the development of neuroblasts from neuroepiderm in *Drosophila* and in the mouse model supports the idea that asymmetric divisions are defined mostly by cell-autonomous (i.e., intrinsic) information. For example, *Drosophila* neuroblasts (NBs) – which are precursors of the central nervous system – arise from polarized epithelial cells during development [48]. The NBs enlarge and delaminate from the ventral ectoderm, forming a subepithelial array of neural stem cells. The plane of cleavage upon division is orientated parallel to their apical-basal axis, resulting in symmetric division. The NBs then undergo a series of asymmetric divisions and, while maintaining the axis of apical-basal polarity, adjust their cleavage plane such that it is perpendicular to this axis, and produce NBs to renew themselves and smaller ganglion mother cells (GMC-1). Asymmetric division is orientated intrinsically and autonomously. GMC production is followed by a single division, generating two post-mitotic neurons or glial precursors. During early embryonic development, asymmetric divisions therefore provide a mechanism for positioning specific cell types at defined sites. An axis of polarity is established in the mother cell, and this coordinates with the general body plan. Cell-fate determinants are distributed asymmetrically along this axis. During mitosis the spindle is also orientated along this axis so that cytokinesis creates two daughter cells containing different concentrations of these determinants [49–53]. In *Drosophila*, for example, homologues of PAR-3, atypical protein kinase C and PAR-6 mediate polarity and they direct epithelial cell polarity and mediate both spindle orientation and localization of cell-fate determinants in NBs [54, 55]. For subsequent development, intracellular or extrinsic mechanisms

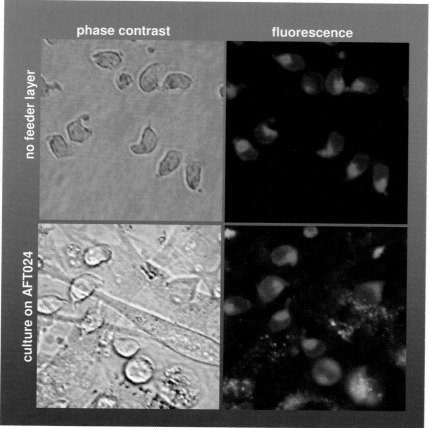

Figure 1.5 Hematopoietic stem cells (HSCs) can be enriched in the CD34+/CD38– fraction. The fluorescent membrane dye PKH26 can be used to label these cells red. The cells can be maintained *in vitro* by direct contact with a supportive cell layer that provides the appropriate microenvironment (e.g., the AFT024 feeder layer; green).

(as a consequence of communication of the daughter cells with each other or with surrounding cells) play then a major role and extrinsic signals might be involved in instructing the asymmetric fates of the daughter cells [56, 57]. Kiger et al. [58] and Tulina and Matunis [59], for example, have defined the molecular nature and spatial organization of the signaling pathway that governs asymmetric divisions of stem cells in the *Drosophila* testis. In the latter, germline cells and SSCs attach to a cluster of support cells called the "hub". Upon division of a germline stem cell, the daughter cell in direct contact with the hub retains the self-renewal potential, whereas the other daughter cell was destined to differentiate into a gonioblast and subsequently into spermatogonia. Evidence was provided that *Unpaired*, a ligand which activates the JAK-STAT signaling cascade and is expressed by the apical hub cells in the testis, causes stem cells to retain their self-renewal potential. Analogous to this finding, the maintenance of mammalian ESCs has been shown to require a similar JAK-STAT signaling, which is counterbalanced by the requirement for MAP kinase activation, and the latter in turn promotes ESC differentiation [60]. In another recent publication, Yamashita et al. demonstrated that germline stem cells were anchored to the hub through localized adherens junctions. Interactions between DE-cadherin on the surface of hub cells and germline stem cells could stabilize a localized binding site for beta-catenin and Apc2 at the germline stem cell (GSC) cortex [61]. The cadherin-catenin and the associated cytoskeletal system seem to be key players in this context.

For HSCs, our group has demonstrated that only contact of primitive CD34+/CD38− cells with a stem cell-supporting microenvironment (AFT024) increased asymmetric divisions of both primitive and committed progenitors by recruiting significant numbers of primitive cells into the cell cycle [62] (Fig. 1.5). This phenomenon of recruitment, as well as the shift in asymmetric division, could not be induced by cytokines [63]. Thus, dormant cells that are usually in G_0 can be recruited to cycle without loss of primitive function after cell–cell contact with AFT024. Only direct contact with cellular elements of the niche could increase the absolute number of cells undergoing asymmetric division. The stem cell niche thus provides the cues to regulate self-renewing divisions and subsequently to control cell numbers.

1.11
Genotype and Expression Profiles of Primitive HSCs

As the slow-dividing fraction (SDF) of HSCs is associated with primitive function and self-renewal, while the fast-dividing fraction (FDF) predominantly proceeds to differentiation, we have separated the CD34$^+$/CD38$^-$ cells according to their divisional kinetics as a functional parameter for the isolation of primitive stem cells [64]. We then performed a genotypic analysis of these two populations (FDF versus SDF) using genome-wide analysis. Genome-wide gene expression analysis of these populations was determined using a Human Transcriptome Microarray containing 51 145 cDNA clones of the Unigene Set-RZPD3 [65]. In addition, gene expression profiles of CD34$^+$/CD38$^-$ cells were compared with those of

CD34$^+$/CD38$^+$ cells. Among the genes showing the highest expression levels in the SDF were the following: CD133, erg, cyclin g2, MDR1, osteopontin, clqr1, ifi16, jak3, fzd3 and hoxa9, a pattern compatible with their primitive function and self-renewal capacity. We have also demonstrated that the SDF of CD34$^+$/CD38$^-$ cells displayed significantly more podia formation and migratory activity as compared to the more committed progenitor cells found among the FDF [65].

Several other attempts have been made to identify the gene expression profiles of stem cells using microarray technology. In most of these studies, the target population was separated from their native stem cell niche before analysis [66–69]. In a meta-analysis of our own data and the data of three other studies on HSCs, we have shown that, despite the use of different starting materials, derivation from different species, applying very different platforms and methods of analysis, an interesting overlap of genes that are overexpressed in the primitive subsets of HSCs was found [65]. This included *fzd6, mdr1*, RNA-binding protein with multiple splicing (*rbpms*), *jak3, and hoxa9*. Other studies have focused on the specific molecular make-up of the HSC niche. Hackney et al. have analyzed the gene expression profiles of AFT024 cells in comparison to other fetal liver-derived lines of varying stem cell support [69]. A number of genes that potentially influenced stem cell function were highly expressed in AFT024 cells, underscoring the hypothesis that many pathways might be involved in supporting stem cell function [70, 71].

1.12
Maintaining Stemness: Interactions between HSCs and the Cellular Microenvironment

Our group, as well as other investigators, has shown that direct contact with the cellular microenvironment was able to maintain the stem cell function of CD34$^+$/CD38$^-$ cells to increase the number of asymmetric divisions, and recruit more CD34$^+$/CD38$^-$ cells into cell cycle compared to those exposed to cytokines alone [63,72–74]. In order to define the essential cellular and molecular mechanisms involved in the interaction between HSCs and the stroma feeder layer, we have studied the impact of cocultivation on the behavior and gene expression of HSCs [75]. We have shown that HSCs developed directed migratory activity towards stroma cells, indicating that HSCs migrated towards signals secreted by the supportive stroma cells [76]. The HSCs subsequently established stable contact to stroma cells by means of a uropod, on which CD44 and CD34 were colocalized. CD44 is known to bind fibronectin and hyaluronic acid, and is essential for the homing and proliferation of HSCs [77–79].

Using a human genome cDNA microarray developed in our group, we have subsequently analyzed the gene expression profiles of CD34$^+$/CD38$^-$ cells upon cultivation with or without stroma for 16, 20, 48, or 72 h. Several genes that play a role in cell adhesion, the re-organization of the cytoskeleton system, the maintenance of methylation patterns, stabilization of DNA during proliferation and repair were up-regulated within the first 72 h upon exposure. The over-expression of genes coding for tubulin α, tubulin β, and ezrin was indicative of the significant role of reorganization of the cytoskeleton system upon interaction with the cellular microenvironment. This was also compatible with the increase in motility and adhesion, as described previously [76]. Other genes that were up-regulated included proliferating cell nuclear antigen (*pcna*), which is involved in the control of DNA replication, and DNA (cytosine-5)-methyltransferase (*dnmt1*), which is responsible for maintaining methylation patterns during embryonic development [80–82]. A few genes characteristic for primitive HPC were again overexpressed, which included the receptor for the complement component molecule C1q (*c1qr1*) and *HLA-DR* [65]. Among the gene sequences that were down-regulated were various hemoglobin genes [76]. Our previous experiments also showed that hemoglobin genes were expressed more highly in the more committed progenitors, and these results indicate that HSCs cultivated without stroma showed an intrinsic propensity to differentiate along the erythrocyte lineage [65].

1.13
Mesenchymal Stem Cells

Mesenchymal stem cells (MSCs) represent another archetype of multipotent SSC that give rise to a variety of cell types including osteocytes, chondrocytes, adipocytes and other kinds of connective tissue cells such as those in tendons. Recent studies have indicated that, given the appropriate microenvironment, MSCs could also differentiate into cardiomyocytes or even cells of nonmesodermal derivation, including hepatocytes and neurons. MSCs have been used within clinical trials for regenerative medicine. These multipotent stem cells might hold promise for the following reasons:

- In contrast to most other SSCs, they can be isolated from a diverse set of tissues that are readily accessible, such as bone marrow, fat tissues and umbilical cord blood.
- These cells could be expanded *in vitro* without losing their "stemness" or self-renewal capacity [83, 84].
- MSCs have been shown to differentiate *in vitro* into bone, cartilage, muscle, tendon, and fat, and possibly also into cardiomyocytes and hepatocytes [85–91].
- In conjunction with HSCs, allogeneic MSCs have been transplanted without graft rejection or major toxicities [92] (Fig. 1.6).

Verfaille and coworkers have described the derivation of multipotent adult progenitor cells (MAPCs) from murine and human bone marrow [93, 94] (see also Chapter 11). These MAPCs were able to differentiate into functional hepatocyte-like cells and were probably related to the MSCs. Similar multipotent pro-

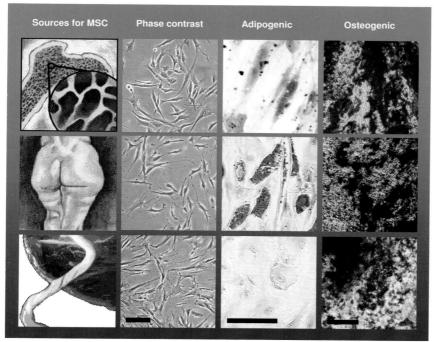

Figure 1.6 Mesenchymal stem cells (MSCs) can be isolated from various tissues, including bone marrow, adipose tissue and from umbilical cord blood. MSCs are plastic adherent with a spindle-shaped morphology. Adipogenic and osteogenic differentiation can be induced by appropriate culture conditions, as examined by Oil Red-O staining or von Kossa staining. Scale bar: 100 μm.

genitors, "unrestricted somatic stem cells" (USSCs), derived from umbilical cord blood, have recently been described by Kögler et al. [95] (see also Chapter 2).

The clinical relevance of all these multipotent stem cells of mesenchymal origin is highly controversial for the following reasons. The prerequisite of prolonged *in-vitro* culture prior to the emergence of MSCs, MAPCs or USSCs raises the question as to whether such cells exist naturally in postnatal tissues. The precise definition of these MSCs, MAPCs or USSCs – and especially their precise cellular and molecular characterization – have remained elusive. Under the culture conditions for the propagation of MSCs, MAPCs or USSCs, these cells might become epigenetically unstable. Further expansion or trans-differentiation for specific maturation pathways *in vitro* might render them more so, and pre-malignant transformation of cells cannot be completely excluded at this juncture. Last, but not least, a sophisticated analysis of self-renewal and differentiation on a single cell basis has to date proved elusive in MSCs, and serial transplantations have not been performed. Current preclinical research on the trans-differentiation potentials of ASCs (including MSCs) has focused mainly on descriptive phenomena such as emergence of differentiation markers, but lacks the solid fundamentals of cell biology. Almost no data exist on the cellular and molecular processes involved in the complex cascade of differentiation into specific pathways such as from HSCs or MSCs into cardiomyocytes or hepatocytes. Further characterization of MSCs requires the development of robust phenotypic and functional markers, and the demonstration of multipotentiality [A, B]. Further basic cell biology research, based especially on precise knowledge of molecular and genetic mechanisms, is urgently needed to provide a safe background for the use of cultured stem cells within the clinical setting, irrespective of their origin – that is, embryonic cells or adult tissues.

1.14
Preliminary Clinical Studies

With the exception of HSCs, the application of stem cell preparations or other cell products thereof as replacement therapy for organ failure, though tantalizing, is yet far from clinical practice. Nevertheless, a few clinical studies have suggested benefits for the use of marrow-derived progenitor cells for cardiovascular diseases, and for the use of liver cell preparations for hepatic failure. Stamm et al., for example, have demonstrated the feasibility and safety of administering progenitor cells derived from autologous bone marrow to the infarcted myocardium of patients with ischemic heart disease who undergo a coronary artery bypass surgery [96] (see also Chapter 14). In an ongoing controlled clinical trial, the same authors have also provided evidence of a pronounced effect of cell therapy on the blood supply to ischemic tissue, associated with an improvement of contractile function. Thus, the scientific basis for the use of ASCs or ESCs for regenerative medicine has remained controversial. As shown by the adverse events associated with gene therapy during the past years [97, 98], clinical trials without any precise scientific foundation might in the long run jeopardize scientific progress and public trust. Issues of concern for the application of stem cell technology in regenerative medicine include the reproducibility for the early trans-differentiation experiments, a definition of the starting cell population and cell products, the standardization of expansion or differentiation processes, and the toxicology and functional properties of the differentiated cell products compared to the target tissue.

1.15
Concluding Remarks and Future Perspectives

At present, it is unclear whether ASCs can match the ESC's capacity to differentiate into cells of almost any organ. Whereas most studies in the past have focused on dramatic changes in long-term fate, such as the conversion into tissues of another germinal derivation, little is known about the mechanisms of the initial steps leading to a different maturation pathway. Neither has the hierarchy of molecular changes involved in switching to another differentiation program been defined. Cross-talk with the microenvironment probably determines the long-term fate, both in terms of the differentiation program as well as in terms of the balance between self-renewal *versus* differentiation.

During the past 40 years, we have learned that stem cell research requires intensive resources and scientific environment that is conducive to innovation. In the case of blood stem cell transplantation, some 20 years of continuous improvements in clinical research has contributed to the establishment of this procedure as a standard and curative treatment for specific diseases. During this time, many groups have attempted to expand HSC use *ex vivo*, though attempts by others to reproduce the initial reports of the expansion of HSCs have not proved successful. Similar concerns also apply to the use of other ASCs that need to be expanded *in vitro*. It is absolutely essential that the initial population is well characterized and the subsequent expansion procedure standardized. Other than morphology, immunophenotyping and alternative methods of characterizing the stem cell population (e.g., division history, molecular markers, genotypic and proteomic analysis) might be required in order to define specific stem cell populations. Importantly, these experiments are currently under way.

Research into the trans-differentiation potentials of ASCs has thus far focused mainly on descriptive phenomena such as the emergence of differentiation markers, but lacks the solid fundaments of cell biology. Very few data exist on the cellular and molecular processes involved in the complex process of trans-differentiation. For example, the molecular mechanism behind the dramatic change in cell fate from HSCs or MSCs into progenitors of cardiomyocytes or hepatocytes is totally unknown. Indeed, the cues and mechanisms governing the decision processes of self-renewal versus differentiation, as well as differentiation along specific pathways are, at best, sketchy. Specific soluble regulatory molecules and direct contact with the cellular microenvironment might play a role in the regulation of self-renewal versus differentiation, as well as the adoption of a specific differentiation program. Today, an understanding of the basic principles which govern stem cell fate is more important than demonstrating dramatic changes therein. Consequently, there is an urgent need for basic cell biological research, especially based on precise knowledge of molecular and genetic mechanisms, in order to provide a safe background for the use of cultured stem cells in the clinical setting, whether of embryonic or adult origin.

Given the present status in stem cell research, it is essential that we keep all options open with regard to investigations into both ESCs and ASCs in order to appreciate the complexity of their differentiation pathways and of their developmental processes. Only with a thorough understanding of the molecular mechanisms involved might we acquire the power to manipulate the destiny of stem cells.

Abbreviations/Acronyms

ASC	adult stem cell
CNS	central nervous system
ESC	embryonic stem cell
FDF	fast-dividing fraction
GFP	green fluorescent protein
GMC	ganglion mother cell
GSC	germline stem cell
HSC	hematopoietic stem cell
LTC-IC	long-term initiating cell
MAPC	multipotent adult progenitor cell
ML-IC	myeloid-lymphoid initiating cell
MSC	mesenchymal stem cell
NB	neuroblast
PNS	peripheral nervous system
SDF	slow-dividing fraction
SSC	somatic stem cell
USSC	unrestricted somatic stem cell

References

1. Wilmut I, Schnieke AE, McWhir J, Kind AJ, Campbell KH (1997). Viable offspring derived from fetal and adult mammalian cells. *Nature* 385: 810.
2. Thomson JA, Itskovitz-Eldor J, Shapiro SS, Waknitz MA, Swiergiel JJ, Marshall VS, Jones JM (1998). Embryonic stem cell lines derived from human blastocysts. *Science* 282: 1145.
3. Fuchs E, Segre JA (2000). Stem cells: a new lease on life. *Cell* 100: 143.
4. Carlson BM (1996). *Patten's Foundations of Embryology.* 6th edn. New York, McGraw-Hill.
5. Sadler TW (2002). *Langman's Medical Embryology,* 8th edn. Philadelphia, Lippincott Williams and Wilkins.
6. Ho AD, Wagner W, Mahlknecht U. (2005). Stem cells and ageing. *EMBO Rep* 6: S35–S38.
7. Stenderup K, Justesen J, Clausen C, Kassem M. (2003). Aging is associated with decreased maximal life span and accelerated senescence of bone marrow stromal cells. *Bone* 33: 919–926.
8. Figueroa R, Lindenmaier H, Hergenhahn M, Nielsen KV, Boukamp P. (2000). Telomere erosion varies during in vitro aging of normal human fibroblasts from young and adult donors. *Cancer Res* 60(11): 2770–2774. Erratum in: *Cancer Res* 60(15): 4301.
9. Thomson JA, Itskovitz-Eldor J, Shapiro SS, Waknitz MA, Swiergiel JJ, Marshall

VS, Jones JM (1998). Embryonic stem cell lines derived from human blastocysts. *Science* 282, 1145–1147.

10. Richards M, Fong CY, Chan WK, Wong PC, Bongso A. (2002). Human feeders support prolonged undifferentiated growth of human inner cell masses and embryonic stem cells. *Nat Biotechnol* 20: 933–936.

11. Amit M, Shariki C, Margulets V, Itskovitz-Eldor J (2004). Feeder layer- and serum-free culture of human embryonic stem cells. *Biol Reprod* 70: 837–845.

12. Gage FH (2000). Mammalian neural stem cells. *Science* 287, 1433–1438.

13. Weissman I L (2000). Translating stem and progenitor cell biology to the clinic: barriers and opportunities. *Science* 287, 1442–1446.

14. Ho AD, Punzel M (2003). Hematopoietic stem cells: can old cells learn new tricks? *J Leukocyte Biol* 73: 547–555.

15. Wagers AJ, Christensen JL, Weissman IL (2002). Cell fate determination from stem cells. *Gene Ther* 9: 606–612.

16. Terada N, Hamazaki T, Oka M, Hoki M, Mastalerz DM, Nakano Y, Meyer EM, Morel L, Petersen BE, Scott EW (2002). Bone marrow cells adopt the phenotype of other cells by spontaneous cell fusion. *Nature* 416: 542–545.

17. Ying QL, Nichols J, Evans EP, Smith AG (2002). Changing potency by spontaneous fusion. *Nature* 416: 545–548.

18. Morshead CM, Benveniste P, Iscove NN, van der Kooy D (2002). Hematopoietic competence is a rare property of neural stem cells that may depend on genetic and epigenetic alterations. *Nat Med* 8: 268–273.

19. Maximow A (1909). Der Lymphozyt als gemeinsame Stammzelle der verschiedenen Blutelemente in der embryonalen Entwicklung und im postfetalen Leber der Säugetiere. *Folia Haematol (Leipz)* 8: 125–141.

20. Siminovitch L, McCulloch EA, Till JE (1963). The distribution of colony-forming cells among spleen colonies. *J Cell Comp Physiol* 62: 327.

21. Becker A, McCulloch EA, Till JE (1963). Cytological demonstration of the clonal nature of spleen colonies derived from transplanted mouse marrow cells. *Nature* 197: 452.

22. Gage FH (1998). Stem cells of the central nervous system. [Review]. *Curr Opin Neurobiol* 8: 671.

23. Block GD, Locker J, Bowen WC, Petersen BE, Katyal S, Strom SC, Riley T, Howard TA, Michalopoulos GK (1996). Population expansion, clonal growth, and specific differentiation patterns in primary cultures of hepatocytes induced by HGF/SF, EGF and TGFa in a chemically defined (HGM) medium. *J Biol Chem* 132: 1133.

24. Bach FH, Albertini RJ, Joo P, Anderson JL, Bortin MM (1968). Bone-marrow transplantation in a patient with the Wiskott-Aldrich syndrome. *Lancet* 2: 1364.

25. Gatti RA, Meuwissen HJ, Allen HD, Hong R, Good RA (1968). Immunological reconstitution of sex-linked lymphopenic immunological deficiency. *Lancet* 2: 1366.

26. de Koning J, van Bekkum DW, Dicke KA, Dooren LJ, Radl J, van Rood JJ (1969). Transplantation of bone-marrow cells and fetal thymus in an infant with lymphopenic immunological deficiency. *Lancet* 1: 1223.

27. Thomas ED, Bryant JI, Buckner CD, et al. (1971). Allogeneic marrow grafting using HL-A matched donor-recipient sibling pairs. *Trans Assoc Am Physicians* 84: 248.

28. Thomas ED, Flournoy N, Buckner CD, et al. (1977). Cure of leukaemia by marrow transplantation. *Leukemia Res* 1: 67–70.

29. Körbling M, Dörken B, Ho AD, Pezzutto A, Hunstein W, Fliedner TM (1986). Autologous transplantation of blood-derived hemopoietic stem cells after myeloablative therapy in a patient with Burkitt's lymphoma. *Blood* 67: 529.

30. Haas R, Ho AD, Bredthauer W, Cayeux S, Egerer G, Knauf W, Hunstein W (1990). Successful autologous transplantation of blood stem cells mobilized with recombinant human granulocyte-macrophage colony-stimulating factor. *Exp Hematol* 18: 94.

31. Lane TA, Law P, Maruyama M, Young D, Burgess J, Mullen M, Mealiffe M, Ter-

stappen LWMM, Hardwick A, Moubayed M, Oldham F, Corringham RET, Ho AD (1995). Harvesting and enrichment of hematopoietic progenitor cells mobilized into the peripheral blood of normal donors by granulocyte macrophage-colony stimulating factor (GM-CSF) or G-CSF: Potential role in allogeneic marrow transplantation. *Blood* 85: 275.

32. Juttner CA, To HO, Ho JQ, Bardy PG, Dyson PG, Haylock DN, Kimber RJ (1988). Early lymphoemotopoietic recovery after autografting using peripheral blood stem cells in acute non-lymphoblastic leukemia. *Transplant Proc* 20: 40.

33. Corringham RET, Ho AD (1995). Rapid and sustained allogeneic transplantation using immunoselected CD34+-selected peripheral blood progenitor cells mobilized by recombinant granulocyte- and granulocyte-macrophage colony-stimulating factors. *Blood* 86: 2052.

34. *International Bone Marrow Transplant Registry*, Annual Report, 2004.

35. Ferrari G, Cusella-De Angelis G, Coletta M, Paolucci E, Stornaiuolo A, Cossu G, Mavilio F (1998). Muscle regeneration by bone marrow-derived myogenic progenitors. *Science* 279: 1528.

36. Eglitis MA, Mezey E (1997). Hematopoietic cells differentiate into both microgliaand macroglia in the brains of adult mice. *Proc Natl Acad Sci USA* 94: 4080.

37. Petersen BE, Bowen WC, Patrene KD, Mars WM, Sullivan AK, Murase N, Boggs SS, Greenberger JS, Goff JP (1999). Bone marrow as a potential source of hepatic oval cells. *Science* 284: 1168.

38. Theise ND, Badve S, Saxena R, Henegariu O, Sell S, Crawford JM, Krause DS (2000). Derivation of hepatocytes from bone marrow cells of mice after radiation-induced myeloablation. *Hepatology* 31: 235.

39. Krause DS, Theise ND, Collector MI, Henegariu SH, Gardner R, Neutzel S, Sharkis SJ (2001). Multi-organ, multilineage engraftment by a single bone marrow-derived stem cell. *Cell* 105: 369.

40. Bjornson CR, Rietze RL, Reynolds BA, Magli MC, Vescovi AL (1999). Turning brain into blood: a hematopoietic fate

adopted by adult neural stem cells *in vivo*. *Science* 283: 534.

41. Clarke DL, Johansson CB, Wilbertz J, Veress B, Nilsson E, Karlstrom H, Lendahl U, Frisen J (2000). Generalized potential of adult neural stem cells. *Science* 288: 1660.

42. Knapp AC, Franke WW (1989). Spontaneous losses of control of cytokeratin gene expression in transformed, non-epithelial human cells occurring at different levels of regulation. *Cell* 59: 67–79.

43. Wagers AJ, Sherwood RI, Christensen JL, Weissman IL (2002). Little evidence for developmental plasticity of adult hematopoietic stem cells. *Science* 297: 2256.

44. Schmittwolf C, Kirchhof N, Jauch A, Dürr M, Harder F, Zenke M, Müller AM (2005). *In vivo* haematopoietic activity is induced in neurosphere cells by chromatin-modifying agents. *EMBO J* 24: 554–566.

45. Almeida-Porada G, Porada, CD, Chamberlain J, Torabi A, Zanjani ED (2004). Formation of human hepatocytes by human hematopoietic stem cells in sheep. *Blood* 104: 2582–2590.

46. Ho AD, Haas R, Champlin R (Eds.) (2000). *Hematopoietic stem cell transplantation.* Marcel Dekker, New York, p. 604.

47. Zanjani ED, Pallavicini MG, Ascensao JL, Flake AW, Langlois RG, Reitsma M, MacKintosh FR, Stutes D, Harrison MR, Tavassoli M (1992). Engraftment and long-term expression of human fetal hematopoietic stem cells in sheep following transplantation in utero. *J Clin Invest* 89: 1178.

48. Lin H, Schagat T (1997). Neuroblasts: a model for the asymmetric division of stem cells. *Trends Genet* 13(1): 33.

49. Hirate J, Nakagoshi H, Nabeshima Y, Matsuzaki F (1995). Asymmetric segregation of the homeodomain protein Prospero during *Drosophila* development. *Nature* 377: 627–630.

50. Knoblich JA, Jan LY, Jan YN (1995). Asymmetric segregation of Numb and Prospero during cell division. *Nature* 377: 624–627.

51. Spana EP, Doe CQ (1995). The Prospero transcription factor is asymmetrically localized to the cell cortex during neuro-

blast mitosis in *Drosophila*. *Development* 121: 3187–3195.

52. Li P, Yang X, Wasser M, Cai Y, Chia W (1997). Inscuteable and Staufen mediate asymmetric localization and segregation of Prospero RNA during *Drosophila* neuroblast cell divisions. *Cell* 90: 437–447.

53. Ikeshima-Kataoka H, Skeath JB, Nabeshima Y, Doe CQ, Matsuzaki F (1997). Miranda directs Prospero to a daughter cell during *Drosophila* asymmetric divisions. *Nature* 390: 625–629.

54. Joberty G, Petersen C, Gao L, Macara IG (2000). The cell polarity protein Par6 links Par3 and atypical protein kinase C to CDc42. *Nature Cell Biol* 2: 531–539.

55. Lin D, et al. (2000). A mammalian PAR-3-PAR-6 complex implicated in CDc42/Rac1 and aPKC signalling and cell polarity. *Nature Cell Biol* 2: 540–547.

56. Sherley JL (2002). Asymmetric cell kindetics genes: The key to expansion of adult stem cells in culture. *Stem Cells* 20: 561–572.

57. Lee, HS, Crane GG, Merok JR, Tunstead JR, Hatch NL, Panchalingam K, Powers MJ, Griffith LG, Sherley JL (2003). Clonal expansion of adult rat hepatic stem cell lines by suppression of asymmetric cell kinetics (Sack). *Wiley Periodicals* 760–771.

58. Kiger AA, Jones DL, Schulz C, Rogers MB, Fuller MT (2001). Stem cell self-renewal specified by JAK-STAT activation in response to a support cell cue. *Science* 294: 2542.

59. Tulina N, Matunis E (2001). Control of stem cell self-renewal in *Drosophila* spermatogenesis by JAK-STAT signaling. *Science* 294: 2546.

60. Matsuda T, Nakamura T, Nakao K, Arai T, Katsuki M, Heike T, Yokota T (1999). STAT3 activation is sufficient to maintain an undifferentiated state of mouse embryonic cells *EMBO J* 18: 4261.

61. Yamashita QM, Jones DL, Fuller MG (2003). Orientation of asymmetric stem cell division by the APC tumor suppressor and centrosome. *Science* 301: 1547–1550.

62. Punzel M, Liu D, Zhang T, Eckstein V, Miesala K, Ho AD (2003). The symmetry of initial divisions of human hematopoietic progenitors is altered only by the cellular microenvironment. *Exp Hematol* 31: 339–347.

63. Huang S, Law P, Francis K, Palsson BO, Ho AD (1999). Symmetry of initial cell divisions among primitive hematopoietic progenitors is independent of ontogenic age and regulatory molecules. *Blood* 94: 2595–2604.

64. Punzel M, Zhang T, Liu D, Eckstein V, Ho AD. Functional analysis of initial cell divisions defines the subsequent fate of individual human CD34+CD38– cells. *Exp Hematol* 30: 464–472.

65. Wagner W, Ansorge A, Wirkner U, Eckstein V, Schwager J, Miesala K, Selig J, Saffrich R, Ansorge W, Ho AD (2004). Molecular evidence for stem cell function of the slow-dividing fraction among human hematopoietic progenitor cells by genome-wide analysis. *Blood* 104: 675–686.

66. Terskikh AV, Miyamoto T, Chang C, et al. (2003). Gene expression analysis of purified hematopoietic stem cells and committed progenitors. *Blood* 102: 94–101.

67. Ivanova NB, Dimos JT, Schaniel C, et al. (2002). A stem cell molecular signature. *Science* 298: 601–604.

68. Ramalho-Santos M, Yoon s, Matsuzaki Y, et al. (2002). 'Stemness': transcriptional profiling of embryonic and adult stem cells. *Science* 298: 597–600.

69. Hackney JA, Charbord P, Brunk BP, et al. (2002). A molecular profile of a hematopoietic stem cell niche. *Proc Natl Acad Sci USA* 99: 13061–13066.

70. Pazianos G, Uqoezwa M, Reya T (2003). The elements of stem cell self-renewal: a genetic perspective. *Biotechniques* 35: 1240–1247.

71. Lauzurica P, Sancho D, Torres M, et al. (2000). Phenotypic and functional characterisics of hematopoietic cell lineages in CD69-deficient mice. *Blood* 95: 2312–2320.

72. Thiemann FT, Moore KA, Smogorzewska EM, et al. (1998). The murine stromal cell line AFT024 acts specifically on human CD34+ CD38– progenitors to maintain primitive function and immunophenotype in vitro. *Exp Hematol* 26: 612–619.

73. Prosper F, Verfaillie CM (2001). Regulation of hematopoiesis through adhesion receptors. *J Leukocyte Biol* 69: 307–316.

74. Punzel M, Gupta P, Verfaillie CM (2002). The microenvironment of AFT024 cells maintains primitive human hematopoiesis by counteracting contact mediated inhibition of proliferation. *Cell Commun Adhes* 9: 149–159.

75. Fruehauf S, Srbic K, Seggewiss R, et al. (2002). Functional characterization of podia formation in normal and malignant hematopoeitic cells. *J Leukocyte Biol* 71: 425–432.

76. Wagner W, Saffrich R, Wirkner U, Eckstein V, Blake J, Ansorge A, Schwager C, Wein F, Miesala K, Ansorge W, Ho AD (2005). Hematopoietic progenitor cells and cellular microenvironment: Behavioral and molecular changes upon interaction. *Stem Cells* 23: 1180–1191.

77. Francis K, Ramakrishna R, Holloway W, et al. (1998). Two new pseudopod morphologies displayed by the human hematopoietic KG1a progenitor cell line and by primary human CD34+ cells. *Blood* 92: 3616–3623.

78. Zhang J, Niu C, Ye L, et al. (2003). Identification of the haematopoietic stem cell niche and control of the niche size. *Nature* 425: 836–841.

79. Calvi LM, Adams GB, Weibrecht KW, et al. (2003). Osteoblastic cell regulate the haematopoietic stem cell niche. *Nature* 425: 841–846.

80. Reik W, Dean W, Walter J (2001). Epigenetic reprogramming in mammalian development. *Science* 293: 1089–1093.

81. Espada J, Ballestar E, Fraga MF, et al. (2004). Human DNA methyltransferase 1 is required for maintenance of the histone H3 modification pattern. *J Biol Chem* 279: 37175–37184.

82. Bakin AV, Curran T (1999). Role of DNA 5-methylcytosine transferase in cell transformation by fos. *Science* 283: 387–390.

83. Caplan AI (2000). Mesenchymal stem cells and gene therapy. *Clin Orthop Rel Res* 379 (Suppl.): S67–S70.

84. Bianco P, Gehron Robey P (2000). Marrow stromal stem cells. *J Clin Invest* 105: 1663–1668.

85. Pereira RF, Halford KW, O'Hara MD, Leeper DB, Sokolov BP, Pollard MD, Bagasra O, Prockop DJ (1995). Cultured adherent cells from marrow can serve as long-lasting precursor cells for bone, cartilage, and lung in irradiated mice. *Proc Natl Acad Sci USA* 92: 4857–4861.

86. Pittenger MF, Mackay AM, Beck SC, Jaiswal RK, Douglas R, Mosca JD, Moorman MA, Simonetti DW, Craig S, Marshak DR (1999). Multilineage potential of adult human mesenchymal stem cells. *Science* 284: 143–147.

87. Conget PA, Minguell JJ (1999). Phenotypical and functional properties of human bone marrow mesenchymal progenitor cells. *J Cell Physiol* 181: 67–73.

88. Dennis JE, Merriam A, Awadallah A, Yoo JU, Johnstone B, Caplan AI (1999). A quadripotential mesenchymal progenitor cell isolated from the marrow of an adult mouse. *J Bone Miner Res* 14: 700–709.

89. Fukuda K (2000). [Generation of cardiomyocytes from mesenchymal stem cells]. *Tanpakushitsu Kakusan Koso* 45: 2078–2084.

90. Lee KD, Kuo TK, Whang-Peng J, Chung YF, Lin CT, Chou SH, Chen JR, Chen YP, Lee OK (2004). In vitro hepatic differentiation of human mesenchymal stem cells. *Hepatology* 40: 1275–1284.

91. Remy-Martin JP, Marandin A, Challier B, Bernard G, Deschaseaux M, Herve P, Wei Y, Tsuji T, Auerbach R, Dennis JE, et al. (1999). Vascular smooth muscle differentiation of murine stroma: a sequential model. *Exp Hematol* 27: 1782–1795.

92. Le Blanc K, Rasmusson I, Sundberg B, Gotherstrom C, Hassan M, Uzunel M, Ringden O (2004). Treatment of severe acute graft-versus-host disease with third party haploidentical mesenchymal stem cells. *Lancet* 363: 1439–1441.

93. Reyes M, Lund T, Lenvik T, Aguiar D, Koodie L, Verfaillie CM (2001). Purification and ex vivo expansion of postnatal human marrow mesodermal progenitor cells. *Blood* 98: 2615–2625.

94. Jiang Y, Jahagirdar BN, Reinhardt RL, Schwartz RE, Keene CD, Ortiz-Gonzalez XR, Reyes M, Lenvik T, Lund T, Blackstad M, et al. (2002). Pluripotency of mesenchymal stem cells derived from adult marrow. *Nature* 418: 41–49.

95. Kogler G, Sensken S, Airey JA, Trapp T, Muschen M, Feldhahn N, Liedtke S, Sorg RV, Fischer J, Rosenbaum C, et al. (2004). A new human somatic stem cell from placental cord blood with intrinsic pluripotent differentiation potential. *J Exp Med* 200: 123–135.

96. Stamm C, Westphal B, Kleine HD, Petzsch M, Kittner C, Klinge H, Schumichen C, Nienaber CA, Freund M, Steinhoff G (2003). Autologous bone-marrow stem-cell transplantation for myocardial regeneration. *Lancet* 361: 45–46.

97. Tiberghien P, Ferrand C, Lioure B, Milpied N, Angonin R, Deconinck E, Certoux JM, Robinet E, Saas P, Petracca B, et al. (2001). Administration of herpes simplex-thymidine kinase-expressing donor T cells with a T-cell-depleted allogeneic marrow graft. *Blood* 97: 63–72.

98. Hacein-Bey-Abina S, von Kalle C, Schmidt M, Le Deist F, Wulffraat N, McIntyre E, Radford I, Villeval JL, Fraser CC, Cavazzana-Calvo M, Fischer A (2003). A serious adverse event after successful gene therapy for X-linked severe combined immunodeficiency. *N Engl J Med* 348: 255–256.

A. Wagner W, Feldman RE, Jr., Seckinger A, Maurer MH, Wein F, Blake J, Krause U, Kalenka A, Burgers HF, Saffrich R, Wuchter P, Kuschinsky W, and Ho AD (2006). The heterogeneity of human mesenchymal stem cell preparations – Evidence from simultaneous analysis of proteomes and transcriptomes. *Exp Hematol* 34:536–548.

B. Wagner W, Wein F, Seckinger A, Frankhauser M, Wirkner U, Krause U, Blake J, Schwager C, Eckstein V, Ansorge W, and Ho AD (2005). Comparative characteristics of mesenchymal stem cells from human bone marrow, adipose tissue, and umbilical cord blood. *Exp Hematol* 33:1402–1416.

2
Alteration of Hematopoietic Stem Cell Fates by Chromatin-Modifying Agents

Nadim Mahmud, Mohammed Milhem, Hiroto Araki, and Ronald Hoffman

Abstract

Epigenetics is defined as modifications of the genome heritable during subsequent cell divisions that do not involve changes in the DNA sequence. DNA methylation plays a critical role in the regulation of genes which define tissue specific gene expression patterns, genome imprinting, and X-chromosome inactivation. DNA methylation has also been shown to be essential for normal mammalian development. DNA methylation and histone deacetylation are components of an epigenetic program that regulates gene expression pattern. Efforts devoted to maintaining the self-renewal and multilineage differentiation potential of human hematopoietic stem cells (HSC) and hematopoietic progenitor cells *in vitro* have met with limited success. The studies discussed here are based upon a hypothesis that conditions previously utilized for *ex-vivo* expansion of HSC result in the epigenetic silencing of genes required for HSC to undergo symmetrical cell divisions resulting in a progressive decline of primitive HSC function. Using pharmacological agents that interfere with DNA methylation and histone deacetylation, we and others have attempted to modify the fate of HSC *in vitro*. In this chapter, we review current advances in the use of chromatin-modifying agents to alter the fate of normal HSC. In particular, the effects of three drugs on normal HSC are described: a hypomethylating drug, 5-aza-2'-deoxycytidine; and two histone deacetylase inhibitors, trichostatin A and a drug traditionally used for treating epilepsy, valproic acid.

2.1
Introduction

Epigenetic modifications of chromatin such as DNA methylation and histone acetylation are important for modifying gene function [1–4]. Silencing of genes has been shown to be accompanied by DNA methylation of a gene's promoter and by histone deacetylation in the region containing the genes of interest,

Stem Cell Transplantation. Biology, Processing, and Therapy.
Edited by Anthony D. Ho, Ronald Hoffman, and Esmail D. Zanjani
Copyright © 2006 WILEY-VCH Verlag GmbH & Co. KGaA, Weinheim
ISBN: 3-527-31018-5

with inhibition of DNA methylation or histone deacetylation reversing the silencing effect. Specific patterns of DNA methylation of genes and histone acetylation are known to be associated with transcriptional regulation and modification of chromatin structure [5]. Histone acetylation has also been shown to have a profound effect on the normal transition from a fetal to an adult hematopoietic cellular differentiation program during ontogeny [6]. Numerous reports have shown that the proliferative and self-renewal capacity of hematopoietic stem cells (HSC) decrease progressively with HSC differentiation [7, 8]. There is evidence that early HSC express a promiscuous set of transcription factors [9] and an open chromatin structure required to maintain their multipotentiality, which is quenched progressively as these cells progress down a particular lineage of differentiation [10]. The mechanisms that govern these stem cell fate decisions are likely under tight control, but remain potentially alterable. Although the global gene expression profiles for HSC have recently been described, very little is known regarding the dynamics of gene expression necessary for HSC fate decisions [11–13]. Interestingly, it has been shown that Oct-4 gene expression is regulated by an epigenetic mechanism, which is essential for the maintenance of pluripotency of embryonic stem cells [14]. The mechanisms underlying the proliferation and differentiation of HSC are incompletely understood. Both extrinsic and intrinsic factors are involved in the regulation of HSC. Although hematopoietic growth factors and an adequate microenvironment are crucial for the survival and proliferation of HSCs, self-renewal/differentiation decisions in HSC appear likely to occur independently of cytokines and are postulated to be determined by the intrinsic properties of HSC.

The role of epigenetic mechanisms in CD34+ cell differentiation and transformation into a leukemic phenotype is emerging. For example, the p15 gene in normal CD34+ cells is unmethylated, becomes methylated after 7 days of *in-vitro* differentiation, and demethylated after prolonged culturing [15]. In general, an HSC has multiple choices once it starts to divide. First of all, it may undergo symmetric or asymmetric cell divisions (Fig. 2.1). The symmetrical stem cell division pattern would lead to expansion of the absolute numbers of HSC, while asymmetric HSC division would lead to maintenance of numbers of HSC. If a HSC population differentiates and both of the daughter cells are committed to the generation of lineage committed progenitor cells, this will ultimately lead to exhaustion of the HSC pool. HSC and hematopoietic progenitor cells (HPC) may also undergo apoptosis, which is essential for homeostasis. One of the drawbacks of *in-vitro* cell culture conditions is that, in general, this results in exhaustion of the stem cells due to a lack of HSC undergoing self-renewal types of divisions. On the other hand, the hematopoietic microenvironment *in vivo* allows sustenance of lifelong hematopoiesis by promoting a self-renewal type of HSC divisions. Following HSC transplantation, the reconstitution of a myeloablated host support the view that a limited number of HSC can divide and reconstitute a myeloablated host and create a HSC pool large enough to sustain long-term donor-derived hematopoiesis [16–19]. While the methylation machinery in normal hematopoietic development is regulated to allow lineage-specific differentiation and control of proliferation, disturbances in methylation leading to gene-specific inactivation by promoter si-

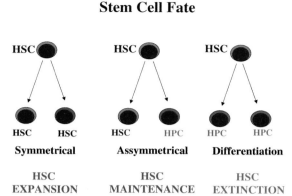

Figure 2.1 The fate of a hematopoietic stem cell (HSC) division: HSC expansion, HSC maintenance and HSC extinction.

lencing have been recognized as a key epigenetic mechanism during malignant transformation [15, 20, 21].

There is mounting evidence that demethylation of target genes by treatment with 5-azacytidine or 5-aza-2′-deoxycitidine (5azaD; decitabine) occurs not only *in vitro* (as shown in virtually hundreds of studies using cell lines), but also in patients with hematological malignancies treated with inhibitors of DNA methylation. Combinations of demethylating agent with histone deacetylase (HDAC) inhibitors have been suggested as a basis for removing epigenetic gene inactivation [15]. Although recently there has been a great deal of interest concerning the epigenetic regulation of genes involved in the development of malignancies and chromatin-modifying agents (5azaD and trichostatin A; TSA) as possible therapeutic agents for the treatment of various cancers, very little is known about the effects of these agents on the fate of normal HSC [22–25].

2.2
Cytotoxicity/Antitumor Activity versus Hypomethylating Effects of 5-Azacytidine and its Analogues

5-Azacytidine and 5azaD – two potent inhibitors of cytosine methylation – have been shown to have strong antileukemic activity in acute myeloid leukemia [26–29]. In addition, both of these agents induce trilineage responses in myelodysplastic syndromes with moderate myelotoxicity and with acceptable nonhematologic toxicity. It has been postulated that the antitumor effect of 5azaD is a consequence of hypomethylation. However, Juttermann et al. have demonstrated in an elegant study that levels of DNA methyl transferase (DNMT) rather than genomic DNA demethylation, mediate 5azaD-induced cytotoxicity [30]. Mutant cells that contained reduced amounts of DNMT were more resistant to the cytotoxic effects of 5azaD than were wild-type cells [30]. Recently, it has been shown that the DNMT genes are constitutively expressed in normal hematopoietic cells (including BM CD34+ cells), but are down-regulated as these cells differentiate. However, the DNMTs were substantially overexpressed in leukemia cells [31]. The higher levels of DNMTs detected in leukemias and other malignant cell lines is likely responsible for greater cytotoxicity following 5azaD treatment of malignant cells in comparison to more limited cytotoxic effects observed in normal cells.

2.3
Treating HSC with 5azaD/TSA can Alter their Fate

5azaD, a cytosine analogue, is a DNMT inhibitor [32] which, when incorporated into DNA, binds covalently to and irreversibly blocks the maintenance of methyltransferase, allowing passive demethylation to take place as a cell replicates its DNA and divides [33]. Trichostatin A is a HDAC inhibitor which exerts its activity by interacting with the catalytic site of the HDAC enzyme [34].

5-Azacytidine and its analogues, as well as HDAC inhibitors, each have dramatic effects on transcriptional regulation [1–4]. These two classes of drugs in combination may be capable of synergistically reactivating developmentally silenced genes [35–37]. HSCs are slowly cycling cells that are capable of exiting G_0/G_1 following exposure to early-acting cytokines including stem cell factor (SCF), FLT-3 ligand, and thrombopoietin (TPO) [38, 39]. There has been very limited success in controlling this process of HSC commitment and differentiation *in vitro* beyond HSCs undergoing a limited number of cell divisions. The model developed by Milhem et al. assesses the role of 5azaD/TSA in HSC/HPC fate decisions (Fig. 2.2) [24, 25]. This group has reported that the *in-vitro* fate of adult bone marrow (BM) CD34$^+$ cells can be altered by adding 5azaD and TSA [24, 25]. Using human BM CD34+ cells, it was demonstrated that following 5azaD/TSA treatment, the methylation status of γ-globin promoter gene was reduced significantly. Such methylation status served as an indicator of the global alteration in the methylation status of CD34+ cells during *in-vitro* culture. More importantly, the CD34+ cells cultured in the absence of 5azaD/TSA in the cytokines alone did not undergo reduction in their methylation status. These results suggest that the hypomethylation status induced by 5azaD/TSA in the culture may promote reactivation of genes critical for maintenance of the HSC phenotype and function in the culture. 5azaD/TSA-treated BM CD34+ cells maintained a primitive CD34+CD90+ cell phenotype (77 %), despite 9 days of culture. On the other hand, CD34+ BM cells cultured in the absence of 5azaD/TSA had only a minor fraction of cells expressing the CD34+CD90+ phenotype (<1 %). Furthermore, the 5azaD/TSA treatment was capable of leading to a 2.5-fold expansion of the numbers of CD34+CD90+ cells in comparison to the input number of cells after 9 days of culture. These expanded CD34+CD90+ cells from 5azaD/TSA cultures possessed properties characteristics of primitive HSC/HPC. For instance, BM CD34+ cells treated with cytokines alone for 9 days contained dramatically reduced numbers of colony-forming cells in comparison to primary BM CD34+

cells. By contrast, unfractionated cells from the culture treated with 5azaD/TSA after 9 days had far greater numbers of colonies than the culture treated with cytokines alone. The plating efficiency of primary CD34+ cells and 5azaD/TSA-treated cells proved to be comparable, whereas the plating efficiency of cells cultured without 5azaD/TSA had a 12-fold lower plating efficiency than cells treated with 5azaD/TSA. Similarly, CD34+ cells cultured in the presence of 5azaD/TSA were characterized by a greater frequency of more primitive HPC, the cobblestone area-forming cells (CAFC), while cells cultured in the absence of 5azaD/TSA experienced a net loss of CAFC in comparison to primary unmanipulated BM CD34+ cells. These data further ensure that the CD34+CD90+ cells expanded in the presence of 5azaD/TSA are enriched for functional progenitors (CFC/CAFC), excluding a discordance between HSC function and phenotype previously observed by many investigators following *ex-vivo* HSC expansion [40, 41]. To further evaluate the differentiation potential of cells generated following 9 days of culture in the presence of cytokines and treatment with 5azaD/TSA, Milhem et al. sorted CD34+ cells into two subsets – CD34+CD90+ and CD34+CD90– cells – and placed in a secondary liquid culture supplemented with 30 % fetal bovine serum (FBS), interleukin (IL)-3, SCF, TPO, granulocyte macrophage colony-stimulating factor (GM-CSF), erythropoietin (EPO), and IL-6 for an additional 7 days [25]. The CD34+CD90+ cells pretreated with 5azaD/TSA still contained 68 % blast-like immature cells, while those cultures exposed to the same cytokine combinations but not pretreated with 5azaD/TSA were composed of almost exclusively terminally differentiated cells [25].

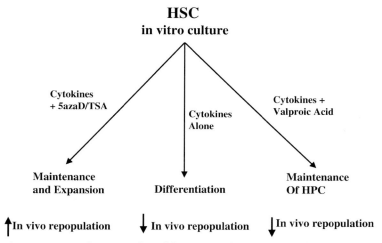

Figure 2.2 Fate of HSC in culture following treatment with cytokines with or without chromatin-modifying agents. HPC includes committed progenitors (colony-forming cells). The cytokine cocktail during the first 48 h of culture consisted of IL-3, thrombopoietin, Flt3-ligand and stem cell factor; IL-3 was removed from the culture after 48 h (for more details, see [25]). TSA: trichostatin A.

2.4
Are the Effects of 5azaD/TSA Due to Cytotoxicity?

Although the effects of 5azaD/TSA could in part be explained by their ability to promote HSC cytotoxicity and thereby select for a primitive subset of quiescent HSC, this explanation does not solely account for these findings. A similar preservation of HSC phenotype, number and function was not achieved by treatment with 5-fluorouracil (5-FU), a cytotoxic agent frequently used to select for primitive progenitor cells and HSCs both *in vivo* and *in vitro* [42]. Nevertheless, the presence of 5azaD in the culture led to the generation of 20-fold fewer total nucleated cells (TNC) as compared to cultures treated with cytokines alone. On the other hand, when TSA was added to the cultures there were only 1.3-fold fewer TNC than the culture containing cytokines alone. Milhem et al. tracked cell fate by means of a membrane dye which was added during the *in-vitro* culture [24, 25]. The results of these studies indicated that the CD34+ cells, in the presence of 5azaD/TSA, divided relatively more slowly compared to the culture containing cytokines alone, thereby leading to fewer HSC cell divisions [24, 25]. However, it could be argued that the cytotoxic effects of 5azaD might have resulted in the selection of a more primitive quiescent HSC subpopulation. This hypothesis is not supported by the data, as the membrane-tracking studies revealed that virtually all cells in the culture divided more than once in the presence of 5azaD/TSA during the 9-day culture period [25]. More importantly, the 5azaD/TSA-treated cells – but not the CD34+ cells exposed to cytokines alone – retained the ability to repopulate immunodeficient mice with human donor cells when transplanted into non-obese diabetic/severely compromised immunodeficient (NOD/SCID) mice [24, 25]. BM CD34+ cells expanded in the presence of 5azaD/TSA retained the ability to differentiate into myeloid and lymphoid cells.

Young et al. have reported that murine HSC treated *in vitro* with HDAC inhibitors resulted in the *ex-vivo* expansion of the numbers of functional HSC, including *in-vivo* repopulating cells [43]. Koh et al. also reported that 5azaD/TSA treatment increased the frequency of CD34+ cells in the culture, but caused cell death which could be overcome by co-culture with mesenchymal stem cells [44]. These authors have speculated that the higher frequency of CD34+ cells in the cultures resulted from treatment with 5azaD/TSA, but that the loss of TNC was due to the relative resistance of CD34+ cells to cytotoxicity as the majority of CD34+ cells were in a quiescent state [44]. However, this statement was not consistent with the findings of Milhem et al. and others since, following 5azaD/TSA treatment, Milhem et al. observed that none of the CD34+ cells remained in a dormant state and that all of the CD34+CD90+ cells had divided more than once [24, 25].

2.5
Treating HSC with Valproic Acid

The potential for epigenetic mechanisms to modify normal HSC fate has also been explored by other groups using an alternate chromatin-modifying agent and HDAC inhibitor, valproic acid [45, 46]. De Felice et al. have shown recently that the alteration in chromatin structure induced by valproic acid can result in an expansion of the numbers of HSC as defined by HSC phenotype [45]. Interestingly, these authors observed that valproic acid-treated cells had higher levels of HoxB4 expression [45, 46]. It has been reported previously that ectopic expression of transcription factor HoxB4 increases the number of transplantable HSCs possessing myeloid and lymphoid differentiation potential after transplantation in an immunodeficient (NOD/SCID) mouse model [45, 46]. However, De Felice et al. did not perform *in-vivo* functional studies to demonstrate the repopulation potential of valproic acid-treated cord blood (CB) cells [45]. In addition, Bug et al. have demonstrated an increase in proliferation and self-renewal potential of murine and human HSC after 10 days of culture in the presence of valproic acid. However, the functional studies *in vivo* were performed after treating cells for only 2 days, which made it difficult to estimate the functional activity of cells expanded for 10 days [46]. Preliminary studies conducted by Araki et al. suggested that treatment with different chromatin-modifying agents was capable of altering the differentiation program of distinct populations of HSC [49]. Treatment of CB CD34+ cells using valproic acid alone or in combination with 5azaD primarily affected short-term colony-forming cells (CFC) and long-term CAFC, but not more primitive *in-vivo* SCID repopulating cells (SRC) [49]. On the other hand, 5azaD/TSA treatment during the *ex-vivo* expansion of CB CD34+ cells targets CAFC and SRC, but not CFC. Further studies using valproic acid are required to demonstrate more clearly whether expansion as defined by the phenotype correlates with the increased frequency of *in-vivo* repopulating HSC. At present, valproic acid appears to appears to affect relatively differentiated HPC rather than HSC [49].

2.6
Ex-vivo Expansion of HSC Using Chromatin-Modifying Agents

Although HSC are known to self-renew *in vivo* [16–19, 50–52], the creation of an *in-vitro* environment, which promotes symmetrical HSC division has largely evaded the efforts of numerous investigators [54–61]. Recently, however, several genetic regulatory programs have been implicated in HSC self-renewal [64–71]. The overwhelming numbers of previous attempts at *in-vitro* HSC expansion have led to a decline or at best maintenance of assayable human HSC [53–63]. A successful strategy for *in-vitro* stem cell expansion should take into account the unique biological properties of HSC. HSC are slowly cycling cells that possess the ability to self-renew and to differentiate into progenitors belonging to multiple hematopoietic lineages [16–19, 52]. Previous attempts at *ex-vivo* expansion of HSC have resulted in a progressive loss of the numbers of marrow repopulating cells and the generation of large numbers of committed hematopoietic progenitor cells [53–61]. Most of the cytokines previously utilized for *in-vitro* HSC expansion favor commitment or rapid HSC cycling [53–61]. It is possible that the retention of HSC function following *in-vitro* cell division would require the preservation of a cellular program identical to that of primary HSC, and that the *ex-vivo* conditions previously employed for HSC expansion result in the loss of this cellular program. Recent studies have provided new understanding of intracellular factors that might regulate HSC self-renewal [64–71]. Proteins that are active during embryonic development, such as members of the Wnt family, bone morphogenetic proteins, sonic hedgehog, and Notch families, induce self-renewal in HSCs [64–71]. In support of this, DeFelice et al. have shown that following valproic acid treatment of HSC, one of the critical candidate transcription factor, HoxB4 implicated to be involved in self-renewal of HSC was up-regulated [45].

The culture conditions utilized to alter BM CD34+ cells in the presence of 5azaD/TSA to alter the fate of BM CD34+ cells resulted in expansion of HSC rather than differentiation in the face of presence of medium in the culture likely to promote terminal differentiation [24, 25]. Araki et al. utilized the same protocol for treating CB HSC with 5azaD/TSA, but chose a more optimal cytokine combination in order to expand HSC present in a human CB [72, 73]. Araki et al. also reported a more than 12-fold expansion of CD34+CD90+ cells, a more than 10-fold expansion of CFC/CAFC, as well as a 10-fold increase in SRC [49, 72, 73]. Taken together, these findings suggest that alteration of chromatin structure by treatment with both hypomethylating agents and the appropriate HDAC inhibitor is most likely capable of preventing the culture-induced decline of HSC activity while expanding the numbers of cells expressing the phenotype of HSC and retaining *in-vivo* hematopoietic repopulation potential. These studies by Araki et al. were among the first successful efforts to expand the numbers of functional HSC [49, 72, 73].

2.7
Reactivation of Gene Expression by Treating Cells with Chromatin-Modifying Agents

Genes critical for early stem cell self-renewal fate decisions are likely activated in primitive cells. Recent studies (including our own) have explored whether global chromatin modifiers might influence stem cell fates. As decitabine is a global hypomethylating agent, treatment with this agent would be anticipated to reactivate hundreds of genes that may favor cellular transformation. However, gene expression studies using microarray analysis of tumor cell lines following 5azaD (decitabine) treatment demonstrated increased expression of only a limited number of genes (0.67 % of 25 940 genes analyzed) [74]. Similarly, genes that are either up- or down-regulated by HDAC inhibitor treatment of tumor cell lines, have been reported to range between 2 and 10 % [75–77].

There are candidate genes that have been suggested to play crucial roles in the balance between symmetrical versus asymmetrical cell divisions in order to maintain the number of HSC during *ex-vivo* culture. Recent studies have identified intracellular factors that might regulate HSC self-renewal [62–69]. To date, overexpression of HOXB4 has been reported to be the most potent stimulator of HSC expansion [64]. Although Hox genes are known for their role in the specifications of AP axis in various bilaterians, it is now becoming clear that Hox genes also play a part in determining cell fate in several other tissues, including blood [78–80].

Whether the CD34$^+$ cells exposed to 5azaD and TSA acquire a unique gene expression pattern characteristic of primitive HSC remains to be determined. It appears likely, however, that the transcriptional program normally associated with *in-vitro* terminal differentiation of CD34$^+$ cells has been interrupted by the reactivation of genes associated with the preservation of phenotypically identifiable HSC populations. This redirection of the fate of primitive HSC is associated with the pharmacological alteration of the methylation status of DNA. Methylation is capable of changing the interactions between proteins and DNA, which leads to alterations in chromatin structure and either a decrease or increase in the rate of transcription [81]. Currently, we are investigating global gene expression pattern in CB cells in the presence or absence of 5azaD/TSA in the culture. This strategy may allow us to distinguish a set of genes the expression of which is altered by the treatment of chromatin-modifying agents, which may serve as a signature for identification of culture of expanded HSC retaining their *in-vivo* repopulation potential. These groups of genes may not only prove to be crucial for the maintenance of HSC function during *ex-vivo* culture but they may also play a role during symmetrical HSC divisions.

2.8
Alteration of Nonhematopoietic Fate by Chromatin-Modifying Agents

Although embryonic stem cells have the broadest differentiation capacity, the potential for their use as cellular therapeutics remains controversial. Several investigators are exploring the nonhematopoietic differentiation potential of hematopoietic cells from BM, peripheral blood (PB) or CB. For example, Makino et al. have isolated a cardiomyogenic cell line (CMG) from murine BM stromal cells immortalized by treatment with 5-azacytidine, and this has resulted in spontaneously beating cardiomyocytes within 2–3 weeks of culture [82]. Recently, 5azaD/TSA has been used to alter the developmental potential of fetal brain-derived cells that form neurospheres [83]. The neurosphere cells, following treatment with 5azaD/TSA, yielded cells with long-term, multilineage and transplantable hematopoietic repopulation potential. Untreated neurosphere cells displayed no detectable hematopoietic repopulation capacity. Over the past years, several studies have shown that BM or peripheral blood cells possess nonhematopoietic differentiation potential [84–89]. Recently, a rare cell from the BM of rodents and humans – called the multipotent adult progenitor cell (MAPC) – was identified and shown to be capable of differentiating *in vitro* into cells belonging to all three germ layers [90, 91]. Furthermore, Kogler al. have identified another pluripotent cell population from human CB, termed the unrestricted somatic stem cells (USSCs) [92]. Since MAPC or USSC have not been detectable in primary hematopoietic tissues, but have been isolated following prolonged *in-vitro* culture, it cannot be ruled out the detection of such cells depend on epigenetic events that occur after prolonged culture.

2.9
Safety/Toxicity of Treating Cells with Chromatin-Modifying Agents

Although several chromatin-modifying agents have been utilized to alter the fate of HSC, some members of these classes of agent, such as butyric acid and sodium valproate, have been used for many years for the therapy of patients with central nervous system disorders and inherited metabolic diseases. There are no reported data available to suggest that these agents cause secondary malignancies [93]. Similarly, 5azaD (decitabine) has been used in clinical trials for over two decades. In patients with myelodysplastic syndrome treated with decitabine, there have been no unusual patterns of cytogenetic instability to suggest that the drug is capable of exacerbating chromosome instability [94]. A brief *in-vitro* exposure of low-dose 5azaD/TSA is unlikely to pose much risk for the development of hematological malignancies. In order to evaluate the long-term safety of the cellular therapy products treated with chromatin-modifying agents, at a minimum the following questions shall be addressed: (i) whether BM/CB cells treated with chromatin-modifying agents for *ex-vivo* expansion are capable of sustaining normal polyclonal hematopoiesis for a prolonged period of time; and (ii) whether treated HSC have a higher preponderance to give rise to myelodysplasia, leukemia or BM failure syndrome.

Before this technology can enter the clinic, several concerns must be addressed. Chromatin-modifying agents are capable of nonspecifically activating genes in normal cells which, in theory, might lead to a neoplastic transformation. Unfortunately, few studies have examined the effects of these agents on normal cells as opposed to cell lines [32]. Although the chromatin-modifying agents might be able to activate silenced oncogenes, these agents can also clearly act as agents to prevent cancer [95]. The clinical use of 5azaD or decitabine in patients with myelodysplastic disorders or sickle cell anemia has not been reported to lead to any increase in chromosome instability [93, 94]. A careful safety assessment of HSC or stem cells belonging to other tissues expanded in the presence of 5azaD/TSA or other chromatin-modifying agents is required before proceeding with clinical development of this technology. However, it is worth pointing out that, in a number of animal models, decitabine has served as a chemopreventive agent rather than as a carcinogenic agent [95].

2.10
Conclusion

The regulation of HSC fate decisions *in vitro* by epigenetic mechanisms offers a
new paradigm by which to elucidate the behavior of a stem cell. Such strategies
may lead to novel approaches resulting in the *ex-vivo* expansion of HSC and the
delivery of foreign genes into HSC, to the modification of cell fates *ex-vivo* prior to
transplantation. Furthermore, knowledge of epigenetic events that govern cellular
fate determination may lead to a better understanding of the basic properties of
maintenance and expansion of stem cells during physiological states, as well as
following malignant transformation.

Abbreviations/Acronyms

5azaD	5-aza-2'-deoxycytidine
5-FU	5-fluorouracil
BM	bone marrow
CAFC	cobblestone area-forming cells
CB	cord blood
CFC	colony-forming cells
CMG	cardiomyogenic
DNMT	DNA methyl transferase
EPO	erythropoeitin
FLT	Fms-like tyrosine kinase
GM-CSF	granulocyte macrophage colony-stimulating factor
HDAC	histone deacetylase
HPC	hematopoietic progenitor cells
HSC	hematopoietic stem cells
MAPC	multipotent adult progenitor cell
NOD/SCID	non-obese diabetic/severe combined immune deficiency
PB	peripheral blood
SCF	stem cell factor
SRC	SCID repopulating cells
TNC	total nucleated cells
TPO	thrombopoietin
TSA	trichostatin A
USSC	unrestricted somatic stem cell

References

1. Reik W, Dean W, Walter J. Epigenetic reprogramming in mammalian development. *Science* **2001**;*293*:1089–1093.

2. Blau HM. Differentiation requires continuous active control. *Annu Dev Biochem* **1992**;*61*:1213–1230.

3. Jones PA, Takai D. The role of DNA methylation in mammalian epigenetics. *Science* **2001**;*293*:1068–1070.

4. Jones PA, Laird PW. Cancer epigenetics comes of age. *Nat Genet* **1999**;*21*:163–167.

5. Marks PA, Richon VM, Rifkind RA. Histone deacetylases inhibitors: inducers of differentiation or apoptosis of transformed cells. *J Natl Cancer Inst* **2000**;*92*:1210–1216.

6. Agata Y, Katakai T, Ye SK, et al. Histone acetylation determines the developmentally regulated accessibility for T cell receptor gamma gene recombination. *J Exp Med* **2001**;*193*:873–879.

7. Jiang Z, Emerson SG. Hematopoietic cytokines, transcription factors and lineage commitment. *Oncogene* **2002**;*21*:3295–3313.

8. Mahmud N, Katayama N, Itoh R, et al. A possible change in doubling time of haemopoietic progenitor cells with stem cell development. *Br J Haematol* **1996**;*94*:242–249.

9. Mahmud N, Rose D, Pang W, Walker R, Patil V, Weich N, Hoffman R. Characterization of primitive marrow CD34+ cells that persist after a sublethal dose of total body irradiation. *Exp Hematol* **2005**;*33*:1388–1401.

10. Akashi K, He X, Chen J, et al.. Transcriptional accessibility for genes of multiple tissues and hematopoietic lineages is hierarchically controlled during early hematopoiesis. *Blood* **2003**;*101*:383–389.

11. Robert LP, Ernest RE, Brunk B, et al. The genetic program of hematopoetic stem cells. *Science* **2000**;*288*:1635–1640.

12. Santos MR, Yoon S, Matsuzaki Y, Mulligan RC, Melton DA. "Stemness": Transcriptional profiling of embryonic and adult stem cells. *Science* **2002**;*298*:597–600.

13. Ivanova NB, Dimos JT, Schaniel C, et al. A stem cell molecular signature. *Science* **2002**;*298*:601–604.

14. Hattori N, Nishino K, Ko YG, Hattori N, Ohgane J, Tanaka S, Shiota K. Epigenetic control of mouse Oct-4 gene expression in embryonic stem cells and trophoblast stem cells. *J Biol Chem* **2004**;*279*:17063–17069.

15. Claus R, Lubbert M. Epigenetic targets in hematopoietic malignancies. *Oncogene* **2003**;*22*:6489–6496.

16. Lemischka IR, Raulet DH, Mulligan RC. Developmental potential and dynamic behavior of hematopoietic stem cells. *Cell* **1986**;*45*:917–927.

17. Abkowitz JL, Persik MT, Shelton GH, et al. Behavior of hematopoietic stem cells in a large animal. *Proc Natl Acad Sci USA* **1995**;*92*:2031–2035.

18. Nash R, Storb R, Neiman P. Polyclonal reconstitution of human marrow after allogeneic bone marrow transplantation. *Blood* **1988**;*72*:2031–2037.

19. Turhan AG, Humphries RK, Phillips GL, Eaves AC, Eaves CJ. Clonal hematopoiesis demonstrated by x-linked DNA polymorphisms after allogeneic bone marrow transplantation. *N Engl J Med* **1989**;*320*:1655–1661.

20. Sakashita K, Koike K, Kinoshita T, Shiohara M, Kamijo T, Taniguchi S, Kubota T. Dynamic DNA methylation change in the CpG island region of p15 during human myeloid development. *J Clin Invest* **2001**;*108*:1195–1204.

21. Daskalakis M, Nguyen TT, Nguyen C, Guldberg P, Kohler G, Wijermans P, Jones PA, Lubbert M. Demethylation of a hypermethylated P15/INK4B gene in patients with myelodysplastic syndrome by 5-Aza-2'-deoxycytidine (decitabine) treatment. *Blood* **2002**;*100*:2957–2964.

22. Saunthararajah Y, Hillery CA, Lavelle D, et al. Effects of 5-aza-2'-deoxycytidine on fetal hemoglobin levels, red cell adhesion, and hematopoietic differentiation in patients with sickle cell disease. *Blood* **2003**;*102*:3865–3870.

23. Gilbert J, Gore SD, Herman JG., Carducci MA. The clinical application of

targeting cancer through histone acetylation and hypomethylation. *Clin Cancer Res* **2004**;*10*:4589–4596.

24. Milhem M, Mahmud N, Lavelle D, Saunthararajah S, DeSimone J, Hoffman R. Reprogramming of primitive hematopoietic stem cell in vitro. *Blood* **2002**;*100*:64a.

25. Milhem M, Mahmud N, Lavelle D, Araki H, DeSimone J, Saunthararajah Y, Hoffman R. Modification of hematopoietic stem cell fate by 5-aza-2'-deoxycytidine and trichostatin A. *Blood* **2004**;*103*:4102–4110.

26. Davidson S, Crowther P, Radley J, Woodcock D. Cytotoxicity of 5-aza-2'-deoxycytidine in a mammalian cell system. *Eur J Cancer* **1992**;*28*:362–368.

27. Jaenisch R, Schnieke A, Harbers K. Treatment of mice with 5-azacytidine efficiently activates silent retroviral genomes in different tissues. *Proc Natl Acad Sci USA* **1985**;*82*(5):1451–1455.

28. Pinto A, Zagonel V. 5-Aza-2'-deoxycytidine (decitabine) and 5-azacytidine in the treatment of acute myeloid leukemias and myelodysplastic syndromes: past, present and future trends. *Leukemia* **1993**;*7*(Suppl.1):51–60.

29. Willemze R, Archimbaud E, Muus P. Preliminary results with 5-aza-2'-deoxycytidine (DAC)-containing chemotherapy in patients with relapsed or refractory acute leukemia. The EORTC Leukemia Cooperative Group. *Leukemia* **1993**;*7*(Suppl.1):49–50.

30. Juttermann R, Li E, Jaenisch R. Toxicity of 5-aza-2'-deoxycytidine to mammalian cells is mediated primarily by covalent trapping of DNA methyltransferase rather than DNA demethylation. *Proc Natl Acad Sci USA* **1994**;*91*:11797–11801.

31. Mizuno S, Chijiwa T, Okamura T, Akashi K, Fukumaki Y, Niho Y, Sasaki H. Expression of DNA methyltransferases DNMT1, 3A, and 3B in normal hematopoiesis and in acute and chronic myelogenous leukemia. *Blood* **2001**;*97*:1172–1179.

32. Issa JP. Decitabine. *Curr Opin Oncol* **2003**;*15*:446–451.

33. Pietrobono R, Pomponi MG, Tabolacci E, Oostra B, Chiurazzi P, Neri G. Quantitative analysis of DNA demethylation and transcriptional reactivation of the FMR1 gene in fragile X cells treated with 5-azadeoxycytidine. *Nucleic Acids Res* **2002**;*30*:3278–3285.

34. Jung M. Inhibitors of histone deacetylase as new anticancer agents. *Curr Med Chem* **2001**;*8*:1505–1511.

35. Kass SU, Pruss D, Wolffe AP. How does methylation repress transcription? *Trends Genet* **1997**;*13*:444–449.

36. Benjamin D, Jost JP. Reversal of methylation mediated repression with short fatty acids: evidence for an additional mechanism to histone deacetylation. *Nucleic Acids Res* **2001**;*29*:3603–3610.

37. Sheikhnejad G, Brank A, Christman JK, et al. Mechanism of inhibition of DNA (cytosine C5)-methyltransferases by oligodexoyribonucleotides containing 5, 6-dihydro-5-azacytosine. *J Mol Biol* **1999**;*285*:2021–2034.

38. Punzel M, Zhang T, Liu D, Eckstein V, Ho AD. Functional analysis of initial cell divisions defines the subsequent fate of individual human CD34⁺CD38⁻ cells. *Exp Hematol* **2002**;*30*:464–472.

39. Srour EF. Proliferative history and hematopoietic function of ex vivo expanded human CD34⁺ cells. *Blood* **2000**;*96*:1609–1612.

40. Danet GH, Lee HW, Luongo JL, Simon MC, Bonnet DA. Dissociation between stem cell phenotype and NOD/SCID repopulating activity in human peripheral blood CD34⁺ cells after ex vivo expansion. *Exp Hematol* **2001**;*29*:1465–1473.

41. Dorrell C, Gan OI, Pereira DS, Hawley RG, Dick JE. Expansion of human cord blood CD34⁺CD38⁻ cells in ex vivo culture during retroviral transduction without a corresponding increase in SCID repopulating cell (SRC) frequency: dissociation of SRC phenotype and function. *Blood* **2000**;*95*:102–110.

42. Berardi AC, Wang A, Levine JD, Lopez P, Scadden DT. Functional isolation and characterization of human hematopoietic stem cells. *Science* **1995**;*267*:104–108.

43. Young JC, Wu S, Hansteen G, Du C, Sambucetti L, Remiszewski S, O'Farrell AM, Hill B, Lavau C, Murray LJ. Inhibitors of histone deacetylases promote he-

matopoietic stem cell self-renewal. *Cytotherapy* **2004**;*6*:328–336.

44. Koh SH, Choi HS, Park ES, Kang HJ, Ahn HS, Shin HY. Co-culture of human CD34+ cells with mesenchymal stem cells increases the survival of CD34+ cells against the 5-aza-deoxycytidine- or trichostatin A-induced cell death. *Biochem Biophys Res Commun* **2005**;*329*:1039–1045.

45. De Felice L, Tatarelli C, Mascolo MG, Gregorj C, Agostini F, Fiorini R, Gelmetti V, Pascale S, Padula F, Petrucci MT, Arcese W, Nervi C. Histone deacetylase inhibitor valproic acid enhances the cytokine-induced expansion of human hematopoietic stem cells. *Cancer Res* **2005**;*65*:1505–1513.

46. Bug G, Gul H, Schwarz K, Pfeifer H, Kampfmann M, Zheng X, Beissert T, Boehrer S, Hoelzer D, Ottmann OG, Ruthardt M. Valproic acid stimulates proliferation and self-renewal of hematopoietic stem cells. *Cancer Res* **2005**;*65*:2537–2541.

47. Sauvageau G, Thorsteinsdottir U, Eaves CJ, et al. Overexpression of HOXB4 in hematopoietic cells causes the selective expansion of more primitive populations in vitro and in vivo. *Genes Dev* **1995**;*9*:1753–1765.

48. Antonchuk J, Sauvageau G, Humphries RK. HOXB4-induced expansion of adult hematopoietic stem cells ex vivo. *Cell* **2002**;*109*:39–45.

49. Araki H, Hoffman R, Mahmud N. Different histone deacetylase inhibitors affect distinct cellular targets within the hierarchy of hematopoietic stem/progenitor cells. *Blood* **2005**:106 (abstract).

50. Ogawa M. Differentiation and proliferation of hematopoietic stem cells. *Blood* **1993**;*81*:2844–2853.

51. Rebel VI, Miller CL, Eaves CJ, Lansdorp PM. The repopulation potential of fetal liver hematopoietic stem cells in mice exceeds that of their adult marrow counterparts. *Blood* **1996**;*87*:3500–3507.

52. Mahmud N, Devine SM, Weller KP, et al. The relative quiescence of hematopoietic stem cells in non-human primates. *Blood* **2001**;*97*:3061–3068.

53. Iscove NN, Nawa K. Hematopoietic stem cells expand during serial transplantation in vivo without apparent exhaustion. *Curr Biol* **1997**;*7*:805–808.

54. Sorrentino BP. Clinical strategies for expansion of haematopoietic stem cells. *Nat Rev Immunol* **2004**;*4*:878–888.

55. McNiece I. Ex vivo expansion of hematopoietic cells. *Exp Hematol* **2004**;*32*:409–410.

56. Devine SM, Lazarus H., Emerson SG. Clinical application of hematopoietic progenitor cell expansion: current status and future prospects. *Bone Marrow Transplant* **2003**;*31*:241–252.

57. Hoffman R. Progress in the development of systems for in vitro expansion of human hematopoietic stem cells. *Curr Opin Hematol* **1999**;*6*:184–191.

58. Rosler E, Brandt J, Chute J, Hoffman R. Cocultivation of umbilical cord blood cells with endothelial cells leads to extensive amplification of competent CD34$^+$CD38$^-$ cells. *Exp Hematol* **2000**;*28*:841–852.

59. Wagner W, Ansorge A, Wirkner U, Eckstein V, Schwager C, Blake J, Miesala K, Selig J, Saffrich R, Ansorge W, Ho AD. Molecular evidence for stem cell function of the slow-dividing fraction among human hematopoietic progenitor cells by genome-wide analysis. *Blood* **2004**;*104*:675–686.

60. Attar EC, Scadden DT. Regulation of hematopoietic stem cell growth. *Leukemia* **2004**;*18*:1760–1768.

61. Williams DA. Ex vivo expansion of hematopoietic stem and progenitor cells – robbing Peter to pay Paul? *Blood* **1993**;*81*:3169–3172.

62. Pecora AL, Stiff P, Jennis A, Goldberg S, Rosenbluth R, Price P, Goltry KL, Douville J, Armstrong RD, Smith AK, Preti RA. Prompt and durable engraftment in two older adult patients with high risk chronic myelogenous leukemia (CML) using ex vivo expanded and unmanipulated unrelated umbilical cord blood. *Bone Marrow Transplant* **2000**;*25*:797–799.

63. Kondo M, Wagers AJ, Manz MG, Prohaska SS, Scherer DC, Beilhack GF, Shizuru JA, Weissman IL. Biology of

hematopoietic stem cells and progenitors: implications for clinical application. *Annu Rev Immunol* **2003**;*21*:759–806.

64. Dick JE. Stem cells: self-renewal writ in blood. *Nature* **2003**;*423*:231–233.

65. Reya T. Regulation of hematopoietic stem cell self-renewal. *Recent Prog Horm Res* **2003**;*58*:283–295.

66. Lessard J, Sauvageau G. Bmi-1 determines the proliferative capacity of normal and leukaemic stem cells. *Nature* **2003**;*423*:255–260.

67. Park IK, Qian D, Kiel M, Becker MW, Pihalja M, Weissman IL, Morrison SJ, Clarke MF. Bmi-1 is required for maintenance of adult self-renewing haematopoietic stem cells. *Nature* **2003**;*423*:302–305.

68. Amsellem S, Pflumio F, Bardinet D, Izac B, Charneau P, Romeo PH, Dubart-Kupperschmitt A, Fichelson S. Ex vivo expansion of human hematopoietic stem cells by direct delivery of the HOXB4 homeoprotein. *Nat Med* **2003**;*9*:1423–1427.

69. Krosl J, Austin P, Beslu N, Kroon E, Humphries RK, Sauvageau G. In vitro expansion of hematopoietic stem cells by recombinant TAT-HOXB4 protein. *Nat Med* **2003**;*9*:1428–1432.

70. Sauvageau G, Thorsteinsdottir U, Eaves CJ, Lawrence HJ, Largman C, Lansdorp PM, Humphries RK. Overexpression of HOXB4 in hematopoietic cells causes the selective expansion of more primitive populations in vitro and in vivo. *Genes Dev* **1995**;*15*(9):1753–1765.

71. Antonchuk J, Sauvageau G, Humphries RK. HOXB4-induced expansion of adult hematopoietic stem cells ex vivo. *Cell* **2002**;*109*:39–45.

72. Araki H, Mahmud N, Milhem M, Patel H, Nunez R, Bruno E, Hoffman R. CD34+CD90+ cord blood cells which have undergone multiple cell divisions retain marrow repopulating potential. *Blood* **2004**:*104*:881a.

73. Araki H, Milhem M, Patel H, Nunez R, Bruno E, Hoffman R, Mahmud N. Extensive self-renewal of engraftable cord blood stem cells induced by chromatin modifying agents. *Cytotherapy* **2005**;*7*(suppl.1):215a.

74. Karpf AR, Lasek AW, Ririe TO, Hanks AN, Grossman D, Jones DA. Limited gene activation in tumor and normal epithelial cells treated with the DNA methyltransferase inhibitor 5-aza-2′-deoxycytidine. *Mol Pharmacol* **2004**;*65*:18–27.

75. Van Lint C, Emiliani S, Verdin E. The expression of a small fraction of cellular genes is changed in response to histone hyperacetylation. *Gene Express* **1996**;*5*:245–253.

76. Chambers AE, Banerjee S, Chaplin T, Dunne J, Debernardi S, Joel SP, Young BD. Histone acetylation-mediated regulation of genes in leukaemic cells. *Eur J Cancer* **2003**;*39*:1165–1175.

77. Glaser KB, Staver MJ, Waring JF, Stender J, Ulrich RG, Davidsen SK. Gene expression profiling of multiple histone deacetylase (HDAC) inhibitors: defining a common gene set produced by HDAC inhibition in T24 and MDA carcinoma cell lines. *Mol Cancer Ther* **2003**;*2*:151–163.

78. Thorsteinsdottir U, Sauvageau G, Humphries RK. Hox homeobox genes as regulators of normal and leukemic hematopoiesis. *Hematol Oncol Clin North Am* **1997**;*11*:1221–1237.

79. Lawrence HJ, Sauvageau G, Humphries RK, Largman C. The role of HOX homeobox genes in normal and leukemic hematopoiesis. *Stem Cells* **1996**;*14*:281–291.

80. Chiba S. Homeobox genes in normal hematopoiesis and leukemogenesis. *Int J Hematol* **1998**;*68*:343–353.

81. Kass SU, Pruss D, Wolffe AP. How does DNA methylation repress transcription? *Trends Genet* **1997**;*13*:444–449.

82. Makino S, Fukuda K, Miyoshi S, Konishi F, Kodama H, Pan J, Sano M, Takahashi T, Hori S, Abe H, Hata J, Umezawa A, Ogawa S. Cardiomyocytes can be generated from marrow stromal cells in vitro. *J Clin Invest* **1999**;*103*:697–705.

83. Schmittwolf C, Kirchhof N, Jauch A, Durr M, Harder F, Zenke M, Muller AM. In vivo haematopoietic activity is induced in neurosphere cells by chromatin-modifying agents. *EMBO J* **2005**;*24*:554–566.

84. Gussoni E, Soneoka Y, Strickland CD, Buzney EA, Khan MK, Flint AF, Kunkel LM, Mulligan RC. Dystrophin expression in the mdx mouse restored by stem cell transplantation. *Nature* **1999**;*401*:390–394.

85. Petersen BE, Bowen WC, Patrene KD, Mars WM, Sullivan AK, Murase N, Boggs SS, Greenberger JS, Goff JB. Bone marrow as a potential source of hepatic oval cells. *Science* **1999**;*284*:1168–1170.

86. Mezey E, Chandross KJ, Harta G, Maki RA, McKercher SR. Turning blood into brain: cells bearing neuronal antigens generated in vivo from bone marrow. *Science* **2000**;*290*:1779–1782.

87. Krause DS, Theise ND, Collector MI, Henegariu O, Hwang S, Gardner R, Neutzel S, Sharkis SJ. Multi-organ, multi-lineage engraftment by a single bone marrow-derived stem cell. *Cell* **2001**;*105*:369–377.

88. Korbling M, Katz RL, Khanna A, Ruifrok AC, Rondon G, Albitar M, Champlin RE, Estrov Z. Hepatocytes and epithelial cells of donor origin in recipients of peripheral-blood stem cells. *N Engl J Med* **2002**;*346*:738–746.

89. Wagers AJ, Sherwood RI, Christensen JL, Weissman IL. Little evidence for developmental plasticity of adult hematopoietic stem cells. *Science* **2002**;*297*:2256–2259.

90. Jiang Y, Vaessen B, Lenvik T, Blackstad M, Reyes M, Verfaillie CM. Multipotent

progenitor cells can be isolated from postnatal murine bone marrow, muscle, and brain. *Exp Hematol* **2002**;*30*:896–904.

91. Reyes M, Lund T, Lenvik T, Aguiar D, Koodie L, Verfaillie CM. Purification and ex vivo expansion of postnatal human marrow mesodermal progenitor cells. *Blood* **2001**;*98*:2615–2625.

92. Kogler G, Sensken S, Airey JA, Trapp T, Muschen M, Feldhahn N, Liedtke S, Sorg RV, Fischer J, Rosenbaum C, Greschat S, Knipper A, Bender J, Degistirici O, Gao J, Caplan AI, Colletti EJ, Almeida-Porada G, Muller HW, Zanjani E, Wernet P. A new human somatic stem cell from placental cord blood with intrinsic pluripotent differentiation potential. *J Exp Med* **2004**;*200*:123–135.

93. Saunthararajah Y, Lavelle D, DeSimone J. DNA hypo-methylating agents and sickle cell disease. *Br J Haematol* **2004**;*126*:629–636.

94. Lubbert M, Wijermans P, Kunzmann R, Verhoef G, Bosly A, Ravoet C, Andre M, Ferrant A. Cytogenetic responses in high-risk myelodysplastic syndrome following low-dose treatment with the DNA methylation inhibitor 5-aza-2′-deoxycytidine. *Br J Haematol* **2001**;*114*:349–357.

95. Belinsky SA, Klinge DM, Stidley CA, Issa JP, Herman JG, March TH, Baylin SB. Inhibition of DNA methylation and histone deacetylation prevents murine lung cancer. *Cancer Res* **2003**;*63*:7089–7093.

3

Increasing Impact of Micro RNAs in Stem Cell Biology and Medicine

Peter Wernet

Abstract

Micro RNAs (miRNAs) are an abundant group of evolutionarily conserved non-coding RNAs of 20–24 nucleotides (nt) in length. Their processing results from double-stranded RNA precursors, and they guide post-transcriptional gene silencing. Members of the RNAseIII enzyme family help in their processing. Gene silencing, however, is mediated by the highly conserved Argonaute family of proteins. Their first discovery was made in *Caenorhabditis elegans*, where miRNAs guide the correct timing of development. To date, hundreds of miRNAs have been detected in entirely different organisms, including plants, flies, fish, and mammals. MiRNAs are expressed specifically during embryonic development, but they are also critical for cell differentiation and cell function control. However, most of the target mRNAs remain elusive. Thus, their stringent identification, together with an understanding of their individual function as well as within-systems network, will enable new knowledge of gene regulation. In time, this might represent a novel molecular platform for the biopathology of, and hitherto descriptive, genetic abnormalities. Of particular relevance within this context would be the interface between stem cells and various forms of malignant tumors.

3.1
Introduction

During the past few years, molecular biologists have been surprised by the discovery of hundreds of genes that encode small RNA molecules, and which became known by the name of "micro RNAs" (miRNAs).

Micro RNAs are an endogenous class of conserved short RNAs, ranging in size from 20 to 24 nucleotides (nt), and with apparently important roles in regulating gene expression (Lagos-Quintana et al., 2001). They are typically excised from a hairpin (fold-back) RNA structure of 60 to 110 nt (named pre-micro RNA) that is transcribed from a larger primary transcript (named pri-micro RNA) (Bartel,

Stem Cell Transplantation. Biology, Processing, and Therapy.
Edited by Anthony D. Ho, Ronald Hoffman, and Esmail D. Zanjani
Copyright © 2006 WILEY-VCH Verlag GmbH & Co. KGaA, Weinheim
ISBN: 3-527-31018-5

2004). They bind to partially complementary sequences within the 3′-untranslated region (UTR) of target mRNAs and suppress their translation. The hitherto few known target mRNAs are regulators of cell differentiation, cell proliferation, or of developmental blueprints.

Recently developed miRNA microarray techniques have helped to accelerate research into these molecules. However, the function of the vast majority in developmental biology, tissue homeostasis and malignant cell proliferation is still poorly understood.

3.2
Biogenesis of miRNAs

Micro RNAs interact with target mRNAs at specific sites to induce cleavage of the message, or to inhibit translation. Target sites on the 3′-UTRs of human gene transcripts for all currently known 320 mammalian miRNAs have been predicted by various computational approaches (Bentwich et al., 2005), though with differing results and interpretations.

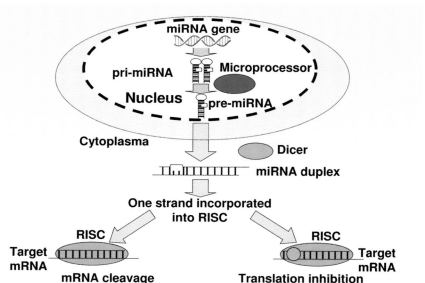

according to P. Melzer, Nature Vol. 435, 9. June 2005

Figure 3.1 The precursor of an miRNA (pri-miRNA) is transcribed in the nucleus. It forms a stem–loop structure that is processed to form another precursor (pre-miRNA) before being exported to the cytoplasm. Further processing of the Dicer protein creates the mature miRNA, one strand of which is incorporated into the RNA-induced silencing complex (RISC). Base pairing between the miRNA and its target directs RISC either to destroy the mRNA or to impede its translation into protein. The initial stem–loop configuration of the primary transcript involves structural clues that have been used to guide searches of genomic sequence for candidate miRNAs. (Modified from Melzer, P.S., *Nature* **2005**, *435*, 745.)

A simplified overview of the production of micro RNAs (Meltzer, 2005) is provided in Figure 3.1.

Micro RNAs frequently are organized into genomic clusters in the cell nucleus, and are primarily transcribed as precursor molecules. The genes for miRNAs are located in exons or introns of noncoding genes (Rodriguez et al., 2004), and also as protein-coding genes (Smalheiser, 2003).

Because splicing and miRNA processing might be coupled, miRNAS and their host mRNAs could be processed at the same time (Baskerville and Bartel, 2005). Certain miRNA genes of mammals are found in repetitive genomic regions, and

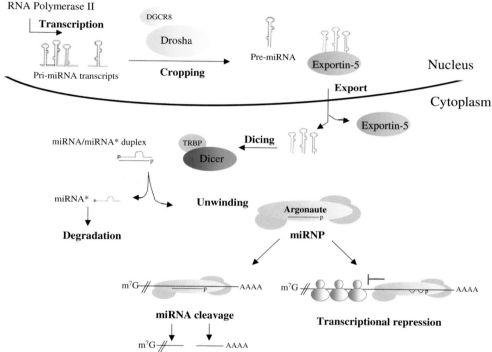

Figure 3.2 Simplified view of post-translational regulation of gene expression, as induced by microRNAs. Primary miRNAs are transcribed by polymerase II, then processed into miRNA precursors by the microprocessor which is a complex consisting of the RNase II enzyme Drosha as well as the RNA-binding protein DGCR8/Pasha. The Exportin 5 receptor transports the miRNA precursor to the cytoplasm. There, the miRNA-precursors are processed further by the Dicer enzyme (an RNase III) into intermediates of miRNA/miRNA duplexes. Dicer interacts with double-stranded RNA-binding protein TRBP for miRNA processing. After unwinding the duplex and degradation of the miRNA, the mature miRNA is incorporated into miRNPs, the main constituents of which are Ago proteins. It is conceivable at this time that miRNPs incorporating Ago2 facilitate the cleavage of complementary target RNAs. In contrast, miRNPs containing other Ago proteins could be involved in translational repression of different target mRNAs. m7G = 7-methyl guanine; AAAA = poly(A) tail. (Modified from Chen, P.Y. and Meister, G., *Biol. Chem.* **2005**, *386*, 1205–1218.)

the transposons might serve as a driving force to create new miRNAs critical for advanced mammalian evolution (Smalheiser and Torvik, 2005).

A more detailed model of the maturation of miRNAs as a stepwise biochemical process is depicted in Figure 3.2.

Micro RNAs are transcribed as long primary transcripts in the nucleus, and are subsequently cleared to produce stem–loop-structured precursor molecules of 70 nt length (pre-miRNAs) by an enzyme named Drosha. This RNase III-type endonuclease contains two RNase III domains and a double-stranded RNA (dsRNA) binding domain; this clears both sites near the base of the stem, leaving a 5'-phosphate and a 2-nt 3' overhang. For efficient processing, Drosha requires nonstructured regions flanking the hairpin (Zeng and Cullen, 2005).

After initial nuclear processing, the pre-miRNA is exported to the cytoplasm by the export receptor Exportin 5. The interaction of Exportin 5 with the miRNA precursor requires a 3' overhang and the stem of the precursor for efficient export. In the cytoplasm, the RNase III enzyme Dicer plays a key role in the further maturation of pre-miRNAS, as well as other sources of long dsRNA. In human cells, Dicer is a cytoplasmic 220-kDa protein that contains two RNase III domains, a dsRNA binding domain, a putative helicase domain, and a PAZ domain. Dicer processes miRNA precursors and generates mature miRNAs.

Dicer recognizes the double-stranded portion and the 2-nt 3' overhang of miRNA precursors, and the cleavage reaction results in a miRNA duplex intermediate comprising the mature miRNA and the opposing strand of the pre-miRNA stem. This short-lived intermediate product is consecutively unwound. As depicted in Figure 3.2, the single-stranded mature miRNA is then incorporated into effector complexes termed miRNPs. This step of unwinding necessitates hydrolysis of ATP, although the Dicer cleavage is ATP-independent.

In contrast to worms and flies, mammals possess only one Dicer enzyme. Consequently, following the Dicer step, miRNAs are incorporated into miRNPs.

The Argonaute protein family is a key component of miRNPs, and has a molecular weight of about 100 kDa. In humans, only one specific Argonaute protein (Argonaute 2) possesses cleavage capacity (Meister et al., 2004).

Further details of the domain structures of proteins involved in human miRNA-guided gene silencing are summarized in Chen and Meister (2005).

3.3
Action Modes of miRNAs

Currently, micro RNA-guided translational regulation is not well understood. Indeed, to date hundreds of miRNAs have been identified, but their mode of action remains elusive.

One of the key characteristics of stem cells is their capacity for self-renewal over long time periods. In this respect, stem cells display similar characteristics to cancer cells, which also are able to evade the "stop" signals of the cell cycle. The critical question, therefore, in stem cell and cancer biology is how cell division is regulated.

The establishment of *in-vitro* translation systems that biochemically recapitulate the mechanism of miRNA-mediated translational control must be developed in order to obtain an understanding of the role of miRNA pathways in regulating the mechanisms that control stem cell division. It could be established that partial complementarity leads to translational repression, whereas fully complementary targets are directed to cleavage.

Thousands of mammalian mRNAs are under selective pressure to maintain 7-nt sites matching miRNAs. These conserved targets are often highly expressed at developmental stages prior to miRNA expression, but their levels fall as the miRNA that targets them begins to accumulate. Nonconserved sites, which outnumber the conserved by ten to one, also mediate repression. As a consequence, genes preferentially expressed at the same time and place as a miRNA have evolved to selectively avoid sites matching the miRNA. This phenomenon of selective avoidance extends to thousands of genes, and enables spatial and temporal specificities of miRNAs to be revealed by finding tissues and developmental stages in which messages with corresponding sites are expressed at lower levels (Farh et al., 2005).

The use of combined bioinformatics and small RNA cloning approaches, together with hundreds of conserved and nonconserved human miRNAs, have recently been reported (Bentwich et al., 2005).

More recently, miRNAs have been found to regulate stem cell division and proliferation (Hatfield et al., 2005) in the model species *Drosophila*. Here, genetic inactivation of Dicer 1 – the RNase III enzyme that processes miRNA precursors to mature miRNAs – results in a significant reduction of stem cell divisions. In dcr $1^{-/-}$ fly mutants the transition from G_1 to S phase is delayed, which suggests the need for miRNA to bypass the normal G_1/S checkpoint.

Analogous miRNA profiling studies have been performed in vertebrates, leading to a variety of miRNAs which act specifically in the developmental stages of the zebrafish (Berezikov and Plasterk, 2005). Mutants carrying a nonfunctional Dicer RNase III enzyme display impaired brain formation and abnormal heart development (Giraldez et al., 2005). Current estimates of the number of mammalian miRNAs range between 250 and 800 (Table 3.1).

Extensive regulation of mammalian miRNAs is not only restricted to embryonic development, but also appears to control differentiation and the specification of most cell types in adult organisms (Krichevsky et al., 2003; Chen et al., 2004; Kuwabara et al., 2004; Suh et al., 2004; Calin et al., 2005; Felli et al., 2005; Förstermann et al., 2005; Hatfield et al., 2005; Li and Carthew, 2005; Lu et al., 2005; Lü et al., 2005; Monticelli et al., 2005; Muljo et al., 2005; O'Donnell et al., 2005; Ramkissoon et al., 2005; Smirnova et al., 2005; Tang et al., 2006). Each of these references provides detailed examples of the enormous biological potential for micro RNAs known to date.

Table 3.1 Estimated numbers of miRNAs in different organisms (Adopted from Chen and Meister, 2005).

Species	miRNAs		References
	Estimated number	Experimentally validated	
Arabidopsis thaliana	379	47	Reinhart et al., 2002; Allen et al., 2004; Jones-Rhoades and Bartel, 2004; Sunkar and Zhu, 2004; Wang et al., 2004
Caenorhabditis elegans	123	114	Ambros et al., 2003; Grad et al., 2003; Lim et al., 2003b
Drosophila melanogaster	78	71	Aravin et al., 2003; Brennecke and Cohen, 2003; Lai et al., 2003
Danio rerio	200	145	Berezikov and Plasterk, 2005; Chen et al., 2005; Giraldez et al., 2005; Wienholds et al., 2005
Xenopus laevis		7	Watanabe et al., 2005
Mus musculus	200-300	214	Lim et al., 2003a, Lim et al., 2005; Bentwich et al., 2005
Homo sapiens	200-800	263	Lim et al., 2003a, Lim et al., 2005; Bentwich et al., 2005

The experimentally validated miRNAs are taken from the miRNA registry (miRBase, Wellcome Trust Sanger Institute; http://microrna.sanger.ac.uk/).

3.4
Potential Function Modes of miRNAs

Following the extensive studies conducted for the cloning of small RNAs, the great relevance of the functional analysis of miRNAs has been recognized, though until now extent of knowledge available regarding the corresponding target RNAs *in vivo* is minimal.

The large majority of these targets were predicted by algorithms based on partial complementarity (Rajewsky and Socci, 2004). By using such bioinformatic calculations, between 2000 and 4000 miRNA targets have been predicted, though certain cellular proteins such as transcription factors, proteins involved in apoptosis and cell cycle regulators, might have been over-represented in these calculations.

Individual miRNAs also may regulate multiple targets (Lin et al., 2003). Furthermore, it is conceivable that individual target mRNAs could be regulated by many different miRNAs, and hence only the interplay of different miRNAs would efficiently inhibit the translation of a target (Krek et al., 2005).

At present, the vast number of predicted mRNA targets available makes it almost impossible to validate specific candidates. Known and experimentally validated miRNA targets are summarized in Table 3.2. Novel experimental approaches will be necessary, aimed at identifying miRNA targets in order to understand the regulatory networks between miRNAs, mRNAs and proteins in cellular networks and intact organisms.

A better understanding of how miRNAs regulate the translation of target-mRNAs will also help to elucidate the involvement of viral miRNAs in health and disease (Pfeffer et al., 2004, 2005; Cai et al., 2005; Sullivan et al., 2005; Omoto and Fuji, 2005).

Table 3.2 Experimentally validated and functionally characterized miRNA targets (from Chen and Meister, 2005).

	miRNA targets		References
	Experimentally confirmed	Function	
Arabidopsis thaliana			
miR-39; miR-171	SCL6 family	Unknown	Llave et al., 2002; Axtell and Bartel, 2005
miR-172	APETALA2; AP2 family	Floral organ identity; floral stem cell proliferation	Chen, 2003; Axtell and Bartel, 2005
miR-JAW; miR-159	TCP4	Leaf development	Palatnik et al., 2003
miR-165; miR-166	IFL1/REV; PHV; PHB	Vascular patterning; organ polarity; leaf polarity	Emery et al., 2003; Juarez et al., 2004; Kidner and Martienssen, 2004; Mallory et al., 2004b; Zhong and Ye, 2004
miR-164	NAC domain gene (CUC1; CUC2)	Embryonic vegetative and floral development	Mallory et al., 2004a
miR-160	ARF10; ARF 16; ARF17	Leaf, flower and root development	Axtell and Bartel, 2005; Mallory et al., 2005
miR-167	ARF6	Unknown	Axtell and Bartel, 2005
Oryza sativa			
OSmiR-408	Planatacynin	Unknown	Sunkar et al., 2005
OSmiR-380	Ser/Thr/Tyr protein kinase	Unknown	Sunkar et al., 2005
OSmiR-444	MADS box transcription factor	Floral organ identity, root development	Sunkar et al., 2005
OSmiR-436	Uncharacterized protein	Unknown	Sunkar et al., 2005
Caenorhabditis elegans			
Lin-4	lin-14; lin-28	Developmental timing	Lee et al., 1993; Wightman et al., 1993
Let-7	lin-41; lin-57; daf-12; pha-4; let-60; hbl-1	Developmental timing	Slack et al., 2000; Lin et al., 2003; Grosshans et al., 2005; Johnson et al., 2005
Lsy-6	cog-1	Neuronal left/right asymmetry	Johnston and Hobert, 2003
miR-273	die-1	Neuronal left/right asymmetry	Chang et al., 2004
Drosophila melanogaster			
bantam	hid	Apoptosis repression	Brennecke et al., 2003
Mus musculus			
miR-1	Hand2; MyoD; Mef2	Cell differentiation/proliferation	Zhao et al., 2005
miR-375	Myotrophin	Glucose-induced insulin secretion	Poy et al., 2004
miR-196	HoxB8; HoxC8; HoxD8; HoxA7	Transcription factors	Yekta et al., 2004
Homo sapiens			
Let-7	Ras	Tumor growth	Johnson et al., 2005
miR-17-5p/miR20a	E2F1	Transcription factor	O'Donnell et al., 2005

With regard to the above-mentioned biological interface between stem cells and cancer, the first results claiming a micro RNA signature associated with prognosis and progression in human chronic lymphocytic leukemia were recently published (Calin et al., 2005). Previously, lymphomas overexpressing miRNAs of the cluster 17-92, as well as c-*myc* oncogene displaying a more severe phenotype of lymphoma compared to tumors over-expressing only c-myc, were reported in a murine model (He et al., 2005; O'Donnell et al., 2005).

Furthermore, it was shown that Ras family proteins – which are frequently overexpressed in lung cancer – carry putative complementary sequences to the miRNA let-7 in their 3'-UTRs (Johnson et al., 2005). The inhibition of let-7 *in vitro* resulted in a dramatic up-regulation of Ras. Moreover, an analysis of lung tumors showed a significant reduction in let-7 expression, suggesting a probable role for let-7 in lung cancer, and possibly also in other types of cancer.

Micro RNAs have also been found to be misregulated in colorectal neoplasias (Michael et al., 2003); thus, it seems feasible that oncogenes are among the critical targets within this context.

3.5
Conclusions

With the recent advent of methods to define miRNAs at the single cell level (Tang et al., 2006), selective miRNA profiling will lead to a clearer understanding of the biology of stem cell pluripotency and molecular control of progenitor fate differentiation. On the other side of the same coin, the diagnostic potential for the assessment of cancer cells and their metastatic appearance will be newly defined through the resultant understanding of components that regulate tumor establishment and growth. Although the knowledge of interactions between the respective miRNAs of embryogenesis, stem cells, viruses and cancer is fairly recent, it is conceivable that multiple blueprints might be developed for novel sequence-specific drugs against miRNA-mediated impairments of affected biological steady states.

Abbreviations

dcr	dicer
dsRNA	double-stranded RNA
miRNA	micro RNA
nt	nucleotide
RISC	RNA-induced silencing complex
UTR	untranslated region

References

Allen, E., Xie, Z., Gustafson, A.M., Sung, G.H., Spatafora, J.W., and Carrington, J.C. (2004). Evolution of microRNA genes by inverted duplication of target gene sequences in *Arabidopsis thaliana*. *Nat. Genet.* 36, 1282–1290.

Ambros, V., Lee, R.C., Lanvaway, A., Williams, P.T., and Jewell, D. (2003). MicroRNAs and other tiny endogenous RNAs in *C. elegans*. *Curr. Biol.* 13, 807–818.

Aravin, A.A., Lagos-Quintana, M., Yalcin, A., Zavola, M., Marks, D., Snyder, B., Gassterland, T., Meyer, J., and Tuschl, T. (2003). The small RNA profile during *Drosophila melanogaster* development. *Dev. Cell* 5, 337–350.

Axtell, M.J. and Bartel, D.P. (2005). Antiquity of microRNAs and their targets in land plants. *Plant Cell* 17, 1658–1673.

Bartel, D.P. (2004). MicroRNAs: genomics, biogenesis, mechanism, and function. *Cell* 116, 281–297.

Baskerville, S. and Bartel, D.P. (2005). Microarray profiling of microRNAs reveals frequent coexpression with neighboring miRNAs and host genes. *RNA* 11, 241–247.

Bentwich, I., Avniel, A., Karov, Y., Aharonov, R., Gilad, S., Barad, O., Barzilai, A., Einat, P., Einav, U., Meiri, E., et al. (2005). Identification of hundreds of conserved and nonconserved human microRNAs. *Nat. Genet.* 37, 766–770

Berezikov, E. and Plasterk, R.H. (2005). Camels and zebrafish, viruses and cancer: a microRNA update. *Hum. Mol. Genet.* 14 Spec. No. 2: R183–R190.

Brennecke, J. and Cohen, S.M. (2003). Towards a complete description of the microRNA complement of animal genomes. *Genome Biol.* 4, 228.

Brennecke, J., Hipfner, D.R., Stark, A., Russell, R.B., and Cohen, S.M. (2003). Bantam encodes a developmentally regulated microRNA that controls cell proliferation and regulates the proapoptotic gene hid in *Drosophila. Cell* 113, 25–36.

Cai, X., Lu, S., Zhang, Z., Gonzalez, C.M., Damania, B., and Cullen, B.R. (2005). Kaposi's sarcoma-associated herpesvirus expresses an array of viral microRNAs in latently infected cells. *Proc. Natl. Acad. Sci. USA* 102, 5570–5575.

Calin, G.A., Ferracin, M., Cimmino, A., Gianpiero, di L., Shimizu, M., Wojcik, S.W., et al. (2005). A MicroRNA signature associated with prognosis and progression in chronic lymphocytic leukemia. *N. Engl. J. Med.* 353, 1793–1801

Chang, S., Johnston, R.J., Jr., Frokjaer-Jensen, C., Locjery, S., and Hobert, O. (2004). MicroRNAs act sequentially and asymmetrically to control chemosensory laterality in the nematode. *Nature* 430, 785–789.

Chen, C.Z., Li, L., Lodish, H.F., and Bartel, D.P. (2004). MicroRNAs modulate hematopoietic lineage differentiation. *Science* 303(5654), 83–86. Epub 2003 December 4.

Chen, P.C., Manninga, H., Slanchev, K., Chien, M., Russo, J.J., Ju, J., Sheridan, R., John, B., Marks, D.S., Gaidatzis, D., et al. (2005). The development miRNA profiles of zebrafish as determined by small RNA cloning. *Genes Dev.* 19, 1288–1293.

Chen, P.Y. and Meister, G. (2005). microRNA-guided postranscriptional gene regulation. *Biol. Chem.* 386, 1205–1218

Chen, X. (2003). A microRNA as a translational repressor of APE-TALA2 in *Arabidopsis* flower development. *Science* 303, 2022–2025.

Emery, J.F., Floyd, S.K., Alvarez, J., Eshed, Y., Hawker, N.P., Izhaki, A., Baum, S.F., and Bowman, J.L. (2003). Radial patterning of *Arabidopsis* shoots by class III HD-ZIP and KANA-DI genes. *Curr. Biol.* 13, 1768–1774.

Farh, K.K., Grimson, A., Jan, C., Lewis, B.P., Johnston, W.K., Lim, L.P., Burge, C.B., and Bartel, D.P. (2005). The widespread impact of mammalian MicroRNAs on mRNA repression and evolution. *Science* 310, 1817–1821. Epub 2005 November 24.

Felli, N., Fontana, L., Pelosi, E., Botta, R., Bonci, D., Facchiano, F., Liuzzi, F., Lulli, V., Morsilli, O., Santoro, S., Valtieri, M., Calin, G.A., Liu, C.G., Sorrentino, A., Croce, C.M., and Peschle, C. (2005). MicroRNAs 221 and 222 inhibit normal erythropoiesis and erythroleukemic cell growth via kit receptor down-modulation. *Proc. Natl. Acad. Sci. USA* 102(50), 18081–18086. Epub 2005 December 5.

Förstemann, K., Tomari, Y., Du, T., Vagin, V.V., Denli, A.M., Bratu, D.P., Klattenhoff, C., Theurkauf, W.E., and Zamore, P.D. (2005). Normal microRNA maturation and germline stem cell maintenance requires Loquacious, a double-stranded RNA-binding domain protein. *PLoS Biol* 3, e236. Epub 2005 May 24.

Giraldez, A.J., Cinalli, R.M., Glasner, M.E., Enright, A.J., Thomson, J.M., Baskerville, S., Hammond, S.M., Bartel, D.P., and Schier, A.F. (2005). MicroRNAs regulate brain morphogenesis in zebrafish. *Science* 308, 833–838.

Grad, Y., Aach, J., Hayes, G.D., Reinhart, B.J., Church, G.M., Ruvkun, G., and Kim, J. (2003). Computational and experimental identification of *C. elegans* microRNAs. *Mol. Cell* 11, 1253–1263.

Grosshans, H., Johnson, T., Reinert, K.L., Gerstein, M., and Slack, F.J. (2005). The temporal patterning micro RNA let-7 regulates several transcription factors at the larval to adult transition in *C. elegans. Dev. Cell* 8, 321–330.

Hatfield, S.D., Shcherbata, H.R., Fischer, K.A., Nakahara, K., Carthew, R.W., and Ruohola-Baker, H. (2005). Stem cell division is regulated by the microRNA pathway. *Nature* 435, 974–978. Epub 2005 June 8.

He, L., Thomson, J.M., Hemann, M.T., Hernando-Monge, E., Mu, D., Goodson, S.,

Powers, S., Cordon-Cardo, C., Lowe, S.W., Hannon, G.J., and Hammond, S.M. (2005). A microRNA polycistron as a potential human oncogene. *Nature* 435, 828–833.

Jones-Rhoades, M.W. and Bartel, D.P. (2004). Computational identification of plant microRNAs and their targets, including a stress-induced miRNA. *Mol. Cell* 14, 787–799.

Johnson, S.M., Grosshans, H., Shingara, J., Borom, M., Jarvis, R., Cheng, A., Labourier, E., Reinert, K.L., Brown, D., and Slack, F.J. (2005). RAS is regulated by the let-7 microRNA family. *Cell* 120, 635–647.

Johnston, R.J. and Hobert, O. (2003). A microRNA controlling left/right neuronal asymmetry in *Caenorhabditis elegans*. *Nature* 426, 845–849.

Juarez, M.T., Kui, J.S., Thomas, J., Heller, B.A., and Timmermans, M.C. (2004). microRNA-mediated repression of rolled leaf1 specifies maize leaf polarity. *Nature* 428, 84–88.

Kidner, C.A. and Martienssen, R.A. (2004). Spatially restricted microRNA directs leaf polarity through Argonaute1. *Nature* 428, 81–84.

Krek, A., Grun, D., Poy, M.N., Wolf, R., Rosenberg, L., Epstein, E.J., MacMenamin, P., da Piedade, I., Gunsalus, K.C., Stoffel, M., and Rajewsky, N. (2005). Combinatorial microRNA target predictions. *Nat. Genet.* 37, 495–500.

Krichevsky, A.M., King, K.S., Donahue, C.P., Khrapko, K., and Kosik, K.S. (2003). A microRNA array reveals extensive regulation of microRNAs during brain development. *RNA* 9, 1274-81. Erratum in: *RNA* (2004) 10(3), 551.

Kuwabara, T., Hsieh, J., Nakashima, K., Taira, K., and Gage, F.H. (2004). A small modulatory dsRNA specifies the fate of adult neural stem cells. *Cell* 116, 779–793.

Lagos-Quintana, M., Rauhut, R., Meyer, J., Borkhardt, A., and Tuschl, T. (2003). New microRNAs from mouse and human. *RNA* 9, 175–179.

Lai, E.C., Tomancak, P., Williams, R.W., and Rubin, G.M. (2003). Computational identification of *Drosophila* microRNA genes. *Genome Biol.* 4, R42.

Lee, R.C., Feinbaum, R.L., and Ambros, V. (1993). The *C. elegans* heterochronic gene lin-4 encodes small RNAs with antisense complementarity to lin-14. *Cell* 75, 843–854.

Li, X. and Carthew, R.W. (2005). A microRNA mediates EGF receptor signaling and promotes photoreceptor differentiation in the *Drosophila* eye. *Cell* 123, 1267–1277.

Lim, L.P., Glasner M.E., Yekta, S., Burge, C.B., Bartel, D.P. (2003a). Vertebrate micro RNA genes. *Science* 299, 1540.

Lim, L.P., Lau, N.C., Weinstein, E.G., Abdelhakim, A., Yetka, S., Rhoades, M.W., Burge, C.B., and Bartel, D.P. (2003b). The microRNAs of *Caenorhabditis elegans*. *Genes Dev.* 17, 991–1008.

Lim, L.P., Lau, N.C., Garrett-Engele, P., Grimson, A., Schelter, J.M., Castle, J., Bartel, D.P., Linsley, P.S., and Johnson, J.M. (2005). Microarray analysis shows that some microRNAs downregulate large numbers of target mRNAs. *Nature* 433, 769–773.

Lin, S.Y., Johnson, S.M., Abraham, M., Vella, M.C., Pasquinelli, A., Gamberi, C., Gottlieb, E., and Slack, F.J. (2003). The *C. elegans* hunchback homolog, hbl-1, controls temporal patterning and is a probable microRNA target. *Dev. Cell* 4, 639–650.

Llave, C., Xie, Z., Kasschau, K.D., and Carrington, J.C. (2002). Cleavage of Scarecrow-like mRNA targets directed by a class of *Arabidopsis* miRNA. *Science* 297, 2053–2056.

Lü, J., Qian, J., Chen, F., Tang, X.Z., Li, C., and Wellington, V.C. (2005). Differential expression of components of the microRNA machinery during mouse organogenesis. *Biochem. Biophys. Res. Commun.* 334, 319–323.

Lu, J., Getz, G., Miska, E.A., Alvarez-Saavedra, E., Lamb, J., Peck, D., Sweet-Cordero, A., Ebert, B.L., Mak, R.H., Ferrando, A.A., Downing, J.R., Jacks, T., Horvitz, H.R., and Golub, T.R. (2005). MicroRNA expression profiles classify human cancers. *Nature* 435, 834–838.

Mallory, A.C., Dugas, D.V., Bartel, D.P., and Bartel, B. (2004a). MicroRNA regulation of NAC-domain targets is required for proper formation and separation of adjacent embryonic, vegetative and floral organs. *Curr. Biol.* 14, 1035–1046.

Mallory, A.C., Reinhart, B.J., Jones-Rhoades, M.W., Tang, G., Zamore, P.D., Barton, M.K., and Bartel, D.P. (2004b). MicroRNA control of PHABULOSA in leaf development: importance of pairing to the microRNA 5 region. *EMBO J.* 23, 3356–3364.

Mallory, A.C., Bartel, D.P., and Bartel, B. (2005). MicroRNA-directed regulation of *Arabidopsis* AUXIN RESPONSE FACTOR 17 is essential for proper development and modulates expression of early auxin response genes. *Plant Cell* 17, 1360–1375.

Meister, G., Landthaler, M., Dorsett, Y., and Tuschl, T. (2004). Sequence-specific inhibition of microRNA- and siRNA-induced RNA silencing. *RNA* 10, 544–550.

Michael, M.Z., O'Connor, S.M., van Holst Pellekaan, N.G., Young, G.P., and James, R.J. (2003). Reduced accumulation of specific microRNAs in colorectal neoplasia. *Mol. Cancer Res.* 1, 882–891.

Monticelli, S., Ansel, K.M., Xiao, C., Socci, N.D., Krichevsky, A.M., Thai, T.H., Rajewsky, N., Marks, D.S., Sander, C., Rajewsky, K., Rao, A., and Kosik, K.S. (2005). MicroRNA profiling of the murine hematopoietic system. *Genome Biol.* 6, R71.

Muljo, S.A., Ansel, K.M., Kanellopoulou, C., Livingston, D.M., Rao, A., and Rajewsky, K. (2005). Aberrant T cell differentiation in the absence of Dicer. *J. Exp. Med.* 202(2), 261–269. Epub 2005 July 11.

O'Donnell, K.A., Wentzel, E.A., Zeller, K.I., Dang, C.V., and Mendell, J.T. (2005). c-Myc-regulated microRNAs modulate E2F1 expression. *Nature* 435, 839–843.

Omoto, S. and Fujii, Y.R. (2005). Regulation of human immunodeficiency virus 1 transcription by nef microRNA. *J. Gen. Virol.* 86, 751–755.

Palatnik, J.F., Allen, E., Wu, X., Schommer, C., Schwab, R., Carrington, J.C., and Weigel, D. (2003). Control of leaf morphogenesis by microRNAs. *Nature* 425, 257–263.

Pfeffer, S., Zavolan, M., Grasser, F.A., Chien, M., Russo, J.J., Ju, J., John, B., Enright, A.J., Marks, D., Sander, C., and Tuschl, T. (2004). Identification of virus-encoded microRNAs. *Science* 304, 734–736.

Pfeffer, S., Sewer, A., Lagos-Quintana, M., Sheridan, R., Sander, C., Grasser, F.A., van Dyk, L.F., Ho, C.K., Shuman, S., Chien, M., et al. (2005). Identification of microRNAs of the herpesvirus family. *Nat. Methods* 2, 269–276.

Poy, M.N., Eliasson, L., Krutzfeldt, J., Kuwajima, S., Ma, X., Macdonald, P.E., Pfeffer, S., Tuschl, T., Rajewsky, N., Rorsman, P., and Stoffel, M. (2004). A pancreatic islet-specific microRNA regulates insulin secretion. *Nature* 432, 226–230.

Rajewsky, N. and Socci, N.D. (2004). Computational identification of microRNA targets. *Dev. Biol.* 267, 529–535.

Ramkissoon, S.H., Mainwaring, L.A., Ogasawara, Y., Keyvanfar, K., Philip McCoy, J., Jr., Sloand, E.M., Kajigaya, S., and Young NS. (2005). Hematopoietic-specific microRNA expression in human cells. *Leuk. Res.* 11, 2278–2283.

Reinhart, B.J., Weinstein, E.G., Rhoades, M.W., Bartel, B., and Bartel, D. (2002). MicroRNAs in plants. *Genes Dev.* 16, 1616–1626.

Rodriguez, A., Griffiths-Jones, S., Ashurst, J.L., and Bradley, A. (2004). Identification of mammalian microRNA host genes and transcription units. *Genome Res.* 14, 1902–1910.

Slack, F.J., Basson, M., Liu, Z., Ambros, V., Horvitz, H.R., and Ruvkun, G. (2000). The lin-41 RBCC gene acts in the *C. elegans* heterochronic pathway between the let-7 regulatory RNA and the LIN-29 transcription factor. *Mol. Cell* 5, 659–669.

Smalheiser, N.R. (2003). EST analyses predict the existence of a population of chimeric microRNA precursor-mRNA transcripts expressed in normal human and mouse tissues. *Genome Biol.* 4, 403.

Smalheiser, N.R. and Torvik, V.I. (2005). Mammalian microRNAs derived from genomic repeats. *Trends Genet.* 21, 322–326.

Smirnova, L., Grafe, A., Seiler, A., Schumacher, S., Nitsch, R., and Wulczyn, F.G. (2005). Regulation of miRNA expression during neural cell specification. *Eur. J. Neurosci.* 21, 1469–1477.

Suh, M.-R., Lee, Y., Kim, K.Y., Kim, S.-K., Moon, S.-H., Lee, J.Y., Cha, K.-Y., Chung, H.M., Yoon, H.S., Moon, S.Y., Kim, V.N., and Kim, S.-S. (2004). Human embryonic stem cells express a unique set of micro RNAs. *Dev. Biol.* 270, 488–498.

Sullivan, C.S., Grundhoff, A.T., Tevethia, S., Pipas, J.M., and Ganem, D. (2005). SV40-encoded microRNAs regulate viral gene

expression and reduce susceptibility to cytotoxic T cells. *Nature* 435, 682–686.

Sunkar, R. and Zhu, J.K. (2004). Novel and
stress-regulated microRNAs and other
small RNAs from *Arabidopsis*. *Plant Cell* 16,
2001–2019.

Sunkar, R., Girke, T., Jain, P.K., and Zhu, J.K.
(2005). Cloning and characterization of microRNAs from rice. *Plant Cell* 17, 1397–
1411.

Tang, F., Hajkova, P., Barton, S.C., Lao, K.,
and Surani, M.A. (2006). MicroRNA expression profiling of single whole embryonic stem cells. *Nucleic Acids Res.* 34(2), e9.

Wang, X.J., Reyes, J.L., Chua, N.H., and Gassterland, T. (2004). Prediction and identification of *Arabidopsis thaliana* microRNAs
and their mRNA targets. *Genome Biol.* 5,
R65.

Watanabe, T., Takeda, A., Mise, K., Okuno, T.,
Suzuki, T., Minami, N., and Imai, H.
(2005). Stage-specific expression of micro-
RNAs during *Xenopus* development. *FEBS
Lett.* 579, 318–324.

Wienholds, E., Kloosterman, W.P., Miska, E.,
Alvarez-Saavedra, E., Berezikov, E., de
Bruijn, E., Horvitz, H.R., Kauppinen, S.

and Plasterk, R.H. (2005). MicroRNA expression in zebrafish embryonic development. *Science* 309, 310–311.

Wightman, B., Ha, I., and Ruvkun, G. (1993).
Posttranscriptional regulation of the heterochronic gene lin-14 by lin-4 mediates
temporal pattern formation in *C. elegans*.
Cell 75, 855–862.

Yekta, S., Shi, I.H., and Bartel, D.P. (2004).
MicroRNA-directed cleavage of HOXB8
mRNA. *Science* 304, 594–596.

Zeng, Y. and Cullen, B.R. (2005). Efficient
processing of primary microRNA hairpins
by Drosha requires flanking non-structured
RNA sequences. *J. Biol. Chem.* 280, 27595–
27603.

Zhao, Y., Samal, E., and Srivastava, D. (2005).
Serum response factor regulates a muscle-
specific microRNA that targets Hand2 during cardiogenesis. *Nature* 436, 214–220.

Zhong, R. and Ye, Z.H. (2004). Amphivasal
vascular bundle 1, a gain-of-function mutation of the IFL1/REV gene, is associated
with alterations in the polarity of leaves,
stems and carpels. *Plant Cell Physiol.* 45,
369–385.

Part II
Standardization and Quality Assurance of Stem Cell Preparations

Stem Cell Transplantation. Biology, Processing, and Therapy.
Edited by Anthony D. Ho, Ronald Hoffman, and Esmail D. Zanjani
Copyright © 2006 WILEY-VCH Verlag GmbH & Co. KGaA, Weinheim
ISBN: 3-527-31018-5

4

Novel Strategies for the Mobilization of Hematopoietic Stem Cells

Stefan Fruehauf, Timon Seeger, and Julian Topaly

Abstract

Stem cell transplantation, whether autologous or allogeneic, improves the outcome of patients with a number of hematological malignancies or solid tumors. A relevant proportion of these patients are excluded from this treatment because sufficient numbers of hematopoietic stem cells cannot be obtained by standard cytokine-assisted mobilization. In this chapter we review the physiology of the blood stem cell mobilization and discuss the role of adhesion molecules such as integrins and selectins, chemokines and their ligands such as SDF-1α and CXCR4, and proteolytic enzymes. Based on this knowledge, several innovative pharmacologic approaches have been proposed to boost the stem cell harvest. Some of these approaches (e.g., SDF-1 peptide analogues, C3a receptor agonist, and GroβT) remain the subject of preclinical development, while others (e.g., chemokine receptor ligand AMD3100) have recently been introduced into clinical trials and have already delivered promising results. It appears possible to successfully harvest peripheral blood progenitor cells (PBPCs) in poor mobilizers and to reduce the number of collections required in the remaining PBPC donors.

4.1
Physiology of Blood Stem Cell Mobilization

Much is known about how hematopoietic stem cells (HSC) traffic from marrow to blood. Granulocyte colony-stimulating-factor (G-CSF) increases the release of granulocytic proteases that cleave adhesion molecules; antibodies to very late antigen-4 (VLA-4) and antagonists of CXCR4 directly interfere with interactions of HSC and microenvironmental cells, as examples. However, little is known about why HSC circulate. Do HSC continuously exit marrow, circulate in blood and return to marrow? Mobilization could protect stem cells from toxic injury. Do HSC exit the marrow when there is no niche – that is – if their survival or persistence is not supported by the marrow microenvironment? Mobilization

Stem Cell Transplantation. Biology, Processing, and Therapy.
Edited by Anthony D. Ho, Ronald Hoffman, and Esmail D. Zanjani
Copyright © 2006 WILEY-VCH Verlag GmbH & Co. KGaA, Weinheim
ISBN: 3-527-31018-5

could be a death pathway, a mechanism that regulates stem cell number. In order to discriminate between these possibilities, the group of Abkowitz tested the transfer of HSC between parabiotic mice of different strains. Their results suggest that marrow homeostasis is maintained by the retention of HSC derived from re-plication events, and not by the engraftment of circulating HSC. Mobilization, therefore, may be a mechanism complementing "asymmetric division" [1]. If a replication event occurs and a niche is not available, one offspring mobilizes or dies. After cytokine exposure, many HSC exit marrow, travel through the blood and then repopulate empty marrow niches. Abkowitz's group went on to quantify the number of HSC (niche capacity) in mouse and cat. Surprisingly, the total number of HSC in mouse and cat was similar (11 200 to 22 400), and conse-quently it was concluded that the number of human HSC may be evolutionarily conserved and comparable.

In the murine system, loci on chromosomes 2 and 11 have been linked to an inter-strain variation in stem cell mobilization proficiency. The locus on chromo-some 2 acts most likely in cells of the bone marrow (BM) microenvironment, whereas the locus on chromosome 11 functions in the mobilized hematopoietic progenitor cells [2]. Candidate target genes for which the expression is regulated by the locus on chromosome 11 were identified by the group of van Zant, and include a novel gene with high homology (94 %) to a member of a recently dis-covered novel human cytokine family, which is predominantly expressed in the vascular epithelium and a protein with an intra-membrane serine protease activ-ity.

4.1.1
Stromal-Derived Factor-1 Alpha (SDF-1α)/CXCR4 Pathway

In the process of mobilization, the SDF-1α(CXLC12) /CXCR4 pathway has been shown to play a key role in the release of HSC (Fig. 4.1). SDF-1α-induced forma-tion of specific podia by hematopoietic cells is associated with their egression from the bone marrow [3]. Recent data imply that G-CSF-induced mobilization also occurs via the SDF-1α/CXCR4 pathway. In contrast to the current opinion of adhesion molecule cleavage by proteases as a major process in HSC mobiliza-tion, the group of Link recently published data that show HSC mobilization by G-CSF was normal in metalloproteinase-9 (MMP-9)-deficient mice, mice with com-bined neutrophil elastase and cathepsin G (NExCG) deficiency, or mice lacking dipeptidyl peptidase I, an enzyme required for the functional activation of many hematopoietic serine proteases. Compared to G-CSF treatment of wild-type mice, the vascular cell adhesion molecule-1 (VCAM-1) expression on BM stromal cells was not decreased in NExCG-deficient mice, indicating that VCAM-1 cleavage is not required for efficient HSC mobilization [4]. However, SDF-1α expression was down-regulated in spite of a lack of proteases. It was pres-ented that on SDF-1α, G-CSF acts primarily through a transcriptional or RNA sta-bility mechanism rather than a proteolytic mechanism [5]. On the other hand, proteases accumulate in the BM and the SDF-1α-induced, increased trans-stromal

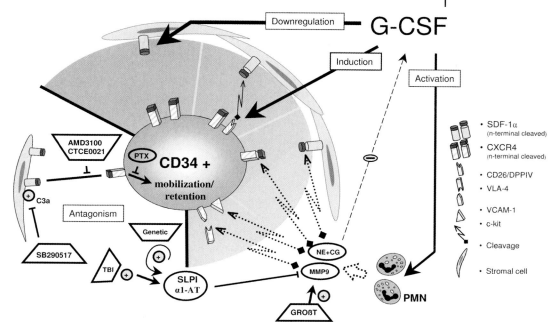

Figure 4.1 Current understanding of the physiology of peripheral blood progenitor cell (PBPC) mobilization and of pharmacological interventions: Role of the SDF-1α–CXCR4 interaction. PBPC are mobilized by: (1) a downregulation of SDF-1α or a decrease in RNA stability; (2) induction of CD26 and subsequent N-terminal cleavage of SDF-1α; (3) neutrophil elastase (NE), cathepsin G (CG) and matrix metalloproteinase-9 (MMP9)-induced cleavage of adhesion molecules. New mobilization agents counteract stem cell retention in the bone marrow: (1) SDF-1α upregulation by C3a is blocked by SB290517; (2) AMD3100, CTCE0021 directly interfere with SDF-1α–CXCR4 interaction. Pertussis toxin (PTX) is used experimentally to block intracellular CXCR4 signaling. TBI = total body irradiation.

migration of MO7e leukemia cells can be reduced by the broad-spectrum MMP inhibitor, GM6001 [6].

Other data supporting a major role of proteases are presented by van Os et al. As previously reported, low-dose total body irradiation (TBI) (0.5 Gy) completely prevents stem cell mobilization induced by interleukin (IL)-8 or by G-CSF [7]. In studying mechanisms underlying this phenomenon, van Os et al. now show a radiation dose-dependent inhibition of elastase activity in BM extracts from irradiated mice in comparison with BM extracts from unirradiated controls. Hence, they concluded that protease inhibitors (i.e., alpha1-antitrypsin (alpha1-AT) and secretory leukocyte protease inhibitor (SLPI) are induced by low-dose irradiation [8]. In another model, the same group established that the number of duplications of the gene for one of these inhibitors, alpha1-AT, inversely correlates with mobilization potential. Strains with five duplications (C57Bl/6, C3H) show less progenitor cell mobilization in response to G-CSF than strains with only three duplications (Balb/c, AKR, DBA). Thus, alpha1-AT and SLPI are candidate modulators of radiation-induced inhibition of stem cell mobilization, and possibly for stem cell mobilization in general. These data suggest a complex model in which both protease-dependent and -independent pathways may contribute to HSC mobilization.

It has been shown that binding of HSC to stroma depends on the presence of heparan sulfate proteoglycan in the extracellular matrix (ECM) secreted by the stromal cells [9]. Heparanase releases ECM-bound molecules to promote cell invasion, adhesion, tissue vascularization and morphogenesis. The primitive, undifferentiated stem and progenitor cell populations in both the BM and the peripheral blood (PB) were increased in transgenic mice overexpressing heparanase compared to control wild-type mice, indicating an involvement of heparanase in steady-state homeostasis. Heparanase-overexpressing cells from transgenic mice, as well as from transfected cell lines, appeared to have increased migration and homing potential [10]. TBI resulted in a higher activity of heparanase in the BM, but this was not due to *de-novo* synthesis. In conformity with this, the stimulation of acute myelogenous leukemia (AML) U937 cells with SDF-1α led to a decline in intracellular heparanase, suggesting that SDF-1α is involved in heparanase translocation and eventual secretion. SDF-1α levels in the BM of heparanase-overexpressing transgenic mice were significantly reduced compared to those on control wild-type mice. These data suggest that there is an important role for the cross-talk between SDF-1α and heparanase in the process of heparanase-mediated cell migration and development.

Data presented by the group of Lapidot provided evidence for a cooperative role of CD44 and its major ligand hyaluronic acid (HA) in the trafficking of human CD34+ stem/progenitor cells to BM. Homing was blocked by anti-CD44 monoclonal antibodies (mAbs) or by soluble HA, and it was significantly impaired after intravenous injection of hyaluronidase. HSC migrating towards a SDF-1α gradient on immobilized HA appeared to have CD44 concentrating at the pseudopodia at the leading edge [11]. CD44 and HA may play a key role in SDF-1α-dependent transendothelial migration.

Further exploration of the signaling of SDF-1α showed that overnight culture of CD34+ cells with simvastatin, a HMG-CoA reductase inhibitor suppressed migration of CD34+ bone marrow cells towards a SDF-1α gradient [12]. In MO7e cells, a hematopoietic growth factor-dependent myeloid cell line, a reduction of SDF-1α-induced activation of Rho, could be demonstrated after pretreatment with simvastatin. Inhibition of chemotaxis was reversed by geranylgeranylpyrophosphate, but not by farnesylpyrophosphate. Both are isoprenoids downstream of mevalonate, the product of HMG-CoA reductase contributing to lipid modification for signaling molecules such as small and heterotrimeric GTP-binding proteins. Statins target geranylgeranylation of Rho and subsequently suppress chemotaxis of hematopoietic cells.

Administration of Flt3 ligand (FL) induces hematopoietic stem cell mobilization *in vivo*, suggesting that FL may enhance HSC motility. Fukuda et al. showed that the FL/Flt3 axis regulates migration of normal and transformed hematopoietic cells, and that there is functional cross-talk between the FL/Flt3 axis and the SDF-1α/CXCR4 pathway that regulates HSC trafficking [13].

It was reported that dextran sulfate (DEX) increased plasma levels of SDF-1α and HSCs in mice and nonhuman primates. Hayakawa et al. examined the effects of preconditioning with SDF-1α and DEX on the homing efficiency of HSCs during BM transplantation. DEX appeared to have multiple functions, including the induction of SDF-1α and other molecules required for homing [14].

The engraftment phenotype is regulated by Rho GTPases Rac1 and Rac2. Rac1, which is a downstream mediator of CXCR4 signals, regulates the engraftment phenotype, likely via the control of both homing and proliferation, while Rac2 is critical for regulation of adhesion and survival, as shown by Cancelas et al. [15]. Thus, Rac1 and Rac2 regulate both specific and overlapping signaling pathways to mediate actin assembly and growth/survival. Rac proteins appear to be critical molecular switches controlling engraftment and mobilization of HSC. For example, compared to wild-type cells, $Rac2^{-/-}$ lineage-negative, Sca-1-positive and c-Kit-negative (LSK) cells seem to have nearly equivalent early engraftment potential but exhibit decreased long-term engraftment. Jansen et al. described defective interactions with hematopoietic microenvironment which also lead to decreased proliferation and an increased rate of apoptosis of $Rac2^{-/-}$ LSK cells in response to stress (four days after 5-fluorouracil treatment) [16].

In a recent study, Papayannopoulou et al. explored the *in-vivo* role of G-protein signaling on the redistribution or mobilization of hematopoietic stem/progenitor cells (HPCs). A single injection of pertussis toxin (Ptx), which blocks G_i-protein signaling, elicited a long-lasting leukocytosis and a progressive increase in circulating hematopoietic progenitor cells in mice. In addition to normal mice, mice genetically deficient in monocyte chemotactic protein 1 (MCP-1), MMP-9, G-CSF receptor, β_2 integrins, or selectins responded to Ptx treatment, suggesting an independence of Ptx-response from the expression of these molecules. As Ptx-mobilized kit+ cells displayed virtually no response to SDF-1α *in vitro*, these data suggest that disruption of CXCR4/SDF-1α signaling may be the underlying mechanism of Ptx-induced mobilization and indirectly reinforce the notion that active signaling through this pathway is required for continuous retention of cells within the bone marrow [17].

Engraftment capacity can be modulated via the SDF-1α/CXCR4 pathway. FTY720, an agonist of the sphingosine 1-phosphate receptor (S1PR) expressed on CD34+ HSC, modulates the effects of SDF-1α [18]. Interestingly, FTY720-incubated hematopoietic cells that express similar amounts of CXCR4 hold a higher SDF-1α-dependent transendothelial migration potential as well as elevated proliferation rates *in vitro*. In NOD/SCID mice, the number of CD34+/CD38– cells that home to the BM after 18 h is significantly raised by pretreatment of animals and cells with FTY720, and this tends to result in improved engraftment. Thus, activation of S1PR by FTY720 might increase CXCR4 function in HSC both *in vitro* and *in vivo*.

The group of Ratajczak proposed a strategy to enhance engraftment which is based on priming of harvested BM or cord blood (CB) HPC for their responsiveness to an SDF-1α gradient. It was reported that both platelet-derived microparticles [19] and the anaphylatoxin C3a [20] enhance the chemotactic responses of CD34+ cells to SDF-1α. Moreover, a similar effect was observed after CD34+ cells were preincubated with supernatants obtained from leukapheresis products of patients mobilized with G-CSF [21] or several other molecules such as soluble VCAM-1, ICAM-1 (intercellular adhesion molecule-1), urokinase receptor (uPAR), thrombin, HA, fibrinogen and fibronectin. Moreover, an inhibitor of cell membrane raft formation (methyl-β-cyclodextrin), by inhibiting the priming effect of the CXCR4 receptor by these factors, indicated that this effect is dependent on CXCR4 being present in lipid rafts. Mice transplanted with cells primed using molecules that promote the incorporation of CXCR4 into membrane lipid rafts engrafted better after transplantation [22].

Adhesion molecules are known to play an exuberant role in the process of mobilization and homing. CD34+ HSC attachment to the BM implies not only cell–cell but also cell–ECM interactions, depending on the rheologic conditions present under static and flow conditions. Exposure of the stromal cells to G-CSF notably increased the expression of VCAM-1 on the cell surface and promoted the activation of p38 MAPK – effects which were paralleled by an augmented adhesion of CD34+ peripheral blood progenitor cells (PBPCs) under flow conditions [23].

Several factors were identified which influenced mobilization and the number of PBPC released from the BM. Flow cytometry served to detect differences in expression of VLA-4, L-selectin, lymphocyte function-associated antigen-1 (LFA-1), platelet-endothelial cell adhesion molecule-1 (PECAM-1), CD44, and G-CSF receptor in CD34+/CD45+ cells among poor mobilizers versus good mobilizers. In conclusion, it was assumed that the extent of expression for VLA-4 and LFA-1 is correlated with the mobilization efficacy in healthy donors with a higher expression in poor mobilizers [24].

More recently, the role of CD26 (DPPIV/dipeptidylpeptidase IV) for mobilization and homing has been studied. CD26 cleaves dipeptides from the N-terminus of polypeptide chains containing the appropriate recognition sequence. CD26 has the ability to cleave the chemokine SDF-1α at its position-2 proline, generating an inactive SDF-1α. The group of Christopherson recently reported that inhibition [25] or loss [26] of CD26 resulted in a reduction in G-CSF-induced mobilization, suggesting that CD26 was a component of G-CSF-induced mobilization and that SDF-1α was a potential downstream target. Based on the theory that CD26 has the ability to regulate the cellular response of HSCs to SDF-1α, these authors hypothesized that endogenous CD26 activity may act as a negative regulator of homing/engraftment. In order to test this hypothesis, they utilized *in-vivo* CD45 congenic mouse models of transplantation to evaluate homing and engraftment efficiency [27] and showed that CD26 negatively regulates short-term homing, long-term engraftment, and competitive repopulation of HSC, as well as survival of mice during HSC transplantation. Suppressing CD26 activity, by using CD26 inhibitors and/or CD26–/– cells, may increase the efficiency of HSC settlement and growth in the transplanted recipient's BM.

A recent study by Selleri et al. addressed the role of urokinase-type plasminogen-activator receptor (uPAR) in cell–cell and cell–ECM adhesion [28]. uPAR has a three-domain structure and is anchored to the cell membrane through a GPI anchor. It lacks a transducing cytoplasmic tail, and probably activates the intracellular signaling pathways by interacting with other surface molecules. It interacts with and regulates the activity of a number of cell adhesion molecules including the integrins LFA-1 and VLA-4 and L-selectin. When uPAR is cleaved at the at the GPI anchor by the endogenous phospholipase D, soluble uPAR (suPAR) is released from the cell membrane. Healthy donors given G-CSF have an up-regulated uPAR on circulating neutrophils and monocytes and show high levels of suPAR in plasma in association with increased suPAR shedding. A cleaved form of suPAR can induce chemotaxis of CD34+ cells and down-regulate CXCR4, which may play an important role in HSC mobilization into the peripheral blood.

4.2
Innovative Agents for PBPC Mobilization

4.2.1
AMD-3100

AMD3100 is a reversible inhibitor of CXCR4/SDF-1α binding which mobilizes CD34+ cells into the PB. In normal volunteers, the administration of AMD-3100 after 4–5 days of G-CSF resulted in a 3- to 3.5-fold increase in circulating CD34+ cells [29].

Since CXCR4 has been implicated in the homing of HSC to the BM, there is concern that AMD3100-mobilized HSC may have reduced engraftment potential. To address this possibility, Devine et al. performed a competitive repopulation assay with AMD3100-mobilized HSC. Kit+/lineage–/Sca+ (KLS) cells of C57BL/6 (Ly5.2) mice treated with either AMD3100, G-CSF or AMD3100+G-CSF, respectively, were infused together with cells from G-CSF treated C57BL/6 (Ly5.1) mice into irradiated syngenic mice. Donor chimerism was assessed at 2 months post-transplantation, and the data suggest that the engraftment potential of AMD3100 mobilized HSC is similar to that of G-CSF mobilized HSC [30].

In order to assess the incidence and severity of acute graft versus host disease (GVHD), T-lymphocytes from the spleens of AMD3100-, G-CSF- and AMD3100+G-CSF-treated, or from naïve unmobilized mice were transplanted into allogeneic recipient mice. GVHD-related weight loss, median survivals, and donor myeloid, T-cell and B-cell engraftment was identical in each group. Thus, these data might set the stage for the use of AMD3100 in the mobilization of allogeneic stem cells in humans.

Experimental transplantation of human CD34+ PBPC into NOD/SCID mice showed a higher frequency of SCID repopulating cells (SRC) in samples mobilized with G-CSF and one injection of AMD3100 (on day 5; 1 in 72 118 ±

10 793 cells (n = 3)), while that in G-CSF-mobilized PB was 1 in 144 277 ± 64 459 cells (n = 3 donors) (P = 0.05) [31]. Chimerism was 8-fold higher in recipients of G-CSF/AMD3100-mobilized cells compared to control cells. *In-vitro* migration assays demonstrated a 7-fold increase in the ability of G-CSF/AMD3100-mobilized CD34+ cells to respond to a chemotactic signal from SDF-1α than G-CSF-mobilized CD34+ cells. Increased expression of CD49d and decreased expression of CD62L (L-selectin) was observed on G-CSF/AMD3100-mobilized cells relative to G-CSF-mobilized cells, which was suggestive of the acquisition of an "engrafting" phenotype. In order to investigate the optimal dose and timing of AMD3100, dose–response studies of CD34+ cell mobilization and pharmacokinetic studies were performed [32]. These investigations indicated the optimal dose of AMD3100 to be 240 μg kg^{-1} body weight (s.c.) for CD34+ mobilization, either alone or as an adjunct to G-CSF, and that the collections could be performed over a broad interval between 4 and 18 h after AMD3100 administration. When analyzing the predicted effect on hematopoietic cell mobilization, there was a linear relationship between mean CD34+ cell peak mobilization and both AMD3100 C_{max} (r^2 = 0.968) and area under the curve (AUC) 0–24 h (r^2 = 0.978) [33, 34]. Pharmacokinetic modeling showed a 4- to 7-h delay between the peak AMD3100 concentration and peak CD34+ cell mobilization. In a Phase I study assessing the safety and clinical effects of AMD3100, 13 patients with multiple myeloma (MM) and non-Hodgkin lymphoma (NHL) were observed. The white blood cell (WBC) and PB CD34+ counts at 4 and 6 h after a single injection with AMD3100 at a dose of either 160 or 240 μg kg^{-1} bodyweight were analyzed [35]. AMD3100 caused a rapid and statistically significant increase in the total WBC and PB CD34+ counts, with higher total CD34+ counts following dosage at 240 μg kg^{-1}. AMD3100 was well tolerated, with only grade 1 toxicities being encountered.

In another study, the aim was to determine if patients mobilized with G-CSF+AMD3100 (G+A) mobilized more CD34+ cells than with G-CSF alone, and whether fewer aphereses were needed to collect 5 × 10^6 CD34+ cells kg^{-1}. An assessment was also made of the pace of engraftment in patients transplanted with products mobilized with G+A [36]. Twelve patients (three MM, nine NHL) in first or second complete or partial remission were been mobilized and transplanted. G+A was found to mobilize more cells per apheresis, irrespective of whether it was the first or second mobilization regimen. AMD3100 toxicities were mild and usually limited to gastrointestinal discomfort/flatulence. No patient required discontinuation of AMD3100 due to drug-related toxicity, and there were no severe adverse events that could be ascribed to AMD3100. Among the 12 patients, 11 engrafted promptly with neutrophil recovery (>500 μL^{-1}) occurring at a median of 10 (range: 9–11) days. None of these patients developed late graft failure, with a median follow-up of 78 (range: 39–189) days. One patient who received 2.3 × 10^6 CD34+ cells kg^{-1} experienced delayed engraftment with absolute nucleated cell (ANC) recovery at day 35, and ultimately died due to bacteremia and fungemia. These data strongly suggest that G+A is a superior mobilizing regimen compared to G-CSF alone, and has the potential to increase the number of CD34+ cells collected and/or reduce the number of aphereses required. Similar results were obtained in a European study with this patient collective [37].

New insights into the mechanisms of stem cell mobilization and genetic manipulation with integrating retroviral vectors in current human gene therapy have been gained by the study of Larochelle et al. AMD3100-mobilized HSC were retrovirally transduced and transplanted into Rhesus macaques. Long-term *in-vivo* marking was observed similar to stem cell factor (SCF)/G-CSF-mobilized PBPC. These data showed that HSCs obtained after AMD3100 application were amenable to genetic manipulation with integrating retroviral vectors, and validated the ability of AMD3100 to induce the mobilization of true, long-term repopulating HSC [38].

CD34+ cell subtyping showed that PBPC mobilized by AMD3100+G-CSF contain more primitive CD34+/CD38– stem cells than G-CSF-mobilized PBPC [39]. Furthermore, CD34+ cells mobilized with the new CXCR4 antagonist AMD3100+G-CSF show increased anti-apoptotic, cell cycle, DNA repair and cell motility-associated gene expression when compared to G-CSF mobilization alone [40].

4.2.2
CTCE0021 and CTCE0214

In a recent report, a small cyclized peptide analogue (31 amino acids) hybrid of the N-terminal and C-terminal regions of SDF-1α linked by a four-glycines linker was shown to have biological function comparable to the native molecule (67 amino acids) [41]. Zhong et al. investigated the effects of SDF-1α analogues (CTCE0021 and CTCE0214) [42]. Enhanced chemotaxis of normal and G-CSF-mobilized hematopoietic cells through transwell inserts was observed with both SDF-1α analogues in a dose-dependent manner. Following an injection of CTCE0214 into mice, an increase in hematopoietic cells was detected as early as 4 h after injection, and became significant at 24 h.

CB CD34+ cells cultured in serum-free medium showed significantly increased survival when incubated with CTCE-0214 alone or synergistically with thrombopoietin (TPO), SCF or Flt-3 ligand [43]. CTCE-0214 also significantly enhanced expansion of CB CD34+ cultured in QBSF-60 containing TPO, SCF and Flt-3L (50 ng mL^{-1} each). After infusion of these cultured cells into NOD/SCID mice, preliminary data showed a trend of more engrafted human cells in the BM of animals receiving cells cultured in CTCE-0214. The results of this study suggest that the SDF-1α peptide analogue CTCE-0214 contributes to the physiologic maintenance/survival of human HSC, and should be explored for the potential use of *ex-vivo* expansion of CB CD34+ cells for transplantation.

4.2.3
C3aR Antagonist SB 290157

In previous studies, the group of Ratajczak showed that C3a, the peptide resulting from the activation of the third component of the complement, C3, primes/sensitizes responses of HSC to the alpha-chemokine SDF-1α [19]. Mice were mobilized in the presence of the C3aR antagonist SB 290157 [44] which, on its own, did not mobilize stem cells, but accelerated their G-CSF-induced mobilization by three days [45]. In view of the fact that the concentration of SDF-1α is reduced in BM during mobilization, and that C3a increases the responsiveness of HSC to SDF-1α, it was suggested that the C3a–C3aR axis might: (i) counteract mobilization; (ii) prevent an uncontrolled release of HSC from the BM into the PB; and (iii) as a result, increase the retention of HSC in the BM. Blockade of the C3a–C3aR axis by a C3aR antagonist (SB 290157) should therefore accelerate/enhance the mobilization of HSC.

4.2.4
GROβT (CXCL2Δ4)

Fukuda et al. reported that GROβT (CXCL2Δ4) rapidly mobilizes HSCs, including long-term repopulating cells (LTRC) into the PB in an equivalent fashion to G-CSF when used alone, and synergizes with G-CSF when used in combination. Interestingly, the accelerated engraftment capability of HSCs mobilized by GROβT compared to HSCs mobilized by G-CSF is not due to increased numbers of transplanted short-term repopulating cells, their homing/migratory potential, or adhesion molecule expression. Enhanced engraftment may result from the selective mobilization of earlier LTRC, or cells with an intrinsic capacity for accelerated engraftment and proliferation [46].

The mobilizing effect of GROβT is hypothesized to be mediated by up-regulation of plasma pro-MMP-9, altering pro-MMP-9:TIMP-1 (tissue inhibitor of metalloproteinase-1) stoichiometry favoring MMP-9 activation [47]. When GROβT is used in combination with G-CSF, a dramatic redistribution in stoichiometry favoring the synergistic activation of MMP-9 is observed, as well as an elevation of active plasma MMP-9 for an extended time frame.

In further investigations, synergistic PBPC mobilization observed when G-CSF and GROβ/GROβT are combined correlates with a synergistic rise in the level of plasma MMP-9, reductions in levels of BM neutrophil elastase (NE), cathepsin G (CG) and MMP-9, and a coincident increase in PB polymorphonuclear neutrophils (PMNs), but a decrease in marrow PMNs compared to G-CSF. Synergistic mobilization is completely blocked by anti-MMP-9 but not NE inhibitor (MeO-Suc-Ala-Ala-Pro-Val-CMK), and is absent in MMP-9-deficient or PMN-depleted mice [48].

These results indicate that PMNs are a common target for G-CSF and GROβ/GROβT-mediated PBPC mobilization and, importantly, that synergistic mobilization by G-CSF plus GROβ/GROβT is mediated by PMN-derived plasma MMP-9.

The developments described here show how stem cell mobilization can be applied more efficiently, as well as effectively. In particular, those patients with a low baseline CD34+ count may profit from these new mobilization strategies, since it is known that this group mobilizes poorly with standard cytokine-based regimens [49, 50].

Acknowledgment

The authors are grateful to A.D. Ho for valuable comments.

Abbreviations/Acronyms

alpha1-AT	alpha1-antitrypsin
AML	acute myelogenous leukemia
ANC	absolute nucleated cell
BM	bone marrow
CB	cord blood
CG	cathepsin G
DEX	dextran sulfate
ECM	extracellular matrix
FL	Flt3 ligand
G-CSF	granulocyte colony-stimulating-factor
GVHD	graft versus host disease
HA	hyaluronic acid
HMG-CoA	hexamethylglutaryl coenzyme A
HPC	hematopoietic progenitor cell
HSC	hematopoietic stem cell
ICAM-1	intercellular adhesion molecule-1
IL	interleukin
KLS	Kit+/lineage–/Sca+
LFA-1	lymphocyte function-associated antigen-1
LTRC	long-term repopulating cells
mAb	monoclonal antibody
MM	multiple myeloma
MMP-9	metalloproteinase-9
NE	neutrophil elastase
NExCG	neutrophil elastase and cathepsin G
NHL	non-Hodgkin lymphoma
NOD/SCID	non-obese diabetic/severely compromised immunodeficient
PB	peripheral blood
PBPC	peripheral blood progenitor cell
PECAM-1	platelet-endothelial cell adhesion molecule-1
PMN	blood polymorphonuclear neutrophil

Ptx   pertussis toxin
S1PR   sphingosine 1-phosphate receptor
SCF   stem cell factor
SLPI   secretory leukocyte protease inhibitor
TBI   total body irradiation
TIMP-1   tissue inhibitor of metalloproteinase-1
TPO   thrombopoietin
uPAR   urokinase-type plasminogen-activator receptor
VCAM-1   vascular cell adhesion molecule-1

References

1. Ho AD. Kinetics and symmetry of divisions of hematopoietic stem cells. *Exp Hematol* 2005;*33*:1–8.

2. Geiger H, Szilvassy SJ, Ragland P, Van Zant G. Genetic analysis of progenitor cell mobilization by granulocyte colony-stimulating factor: verification and mechanisms for loci on murine chromosomes 2 and 11. *Exp Hematol* 2004;*32*:60–67.

3. Fruehauf S, Srbic K, Seggewiss R, Topaly J, Ho AD. Functional characterization of podia formation in normal and malignant hematopoietic cells. *J Leukocyte Biol* 2002;*71*:425–432.

4. Levesque JP, Liu F, Simmons PJ, Betsuyaku T, Senior RM, Pham C, Link DC. Characterization of hematopoietic progenitor mobilization in protease-deficient mice. *Blood* 2004;*104*:65–72.

5. Semerad CL, Christopher MJ, Liu F, Short B, Simmons PJ, Winkler I, Levesque JP, Chappel J, Ross FP, Link DC. G-CSF potently inhibits osteoblast activity and CXCL12 mRNA expression in the bone marrow. *Blood* 2005; *106*:3020–3027 (Epub July 21).

6. Lopez A, Carion A, Herault O, Binet C, Charbord P, Domenech J. In vitro transstromal migration of hematopoietic cells is increased after G-CSF-stimulation of human stromal cells [abstract]. *Blood* 2003;*102*:837a–838a; abstract 3110.

7. van Os R, Konings AW, Down JD. Compromising effect of low dose-rate total body irradiation on allogeneic bone marrow engraftment. *Int J Radiat Biol* 1993;*64*:761–770.

8. van Os R, Velders GA, Hagoort H, van Pel M, van Schie M, Dethmers-Ausema B, de Haan G, Willemze R, Fibbe RE. Protease inhibitors are involved in inhibition of stem cell mobilization induced by low dose irradiation [abstract]. *Blood* 2003;*102*:397b; abstract 5317.

9. Siczkowski M, Clarke D, Gordon MY. Binding of primitive hematopoietic progenitor cells to marrow stromal cells involves heparan sulfate. *Blood* 1992;*80*:912–919.

10. Spiegel A, Zcharia E, Goldshmidt O, Kalinkovich A, Bitan M, Nagler A, Vlodavsky I, Lapidot T. Heparanase facilitates development and SDF-1 induced migration of hematopoietic stem and progenitor cells [abstract]. *Blood* 2003;*102*:825a–826a; abstract 3056.

11. Avigdor A, Goichberg P, Shivtiel S, Dar A, Peled A, Samira S, Kollet O, Hershkoviz R, Alon R, Hardan I, Ben-Hur H, Naor D, Nagler A, Lapidot T. CD44 and hyaluronic acid cooperate with SDF-1 in the trafficking of human CD34+ stem/progenitor cells to bone marrow. *Blood* 2004;*103*:2981–2989.

12. Gotoh A, Ikebuchi K, Miyazawa K, Ohyashiki K. HMG-CoA reductase inhibitors (statins) suppress activation of rho and hematopoietic cell migration toward SDF-1 via suppression of geranylgeranylation [abstract]. *Blood* 2003;*102*:827a; abstract 3063.

13. Fukuda S, Broxmeyer HE, Pelus LM. Flt3 ligand and the Flt3 receptor regulate hematopoietic cell migration by modulating

the SDF-1alpha(CXCL12)/CXCR4 axis. *Blood* 2005;*105*:3117–3126.

14. Hayakawa J, Migita M, Fukazawa R, Adachi K, Hayakawa M, Shimada T, Fukunaga Y. Pretreatment of donor mice with dextran sulfate enhances homing of hematopoietic stem cells [abstract]. *Blood* 2003;*102*:40a; abstract 128.

15. Gu Y, Filippi MD, Cancelas JA, Siefring JE, Williams EP, Jasti AC, Harris CE, Lee AW, Prabhakar R, Atkinson SJ, Kwiatkowski DJ, Williams DA. Hematopoietic cell regulation by Rac1 and Rac2 guanosine triphosphatases. *Science* 2003;*302*:445–449.

16. Jansen M, Yang FC, Cancelas JA, Bailey JR, Williams DA. Rac2-deficient hematopoietic stem cells show defective interaction with the hematopoietic microenvironment and long-term engraftment failure. *Stem Cells* 2005;*23*:335–346.

17. Papayannopoulou T, Priestley GV, Bonig H, Nakamoto B. The role of G-protein signaling in hematopoietic stem/progenitor cell mobilization. *Blood* 2003;*101*:4739–4747.

18. Kimura T, Boehmler AM, Seitz G, et al. The sphingosine 1-phosphate receptor agonist FTY720 supports CXCR4-dependent migration and bone marrow homing of human CD34+ progenitor cells. *Blood* 2004;*103*:4478–4486.

19. Janowska-Wieczorek A, Majka M, Kijowski J, Baj-Krzyworzeka M, Reca R, Turner AR, Ratajczak J, Emerson SG, Kowalska MA, Ratajczak MZ. Platelet-derived microparticles bind to hematopoietic stem/progenitor cells and enhance their engraftment. *Blood* 2001;*98*:3143–3149.

20. Reca R, Mastellos D, Majka M, Marquez L, Ratajczak J, Franchini S, Glodek A, Honczarenko M, Spruce LA, Janowska-Wieczorek A, Lambris JD, Ratajczak MZ. Functional receptor for C3a anaphylatoxin is expressed by normal hematopoietic stem/progenitor cells, and C3a enhances their homing-related responses to SDF-1. *Blood* 2003;*101*:3784–3793.

21. Reca R, Kucia M, Wysoczynski M, Ratajczak J, Sirvaikar N, Janowska-Wieczorek A, Ratajczak MZ. Because mobilized peripheral blood stem/progenitor cells are primed by various inflammatory molecules present in supernatants from leukapheresis products for their chemotactic responses to SDF-1 they engraft faster than bone marrow cells after transplantation [abstract]. *Blood* 2003;*102*:115a; abstract 392.

22. Wysoczynski M, Reca R, Ratajczak J, Kucia M, Shirvaikar N, Honczarenko M, Mills M, Wanzeck J, Janowska-Wieczorek A, Ratajczak MZ. Incorporation of CXCR4 into membrane lipid rafts primes homing-related responses of hematopoietic stem/progenitor cells to an SDF-1 gradient. *Blood* 2005;*105*:40–48.

23. Fuste B, Escolar G, Marin P, Mazzara R, Ordinas A, Diaz-Ricart M. G-CSF increases the expression of VCAM-1 on stromal cells promoting the adhesion of CD34+ hematopoietic cells: studies under flow conditions. *Exp Hematol* 2004;*32*:765–772.

24. Oelschlaegel U, Ehninger G, Kroschinsky F. Differences in expression of adhesion molecules in correlation to mobilisation efficacy in healthy stem cell donors investigated by flow cytometry [abstract]. *Bone Marrow Transplant* 2004;*33*:Abstract:P560.

25. Christopherson KW, II, Cooper S, Broxmeyer HE. Cell surface peptidase CD26/DPPIV mediates G-CSF mobilization of mouse progenitor cells. *Blood* 2003;*101*:4680–4686.

26. Christopherson KW, Cooper S, Hangoc G, Broxmeyer HE. CD26 is essential for normal G-CSF-induced progenitor cell mobilization as determined by CD26-/- mice. *Exp Hematol* 2003;*31*:1126–1134.

27. Christopherson KW, Hangoc G, Broxmeyer HE. Suppression or deletion of CD26 (DPPIV) activity on donor cells greatly enhances the efficiency of mouse hematopoietic stem & progenitor cell homing and engraftment in vivo [abstract]. *Blood* 2003;*102*:38a;abstract 122.

28. Selleri C, Montuori N, Ricci P, Visconte V, Carriero MV, Sidenius N, Serio B, Blasi F, Rotoli B, Rossi G, Ragno P. Involvement of the urokinase-type plasminogen activator receptor in hematopoietic stem cell mobilization. *Blood* 2005;*105*:2198–205.

29. Liles WC, Broxmeyer HE, Rodger E, et al. Mobilization of hematopoietic progenitor cells in healthy volunteers by AMD3100, a CXCR4 antagonist. *Blood* 2003;*102*:2728–2730.

30. Devine SM, Liu F, Holt M, Link D, DiPersio JF. AMD3100-mobilized murine hematopoietic stem cells and T-lymphocytes have identical capacity to induce multilineage stem cell engraftment, donor T-cell chimerism and GVHD in mice compared with G-CSF mobilized cells [abstract]. *Blood* 2003;*102*:938a–939a;abstract 3495.

31. Broxmeyer HE, Orschell CM, Clapp DW, Hangoc G, Cooper S, Plett PA, Liles WC, Li X, Graham-Evans B, Campbell TB, Calandra G, Bridger G, Dale DC, Srour EF. Rapid mobilization of murine and human hematopoietic stem and progenitor cells with AMD3100, a CXCR4 antagonist. *J Exp Med* 2005;*201*:1307–1318.

32. Liles WC, Rodger E, Broxmeyer HE, Dehner C, Badel K, Calandra G, Christensen J, Wood B, Price TH, Dale DC. Augmented mobilization and collection of CD34+ hematopoietic cells from normal human volunteers stimulated with granulocyte-colony-stimulating factor by single-dose administration of AMD3100, a CXCR4 antagonist. *Transfusion* 2005;*45*:295–300.

33. Lack NA, Dale DC, Calandra GB, MacFarland RT, Badel K, Liles WC. A pharmacokinetic/pharmacodynamic profile of the CXCR4 antagonist AMD3100 for the mobilization of hematopoietic progenitor cells [abstract]. *Blood* 2003;*102*:115a;abstract 391.

34. Lack NA, Green B, Dale DC, Calandra GB, Lee H, MacFarland RT, Badel K, Liles WC, Bridger G. A pharmacokinetic-pharmacodynamic model for the mobilization of CD34+ hematopoietic progenitor cells by AMD3100. *Clin Pharmacol Ther* 2005;*77*(5):427–436.

35. Devine SM, Flomenberg N, Vesole DH, Liesveld J, Weisdorf D, Badel K, Calandra G, DiPersio JF. Rapid mobilization of CD34+ cells following administration of the CXCR4 antagonist AMD3100 to patients with multiple myeloma and non-Hodgkin's lymphoma. *J Clin Oncol* 2004;*22*:1095–1102.

36. Flomenberg N, Devine SM, Dipersio JF, Liesveld JL, McCarty JM, Rowley SD, Vesole DH, Badel K, Calandra G. The Use of AMD3100 plus G-CSF for autologous hematopoietic progenitor cell mobilization is superior to G-CSF alone. *Blood* 2005;*106*:1867–1874 (Epub May 12).

37. Fruehauf S, Seeger T, Topaly J, Herrmann D, Calandra G, Laufs S, et al. AMD3100 augments the number of mobilized peripheral blood progenitor cells (PBPC) when added to a G-CSF standard mobilization regime and AMD3100-mobilized PBPC result in rapid hematopoietic reconstitution after autologous transplantation. EBMT meeting Prague February 2005; abstract 1035.

38. Larochelle A, Krouse A, Orlic D, Donahue R, Dunbar C, Hematti P. The use of AMD3100 mobilized peripheral blood stem cell CD34+ targets for retroviral transduction results in high-level in vivo marking in rhesus macaques [abstract]. *Blood* 2003;*102*:396b–397b;abstract 5315.

39. Fruehauf S, Seeger T, Calandra G, Goldschmidt H, Ho AD. The CXCR4 antagonist AMD3100 mobilizes a more primitive subset of CD34+ cells than G-CSF (abstract). Annual Meeting of the German Society for Hematology and Oncology (DGHO), 2005.

40. Fruehauf S, Seeger T, Maier P, Laufs S, Bridger G, Calandra G, Goldschmidt H, Ho AD, CD34+ cells mobilized with the new CXCR4 antagonist AMD3100 + G-CSF show increased anti-apoptotic, cell cycle, DNA repair and cell motility-associated gene expression when compared to G-CSF mobilization alone. EBMT Meeting Hamburg, March 2006, abstract 805.

41. Tudan C, Willick GE, Chahal S, Arab L, Law P, Salari H, Merzouk A. C-terminal cyclization of an SDF-1 small peptide analogue dramatically increases receptor affinity and activation of the CXCR4 receptor. *J Med Chem* 2002;*45*:2024–2031.

42. Zhong R, Law P, Wong D, Merzouk A, Salari H, Ball ED. Small peptide analogs to stromal derived factor-1 enhance chemotactic migration of human and mouse hematopoietic cells. *Exp Hematol* **2004**;*32*:470–475.

43. Li KKH, Chuen CKY, Law P, Wong D, Merzouk A, Salari H, Li CK, Fok TF, Yuen PMP. A small peptide analog of stromal cell-derived factor-1 (SDF-1) enhances survival and expansion of cord blood (CB) CD34+ cells in synergy with cytokines [abstract]. *Blood* **2003**;*102*:960a;abstract 3576.

44. Ames RS, Lee D, Foley JJ, Jurewicz AJ, Tornetta MA, Bautsch W, Settmacher B, Klos A, Erhard KF, Cousins RD, Sulpizio AC, Hieble JP, McCafferty G, Ward KW, Adams JL, Bondinell WE, Underwood DC, Osborn RR, Badger AM, Sarau HM. Identification of a selective nonpeptide antagonist of the anaphylatoxin C3a receptor that demonstrates antiinflammatory activity in animal models. *J Immunol* **2001**;*166*:6341–6348.

45. Ratajczak J, Reca R, Kucia M, Majka M, Allendorf DJ, Baran JT, Janowska-Wieczorek A, Wetsel RA, Ross GD, Ratajczak MZ. Mobilization studies in mice deficient in either C3 or C3a receptor (C3aR) reveal a novel role for complement in retention of hematopoietic stem/progenitor cells in bone marrow. *Blood* **2004**;*103*:2071–2078.

46. Fukuda S, Bian H, King AG, Pelus LM. The accelerated engraftment of peripheral blood cell counts following transplantation with hematopoietic stem and progenitor cells (HSCs) mobilized by the CXC chemokine GROβT (CXCL2Δ4) is independent of homing to recipient bone marrow and the SDF-1α (CXCL12):CXCR4 migration axis [abstract]. *Blood* **2003**;*102*:113a;abstract 385.

47. Pelus LM, Bian H, King AG. Alterations in the stoichiometry and kinetics between matrix metalloproteinase-9 (MMP-9) and tissue inhibitor of metalloproteinase-1 (TIMP-1) are responsible for the synergistic mobilization of peripheral blood hematopoietic stem cells (HSC) by the CXC chemokine GROβT/CXCL2Δ4 plus G-CSF [abstract]. *Blood* **2003**;*102*:694a–695a;abstract 2572.

48. Pelus LM, Bian H, King AG, Fukuda S. Neutrophil-derived MMP-9 mediates synergistic mobilization of hematopoietic stem and progenitor cells by the combination of G-CSF and the chemokines GRObeta/CXCL2 and GRObetaT/CXCL2delta4. *Blood* **2004**;*103*:110–119.

49. Fruehauf S, Haas R, Conradt C, Murea S, Witt B, Mohle R, Hunstein W. Peripheral blood progenitor cell (PBPC) counts during steady-state hematopoiesis allow to estimate the yield of mobilized PBPC after filgrastim (R-metHuG-CSF)-supported cytotoxic chemotherapy. *Blood* **1995**;*85*:2619–2626.

50. Fruehauf S, Schmitt K, Veldwijk MR, Topaly J, Benner A, Zeller WJ, Ho AD, Haas R. Peripheral blood progenitor cell (PBPC) counts during steady-state haemopoiesis enable the estimation of the yield of mobilized PBPC after granulocyte colony-stimulating factor supported cytotoxic chemotherapy: an update on 100 patients. *Br J Haematol* **1999**;*105*:786–794.

5
Pluripotent Stem Cells from Umbilical Cord Blood

Gesine Kögler and Peter Wernet

5.1
Biological Advantages of Cord Blood as a Stem Cell Resource

The treatment of a broad array of malignant and nonmalignant hematopoietic diseases in children using allogeneic HLA-identical sibling cord blood donors or an unrelated cord blood donor has been accomplished for several years [1–4]. Recent experience has shown that it is possible also to reconstitute hematopoiesis in adults after myeloablative conditioning by applying one large cord blood (CB) unit or by combining two CB units [5–7]. It has been shown, that in contrast to its adult bone marrow (BM) counterpart, the stem cell compartment in CB is less mature. This has been documented extensively for the hematopoietic system, including a higher proliferative potential *in vitro* as well as *in vivo*, which is associated with an extended life span of the stem cells and longer telomeres [8–11]. In addition, the frequency of these stem cells is also higher in CB than in adult BM, most likely due to the very rapid growth and remodeling of tissue that is required during the fetal period. Hematopoietic stem cells, for example, proliferate and differentiate at a number of distinct sites such as in yolk sac, fetal liver, thymus, spleen and BM, indicating migration through the circulation [10, 11]. The immaturity and increased frequency of hematopoietic stem cells suggested that CB might also be an attractive source for nonhematopoietic stem cells. Besides this biological advantage, CB is abundantly available, is routinely harvested without risk to the donor, and infectious agents such as cytomegalovirus (CMV), Epstein–Barr virus (EBV) and other viruses are rare exceptions, which in regenerative medicine is advantageous over the use of adult stem cell sources [3, 12]. To date, very few reports have been made indicating that so-called unrestricted somatic stem cells (USCC) [13–15] or mesenchymal stem cells (MSC) from CB [16–19] can be differentiated into nonhematopoietic tissue. In addition to these reports, observations have been made by Ende and colleagues [20, 21] that so-called pluripotent "Berashi cells" were responsible for a clinical response observed in SOD1 mice, Huntington disease mice and Alzheimer mice [21]. However, the responsible cell population could never be detected, enriched, or expanded [21]. Likewise,

Stem Cell Transplantation. Biology, Processing, and Therapy.
Edited by Anthony D. Ho, Ronald Hoffman, and Esmail D. Zanjani
Copyright © 2006 WILEY-VCH Verlag GmbH & Co. KGaA, Weinheim
ISBN: 3-527-31018-5

intravenously administered CB has been shown to reduce behavioral deficits following stroke in rats [22], while large doses of human CB cells have been reported to improve survival in a mouse with amyotrophic lateral sclerosis [21]. Recently, a rare, CD45-negative USSC was detected in CB and established routinely as having the ability to be differentiated into osteoblasts, chondrocytes, adipocytes, and neural progenitors as well as to differentiate *in vivo* into bone, cartilage, hematopoietic cells, neural, liver, and heart tissue [13, 15]. Since USSC seem to be an immature mesodermal progenitor for MSC, and are also an easily accessible source of cells, they should be viewed as a potential candidate for supportive therapy.

5.2
The Generation and Expansion of Pluripotent Cells (USSC) from Cord Blood

5.2.1
Generation and Expansion of USSC from Fresh CB

In the obstetric department, placental CB is collected only after having obtained informed consent from the mother. Following delivery of the baby, the CB was collected from the umbilical cord vein into special bags containing citrate-phosphate-dextrose. USSC cultures were initiated from 325 placental CB samples, with a total generation frequency of 39 %. After 6 to 25 days, between one and 11 USSC colonies of CB were detected, and these grew into monolayers within 2 to 3 weeks (Fig. 5.1).

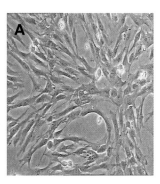

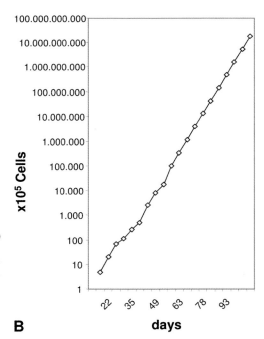

Figure 5.1 Characteristics of unrestricted somatic stem cells (USSC). (A) Spindle-shaped USSC plated at low density after 32 population doublings (original magnification, ×20). (B) Expansion kinetics of USSC for 20 passages, equivalent to 46 population doublings.

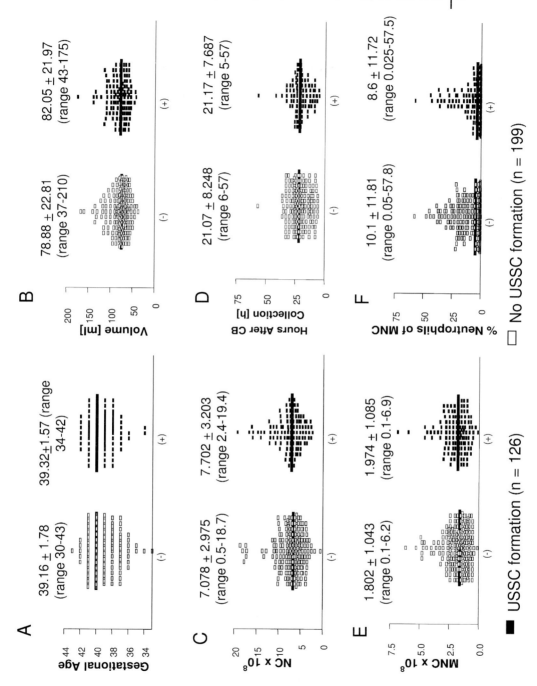

◀ **Figure 5.2** (A–F) Generation of USSC. Note the lack of correlation between: (A) gestational age (>30 weeks); (B) volume including citrate-phosphate-dextrose (CPD); (C) nucleated cells/cord blood collected; (D) hours after cord blood collection; (E) MNC fraction isolated; and (F) % neutrophils in the MNC in the mononuclear fraction. Data were analyzed statistically using the Welsh-corrected Student's *t*-test.

Although USSC could not be generated from all CB samples, if colonies were obtained then it was possible to expand them up to 10^{15} cells in 20 passages. In the presence of Myelocult medium/10^{-7} M dexamethasone, USSC cultures were initiated from 57.1 % of CB samples, whereas in low-glucose DMEM-medium/10^{-7} M dexamethasone/30 % fetal calf serum (FCS) the frequency was 31.6 % [15]. No correlation was observed between gestational age, CB volume, the number of nucleocytes (NC) in the CB collections, hours elapsed after CB collection, the number of mononucleocytes (MNC) in the CB after gradient separation, the number of contaminating granulocytes in the MNC fraction, and the successful initiation of USSC cultures (Fig. 5.2). Although the generation frequency was higher in Myelocult medium, subsequent growth kinetics as well as differentiation assays revealed that low-glucose DMEM medium/pretested FCS is ideal to keep USSC in an undifferentiated state [15]. In contrast, in the presence of Myelocult medium, USSC, which stained positive with Alizarin red, may develop into bone nodules. Simply, by morphology (bone nodule formation) it can be determined whether the cells stay undifferentiated or are triggered into the mesenchymal lineage associated with nodule formation. So far, any attempt to directly isolate this CD45-negative cell population from fresh CB by immunomagnetic selection (data not shown) has failed, most likely due to the very low frequency of this cell population in the MNC fraction of CB [15].

5.2.2
Generation and Expansion of USSC from Cryoconserved CB

The generation of USSC from cryopreserved and thawed CB samples performed according to the Rubinstein thawing procedure, was associated with difficulties as expected, because in thawed CB aggregate formation occurs which results in cell losses. The first experiments (n = 15) with unseparated frozen CB showed the MNC output after Ficoll isolation and erythrocyte lysis to be very low, and the cells did not adhere to the culture flasks. Thus, subsequent experiments were performed with volume-reduced CB units (n = 66), where the majority of erythrocytes were already depleted at the time of cryopreservation. Only one experimental condition, namely the MNC isolation with erythrocyte lysis and cultivation in the presence of Myelocult/10^{-7} M dexamethasone, revealed successful generation of USSC. USSC were generated from seven out of 36 CB samples (generation frequency 19.5 %). Among all seven CB specimens, only one primary USSC colony was detected. In the presence of DMEM/10^{-7} M dexamethasone (n = 10) no USSC were generated [15].

5.2.3
Immunophenotype of USSC Obtained from Fresh and Cryopreserved CB Specimens

USSC obtained from fresh and cryopreserved CB had the same immunophenotype: they were negative for CD4, CD8, CD11a, CD11b, CD14, CD15, CD16, CD25, CD31, CD33, CD34, CD45, CD49b, CD49d, CD49f, CD50, CD62E, CD62L, CD62P, CD80, CD86, CD106, CD117, cadherin V, glycophorin A, and HLA-class II, but positive for CD13, CD29, CD44, CD49e, CD54, CD58, CD71, CD73, CD90, CD105, CD166, vimentin, CD146, cytokeratin 8 and 18, CD10, and von Willebrand factor, and showed weaker expression of HLA class I and CD123 as well as variable expression of FLK1 (KDR) and CD133/1, CD133/2. USSC showed strong staining for the monoclonal antibody (mAb) CD146 (P1H12), expressed on circulating endothelial cells, and stained positive for FLK-1 (KDR) as well as weakly for CD133/1. Other endothelial specific markers such as CD31 and cadherin V were negative [13, 15].

5.2.4
Expansion of USSC

USSC can be cultured for more than 20 passages (equivalent to more than 40 population doublings) without any spontaneous differentiation (see Fig. 5.1). The USSC karyotype was normal 46XX or 46XY as analyzed for six individual USSC specimens for passages 5 (21 population doublings) to 19 (45 population doublings). The average telomere length of USSC obtained after both 21 (passage 4) and 25 population doublings (passage 6) was 8.93 kbp, and after 36 population doublings was 8.60 kbp (passage 13). This is significantly longer than the telomere length of MSC generated from BM donors with 7.27 kbp (passage 4) and 7.11 kbp (passage 9) at 19 and 27 population doublings, respectively [13].

5.2.5
The Differentiation Potential of USSC

5.2.5.1 *In-vitro* and *In-vivo* Differentiation of USSC into Mesenchymal Cell Lineages

In-vitro differentiation of USSC into bone, cartilage, and adipocytes
The USSC tested were capable of differentiating along the osteogenic and chondrogenic and adipogenic lineage. Differentiation into osteoblasts could be induced by dexamethasone, ascorbic acid and β-glycerol phosphate, with the cells showing initial calcium phosphate deposits after only 5 days. Bone-specific alkaline phosphatase (ALP) activity was detected, and a continuous increase in Ca^{2+} release was documented. Osteogenic differentiation was confirmed by the expression of ALP, osteocalcin, osteopontin, bone sialoprotein and collagen type I, all of which were detected using RT-PCR. A pellet culture technique was employed to trigger USSC towards the chondrogenic lineage. The chondrogenic nature of differentiated cells was assessed by Alcian blue staining and by expression of the car-

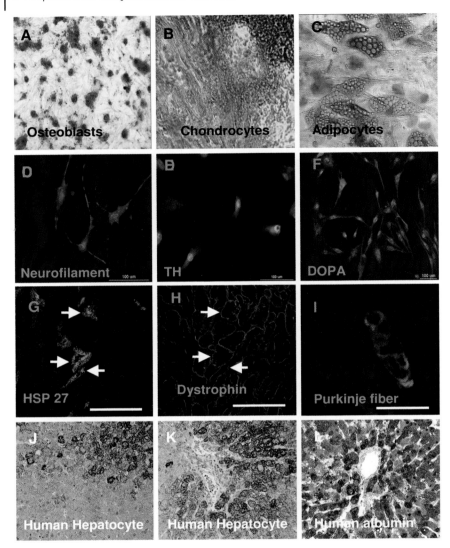

tilage extracellular protein type II collagen. Chondrogenesis was further confirmed by RT-PCR for the cartilage-specific mRNAs encoding Cart-1, collagen type II and chondroadherin. For the induction of adipogenic differentiation, the USSC were cultured with dexamethasone, insulin, 3-isobutyl-1-methylxanthine (IBMX) and indomethacin. Adipogenic differentiation was demonstrated by Oil Red O staining of intracellular lipid vacuoles (Fig. 5.3) [13, 15].

◀ **Figure 5.3** *In-vitro* differentiation of USSC into osteoblasts, chondroblasts, and adipocytes. For osteoblast differentiation, cells were cultured at 8000 cells cm^{-2} in 24-well plates in DMEM, 30% FCS, 10^{-7} M dexamethasone, 50 μM ascorbic acid-2 phosphate and 10 mM β-glycerol phosphate (DAG). (A) Differentiation to osteoblasts is shown by Alizarin red staining to determine calcium deposition (original magnification, ×10). For chondrogenic differentiation a micromass culture system was used; 2 ×10^5 USSC were cultured in a high-glucose DMEM medium containing 100 U mL^{-1} penicillin and 0.1 mg mL^{-1} streptomycin, 100 nM dexamethasone, 35 μg mL^{-1} ascorbic acid-2-phosphate, 1 mM sodium pyruvate, ITS-premix (dilution 1:100) and 10 ng mL^{-1} TGFβ1. Cell aggregates were cultured for 21 days. Collagen type II staining of pellet microsections (B) were analyzed by fluorescence microscopy; nuclei were stained with DAPI (original magnification, ×40). For adipogeneic differentiation, USSC were induced in the presence of 1 μM dexamethasone, 10 μg mL^{-1} insulin, 0.5 mM isobutylmethylxanthine and 100 μM indomethacin. Oil red-O staining of the lipid vesicles performed 2 weeks after stimulation demonstrated an ongoing adipo-genesis (C) (original magnification, ×40). (D–F) *In-vitro* differentiation of USSC into neural cells: Differentiated cells showed positive immunoreactivity for the neuron-specific markers neurofilament (D), the enzyme tyrosine hydroxylase (E) and DOPA-decarboxylase (F). Cell nuclei show a blue color due to DAPI staining. (G–I) *In-vivo* differentiation of USSC to cardiomyocytes and Purkinje fibers: (G) and (H) are serial sections of the right ventricle. (G) is labeled with a human-specific anti-HSP27 mAb, and (H) with an anti-dystrophin mAb with broad species specificity. Arrows indicate the same cells. (I) A section of a Purkinje fiber. Scale bars: (A–E) = 50 μm; (F) = 100 μm. (J,K) *In-vivo* differentiation of USSC from CB into parenchymal liver cells in the preimmune fetal sheep model. Photomicrographs show staining for anti-human hepatocyte antibody in liver sections of animals transplanted with USSC: in close association with the portal veins >80% of cells stained positive, but areas within the tissue, outside the portal spaces, also stained positive for the human hepatocyte antibody. (L) Anti-human albumin staining of the liver (original magnification, ×20).

In-vivo differentiation of USSC into bone and cartilage

To determine the *in-vivo* regeneration capacity for bone, the repair of critical size bone defects with USSC-loaded calcium phosphate ceramic cylinders was demonstrated. USSC were loaded into porous ceramic cylinders of 5 mm length that were in turn implanted into femur critical size bone defects of athymic Harlan Nude rats. At 4 weeks after transplantation, the human cells were still alive within the defect bone. Images of longitudinal and cross-sections demonstrated bony healing between the cell-loaded implant and the host bone. Bony integration was established in the form of cancellous bone as detected by toluidine blue staining. After 12 weeks a clear bony reconstitution was observed. The *in-vivo* chondrogenic potential was assayed by loading USSC into gelatin sponges; these were cultured in chondrogenic medium with supplementation of transforming growth factor-beta (TGF-β) for up to 2 weeks, and then implanted subcutaneously into nude mice. After an additional 3-week period, the implanted cells demonstrated strong chondrogenic differentiation [13].

5.2.5.2 Differentiation of USSC into Neural Cells *In Vitro* and *In Vivo*

USSC could be differentiated in the presence of NGF, bFGF, dibutyryl cAMP, IBMX and retinoic acid into neural cells. The number of neurofilament (NF)-positive cells increased, starting with approximately 30% after one week and rising to >70% homogeneity after 4 weeks. Double immunostaining revealed co-localization of NF and sodium-channel protein in a small proportion of cells. Expression of synaptophysin was detected after 4 weeks, indicating a more mature phenotype of neurons. Consistently, about 90% of the cells stained positive for the inhibitory neurotransmitter, gamma-aminobutyric acid (GABA). USSC-derived neurons could be identified which stained positive for tyrosine hydroxylase, the key enzyme of the dopaminergic pathway. Choline acetyltransferase, the main enzyme of the cholinergic pathway, was detected in about 50% of the cells. Since 90% of USSC showed positive immunoreactivity for GABA *in vitro*, coexpression with the enzymes of other neurotransmitter pathways was observed. These findings could be explained by the function of GABA during very early embryogenesis, where it is considered to act as a signal molecule in neuronal development (see Fig. 5.3). Extensive analyses using patch–clamp recording of different USSC batches have not yet revealed any voltage-activated, fast-inactivating Na^+ current, which is typical of terminally differentiated neurons. This finding suggests that USSC differentiate to a precursor-like phenotype *in vitro*, in which distinct neuron-specific proteins are expressed, but a fully functional neuronal phenotype has not yet developed. This may be explained by the lack of a proper microenvironment with special cytokines and/or molecular signals, which are not present in the *in-vitro* culture system. Moreover, USSC differentiated into astrocytes expressing glial fibrillary acidic protein (GFAP), an astroglial intermediate filament, reaching a transient maximum level of approximately 45% at 2 weeks. During early differentiation the coexpression of both GFAP and NF could be observed, indicating a common progenitor cell type. In order to analyze the potency of the USSC to migrate, integrate and differentiate into neuronal-like cells *in vivo*, the cells were labeled with pKH26 and transplanted stereotactically into the hippocampal region of an intact adult rat brain. At 3 months post-implantation, USSC expressing human Tau protein could be identified widely distributed throughout the brain, indicating a high migratory activity of USSC *in vivo*. Human Tau-immune staining further revealed a neuron-like, highly differentiated morphology of implanted USSC in different ipsilateral and contralateral regions of the adult brain, including the neocortex [13].

5.2.5.3 Application of USSC to the Fetal Sheep Xenograft Model to Study *In-vivo* Hematopoiesis

Between days 50 and 60 during the sheep gestational period, the hematopoietic system is known to expand rapidly. Zanjani's group showed that, during this period, hematopoietic engraftment was facilitated and that the infusion of human cells resulted in a tolerance to human antigens [23]. Therefore, USSC were transplanted in an intrauterine location before immunological maturity occurred.

USSC originating from one expanded colony, were injected intraperitoneally into the preimmune fetal sheep. Multilineage hematopoietic engraftment was observed, which included cells of erythroid (glycophorin A), myeloid (CD13, CD33) and lymphoid (CD3, CD7, CD10, CD20) lineages. The level of these human cells in the BM of these sheep was maintained throughout a 12-month period following transplantation [13].

5.2.5.4 *In-vivo* Differentiation of USSC into Myocardial Cells and Purkinje Fibers in the Preimmune Fetal Sheep

The differentiation of USSC into human cardiomyocytes was analyzed 8 months after transplantation, by using positive staining with the human-specific anti-HSP27 antibody in both atria, both ventricles and the septum of the heart of *in utero* in USSC-transplanted sheep. In order to show that the cells labeled with the human-specific HSP27 Ab were mature cardiomyocytes, tissue sections were probed with several antibodies that have a distinctive staining in cardiac cells. The antibodies used (anti-ryanodine receptor, anti-myosin heavy chain, anti-dystrophin) had a broad specificity so that they could be used to confirm that this staining pattern was identical to that of cells derived from the USSC, as well as from the surrounding cardiac sheep cells. In this study, engraftment of USSC in the Purkinje fiber system was also detected. Identification was made by characteristic morphology and staining with the Purkinje fiber marker PGP 9.5. One major difference between USSC and human MSC from bone marrow is the distribution of USSC-derived cells in the heart. In previous studies using the preimmune fetal sheep model, the human MSC were shown to engraft predominantly in the Purkinje fiber system, with very few ventricular or atrial cardiomyocytes of human origin present [24]. In contrast, USSC-derived cells formed both Purkinje fiber cells and cardiomyocytes (see Fig. 5.3). This would indicate that the USSC were of an earlier cell type than multipotent bone marrow MSC, possibly representing also the precursor cell for MSC. However, further studies should address the question of whether USSC engraft in other cell types in the heart.

5.2.5.5 *In-vivo* Differentiation of USSC into Hepatic Cells in the Preimmune Fetal Sheep

In order to examine human hepatocyte development, the livers of sheep were removed 14 months after USSC transplantation. Donor parenchymal liver cells generated from the USSC, located in close association with the portal veins, represented the majority of the total hepatic cells (see Fig. 5.3). The counting of representative areas of tissue sections of chimeric sheep liver revealed that 21.1 $\pm$ 3.2 % of the total liver cells were stained positive by the antibody (Ab) that specifically recognizes human hepatocytes; more than 80 % of those hepatocytes in association with portal veins were stained positive. In order to use a different human-specific Ab, the mAb HSA-11 recognizing human albumin was applied, and showed identical results. The results were confirmed by Western blotting, which demonstrated the functional production of this human protein *in vivo* in the serum of sheep, obtained at 17 months after transplantation [13].

Cell fusion does not account for liver cell-specific differentiation of USSC

In order to determine whether the human cells which were integrated into the sheep liver parenchyma acquired an organ-specific differentiated phenotype through cell fusion with indigenous ovine liver parenchymal cells or by liver cell differentiation, single microdissected liver parenchymal cells from chimeric liver tissue were monitored for the coexistence of human and ovine genomes in these cells. Single liver parenchymal cells, which were either devoid of human proteins or in which expression of human proteins could be detected, were micromanipulated from chimeric liver sections and separately analyzed by single-cell PCR for the human *IGH* or *TCRB* loci and the ovine *IGH* and *TCRD* loci. In human cells derived from the chimeric liver paraffin sections, amplification gave rise to a PCR product for human DNA, whereas not a single sheep-specific PCR product was detected [13]. Fusion events, if they occur at all under physiological non-injury conditions, could at best account for a very low frequency of the differentiated liver cell-specific phenotype of the USSC-derived hepatocytes integrated into the chimeric liver tissue.

5.2.6
Cytokine Production and Hematopoiesis Supporting Activity of CB-Derived Unrestricted Somatic Stem Cells

5.2.6.1 Rationale for Application of USSC to Support Hematopoiesis

The major disadvantage of CB compared to other stem cell sources remains the comparatively low cell dose available for adult patients, as this results in a longer time to engraftment, higher rates of graft failure, and transplantation-related mortality. Several groups have attempted, in both preclinical and clinical settings, to provide a cytokine-driven *ex-vivo* expansion of the CB product in order to increase hematopoietic progenitor and granulocyte numbers, and also to reduce the duration of post-transplant neutropenia. However, the implementation of such protocols has been complicated by the following facts:

- In the majority of tissue banks the CB transplants are frozen in a single bag. The clinical trials were performed with only a fraction of a CB unit being expanded *ex vivo*, while the larger portion was infused unmanipulated. Thus, the expanded product usually cannot be infused until 10–14 days after transplantation.
- Clinical-grade growth factors are only available for a limited number of cytokines, and are expensive.
- None of the clinical experiences could unequivocally document a clear benefit of infusion of such *ex-vivo* cytokine-expanded components.

Since cytokine-driven *ex-vivo* expansion of CD34+ cells from CB remains the subject of much controversy, other approaches to improve the time to hematopoietic engraftment and reconstitution after CB transplantation are currently being explored, including multiple CB transplants and the co-transplantation of a single CB unit together with highly purified CD34+-mobilized peripheral blood stem cells from a haploidentical, related donor. The specialized BM stroma microenvironment, which consists of extracellular matrix and stromal cells, has been shown to be crucial for hematopoietic regeneration after stem cell transplantation. Furthermore, MSC from BM or lung fibroblasts have been shown in preimmune fetal sheep or in NOD/SCID mice to promote engraftment of CB-derived CD34+ cells. Therefore, the co-transplantation of CB with stromal elements may improve engraftment.

5.2.6.2 Qualitative and Quantitative Assessment of Cytokines Produced by USSC or Bone Marrow MSC

The production of soluble cytokines by unstimulated USSC was determined using a variety of different approaches. Cytokines were identified either by applying the RayBio Human Cytokine Array for qualitative screening, or the Luminex Multiplex cytokine bead kit and ELISA assays for quantitative screening. USSC constitutively produced SCF, LIF, TGF-1β, M-CSF, GM-CSF, VEGF, IL-1β, IL-6, IL-8, IL-11, IL-12, IL-15, SDF-1α, and hepatocyte growth factor (HGF) as determined quantitatively. In addition, angiogenin (sensitivity >10 pg mL^{-1}) as well as MCP-1 (sensitivity >1 pg mL^{-1}) were positive in the cytokine array.

The following cytokines were not detected within the detection limit indicated:
- <1 pg mL^{-1}: IL-4, IL-5, MIG (monokine induced by gamma interferon), exotaxin;
- <7.4 pg mL^{-1}: IL-3;
- <10 pg mL^{-1}: TNFα, BLC (B-lymphocyte chemoattractant), IGF-1;
- <15 pg mL^{-1}: INFα;
- <25 pg mL^{-1}: IL-2;
- <100 pg mL^{-1}: MCP-4 (monocyte chemoattractant protein), GCP-2 (granulocyte chemotactic protein).

Release of Flt3 ligand (<7 pg mL^{-1}) and thrombopoietin (TPO) (<7.4 pg mL^{-1}) was not detected in unstimulated USSC or in IL-1β-induced USSC, though production was observed in their cell lysates for TPO (data not shown). Since only low constitutive production of G-CSF was observed (0.2 ± 0.1 pg mL^{-1}), USSC were stimulated with IL-1β for 24 h, and this resulted in a mean G-CSF production of 1959 ± 443.3 pg mL^{-1} (range: 1516 to 2402 pg mL^{-1}). For unstimulated BM MSC, 1.2 ± 0.5 pg mL^{-1} G-CSF was produced upon stimulation with IL-1β (79.7 ± 19.5 pg mL^{-1}). This was significantly lower than for USSC ($p = 0.0095$). A significantly higher production of stem cell factor (SCF) ($p = 0.0104$) and LIF ($p = 0.046$) was observed in USSC than in BM MSC [15]. SDF-1α production was higher in USSC (up to 3087.8 pg mL^{-1}), but due to the high variability of

SDF-1α production this value was not statistically different from production in BM MSC.

5.2.6.3 Hematopoiesis Supporting Stromal Activity of USSC in Comparison to BM MSC

In order to determine the hematopoiesis-supporting stromal activity of USSC compared to BM MSC, CB CD34+ cells were expanded in co-cultures with USSC and BM MSC. At 1, 2, 3, and 4 weeks, co-cultivation of CD34+ cells on the USSC layer resulted in 14-fold, 110-fold, 151-fold and 183.6-fold amplifications respectively of the total cells, and in 30-fold, 101-fold, 64-fold and 29-fold amplifications of colony-forming cells (CFC), respectively (Fig. 5.4). LTC-IC expansion at 1 and 2 weeks was 2-fold and 2.5-fold significantly higher for USSC than BM MSC (1-fold and 1-fold), but declined after day 21 [15]. Transwell co-cultures of USSC did not significantly alter total cell or CFC expansion. In summary, USSC produce functionally significant amounts of hematopoiesis-supporting cy-

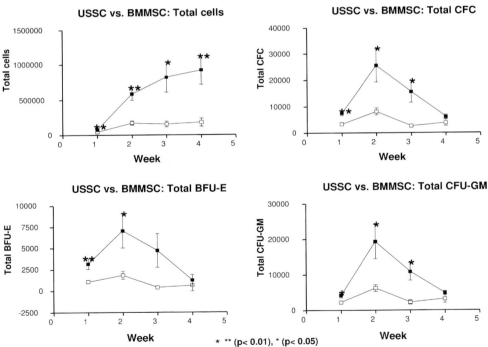

Figure 5.4 Kinetics of *ex-vivo* expansion of CD34+ CB cells on USSC and BMMSC over 28 days. CD34+-enriched CB cells were plated on irradiated layers of USSC (■) and BMMSC (□). Cell and CFC counts were performed from the expanding cultures at days 0, 7, 14, 21, and 28. Results were expressed as mean ± SD for total cells (upper left), CFC (upper right), BFU-E/CFU-E (lower left), and CFU-GM (lower right). **, p ≤0.01; *, p ≤0.05.

tokines, and are superior to BM MSC in the expansion of CD34+ cells from CB [15]. USSC is therefore a suitable candidate for stroma-driven *ex-vivo* expansion of hematopoietic CB cells for short-term reconstitution.

5.3
Other Multipotent Nonhematopoietic Stem Cells: Mesenchymal Cells in CB and CB Tissue

The presence of stromal or mesenchymal progenitor cells in CB remains a subject of discussion. In this regard, several groups have been unable to generate MSC [25, 26], or were able to generate cells in only a limited number of CB specimens [16, 17] or only with a restricted differentiation potential [16–19]. Recently, another potential source of mesenchymal cell, the so-called human umbilical cord perivascular cell (HUCPV), originating from Wharton's jelly of the umbilical cord was reported [27]. Saraguser and coworkers [27] described a nonhematopoietic MHC class I and II negative, alpha-actin, desmin, vimentin and pericyte marker-positive cell with a differentiation capacity into an osteogenic phenotype and the formation of bone nodules. These studies were based on an old description by Thomas Wharton [28], as well as more recent reports by McElreavey et al. [29], Naughton et al. [30], Kadner et al. [31], and Romanov et al. [32]. In summarizing these data, Wharton's jelly contains cells with a myofibroblast phenotype [33], a smooth muscle cell phenotype, and can be induced to differentiate into "neural-like" cells expressing neuron-specific enolase and other neural markers [34]. Whether these described cell populations are different precursor phenotypes of one cell entity or originating from different cells needs to be evaluated.

5.4
Conclusion: Future Efforts Towards the Regenerative Capacity of CB Nonhematopoietic Cells

Ultimately, it would be of great scientific value to know whether cells referred to as USSC can be generated and differentiated on a clonal level, with the same unrestricted differentiation potential. Furthermore, it would be helpful to define antigens, which would allow this cell population to be enriched from each CB collection, and to define more precisely the distinct differentiation pathways with a read-out in selected animal models. In addition, it is apparent that the immunogenicity of USSC has not yet been evaluated, though this issue is currently under investigation. On the basis of their pluripotency and expansion under GMP conditions into large quantities, the USSC or other nonhematopoietic cells from CB, when pretested for infectious agents and matched for the major transplantation antigens, may serve as a universal allogeneic stem cell source for the future development of cellular therapy for tissue repair and regeneration.

Acknowledgments

The authors thank their co-authors of jointly published papers cited in this chapter. Studies conducted in the authors' laboratory and relating to hematopoiesis-supporting activity were funded by the German José Carreras Leukemia Foundation grant DJCLS-R03/06 and the EUROCORD III grant QLRT-2001-01918.

Abbreviations/Acronyms

ALP	alkaline phosphatase
bFGF	basic fibroblast growth factor
BLC	B-lymphocyte chemoattractant
BM	bone marrow
CB	cord blood
CFC	colony-forming cell
CMV	cytomegalovirus
EBV	Epstein–Barr virus
GABA	gamma-aminobutyric acid
GFAP	glial fibrillary acidic protein
GCP	granulocyte chemotactic protein
GM-CSF	granulocyte macrophage colony-stimulating factor
HGF	hepatocyte growth factor
HUCPV	human umbilical cord perivascular
IBMX	3-isobutyl-1-methylxanthine
IGF	Insulin-like growth factor
IL	interleukin
INF	interferon
LIF	leukemia inhibitory factor
MCP	monocyte chemoattractant protein
M-CSF	macrophage colony-stimulating factor
MIG	monokine induced by gamma interferon
MNC	mononucleocyte
MSC	mesenchymal stem cell
NC	nucleocyte
NF	neurofilament
NGF	nerve growth factor
NOD/SCID	non-obese diabetic/severely compromised immunodeficient
RT-PCR	reverse transcription-polymerase chain reaction
SDF	stromal cell-derived factor
TGF-β	transforming growth factor-beta
TNF	tumor necrosis factor
USCC	unrestricted somatic stem cell
VEGF	vascular endothelial growth factor

References

1. E. Gluckman, V. Rocha, A. Boyer-Chammard, F. Locatelli, W. Arcese, R. Pasquini, J. Ortega, G. Souillet, E. Ferreira, J.P. Laporte, M. Fernandez, C. Chastang, Outcome of cord-blood transplantation from related and unrelated donors. Eurocord Transplant Group and the European Blood and Marrow Transplantation Group. *N Engl J Med* **1997**, *337*, 373–381.

2. E. Gluckman, V. Rocha, W. Arcese, G. Michel, G. Sanz, K.W. Chan, T.A. Takahashi, J. Ortega, A. Filipovich, F. Locatelli, S. Asano, F. Fagioli, M. Vowels, A. Sirvent, J.P. Laporte, K. Tiedemann, S. Amadori, M. Abecassis, P. Bordigoni, B. Diez, P.J. Shaw, A. Vora, M. Caniglia, F. Garnier, I. Ionescu, J. Garcia, G. Koegler, P. Rebulla, S. Chevret, Factors associated with outcomes of unrelated cord blood transplant: guidelines for donor choice. *Exp Hematol* **2004**, *32*, 397–407.

3. P. Rubinstein, C. Carrier, A. Scaradavou, J. Kurtzberg, J. Adamson, A.R. Migliaccio, R.L. Berkowitz, M. Cabbad, N.L. Dobrila, P.E. Taylor, R.E. Rosenfield, C.E. Stevens, Outcomes among 562 recipients of placental-blood transplants from unrelated donors. *N Engl J Med* **1998**, *339*, 1565–1577.

4. V. Rocha, J. Cornish, E.L. Sievers, A. Filipovich, F. Locatelli, C. Peters, M. Remberger, G. Michel, W. Arcese, S. Dallorso, K. Tiedemann, A. Busca, K.W. Chan, S. Kato, J. Ortega, M. Vowels, A. Zander, G. Souillet, A. Oakill, A. Woolfrey, A.L. Pay, A. Green, F. Garnier, I. Ionescu, P. Wernet, G. Sirchia, P. Rubinstein, S. Chevret, E. Gluckman, Comparison of outcomes of unrelated bone marrow and umbilical cord blood transplants in children with acute leukemia. *Blood* **2001**, *97*, 2962–2971.

5. V. Rocha, M. Labopin, G. Sanz, W. Arcese, R. Schwerdtfeger, A. Bosi, N. Jacobsen, T. Ruutu, M. de Lima, J. Finke, F. Frassoni, E. Gluckman, Transplants of umbilical-cord blood or bone marrow from unrelated donors in adults with acute leukemia. *N Engl J Med* **2004**, *351*, 2276–2285.

6. M.J. Laughlin, M. Eapen, P. Rubinstein, J.E. Wagner, M.J. Zhang, R.E. Champlin, C. Stevens, J.N. Barker, R.P. Gale, H.M. Lazarus, D.I. Marks, J.J. van Rood, A. Scaradavou, M.M. Horowitz, Outcomes after transplantation of cord blood or bone marrow from unrelated donors in adults with leukemia. *N Engl J Med* **2004**, *351*, 2265–2275.

7. J.N. Barker, D.J. Weisdorf, T.E. DeFor, B.R. Blazar, P.B. McGlave, J.S. Miller, C.M. Verfaillie, J.E. Wagner, Transplantation of 2 partially HLA-matched umbilical cord blood units to enhance engraftment in adults with hematologic malignancy. *Blood* **2005**, *105*, 1343–1347.

8. S.J. Szilvassy, T.E. Meyerose, P.L. Ragland, B. Grimes, Differential homing and engraftment properties of hematopoietic progenitor cells from murine bone marrow, mobilized peripheral blood, and fetal liver. *Blood* **2001**, *98*, 2108–2115.

9. H. Vaziri, W. Dragowska, R.C. Allsopp, T.E. Thomas, C.B. Harley, P.M. Lansdorp, Evidence for a mitotic clock in human hematopoietic stem cells: loss of telomeric DNA with age. *Proc Natl Acad Sci USA* **1994**, *91*, 9857–9860.

10. G. Migliaccio, A.R. Migliaccio, S. Petti, F. Mavilio, G. Russo, D. Lazzaro, U. Testa, M. Marinucci, C. Peschle, Human embryonic hemopoiesis. Kinetics of progenitors and precursors underlying the yolk sac – liver transition. *J Clin Invest* **1986**, *78*, 51–60.

11. E.D. Zanjani, J.L. Ascensao, M. Tavassoli, Liver-derived fetal hematopoietic stem cells selectively and preferentially home to the fetal bone marrow. *Blood* **1993**, *81*, 399–404.

12. G. Kogler, J. Callejas, P. Hakenberg, J. Enczmann, O. Adams, W. Daubener, C. Krempe, U. Gobel, T. Somville, P. Wernet, Hematopoietic transplant potential of unrelated cord blood: critical issues. *J Hematother* **1996**, *5*, 105–116.

13. G. Kogler, S. Sensken, J.A. Airey, T. Trapp, M. Muschen, N. Feldhahn, S. Liedtke, R.V. Sorg, J. Fischer, C. Rosenbaum, S. Greschat, A. Knipper, J. Bender, O. Degistirici, J. Gao, A.I. Caplan, E.J.

Colletti, G. Almeida-Porada, H.W. Muller, E. Zanjani, P. Wernet, A new human somatic stem cell from placental cord blood with intrinsic pluripotent differentiation potential. *J Exp Med* **2004**, *200*, 123–135.

14. P. Wernet, J. Fischer, A. Caplan, E. Zanjani, H. Muller Hw, A. Knipper, G. Kogler, Isolation of non-hematopoietic stem cells from umbilical cord blood. *Biol Blood Marrow Transplant* **2004**, *10*, 738–739.

15. G. Kögler, T. Radke, A. Lefort, S. Sensken, J. Fischer, R.V. Sorg, P. Wernet, Cytokine production and hematopoiesis supporting activity of cord blood derived unrestricted somatic stem cells. *Exp Hematol* **2005**, *33*, 573–583.

16. A. Erices, P. Conget, J.J. Minguell, Mesenchymal progenitor cells in human umbilical cord blood. *Br J Haematol* **2000**, *109*, 235–242.

17. H.S. Goodwin, A.R. Bicknese, S.N. Chien, B.D. Bogucki, C.O. Quinn, D.A. Wall, Multilineage differentiation activity by cells isolated from umbilical cord blood: expression of bone, fat, and neural markers. *Biol Blood Marrow Transplant* **2001**, *7*, 581–588.

18. K. Bieback, S. Kern, H. Kluter, H. Eichler, Critical parameters for the isolation of mesenchymal stem cells from umbilical cord blood. *Stem Cells* **2004**, *22*, 625–634.

19. E.J. Gang, J.A. Jeong, S.H. Hong, S.H. Hwang, S.W. Kim, I.H. Yang, C. Ahn, H. Han, H. Kim, Skeletal myogenic differentiation of mesenchymal stem cells isolated from human umbilical cord blood. *Stem Cells* **2004**, *22*, 617–624.

20. N. Ende, The Berashis cell: a review – is it similar to the embryonic stem cell? *J Med* **2000**, *31*, 113–130.

21. N. Ende, F. Weinstein, R. Chen, M. Ende, Human umbilical cord blood effect on sod mice (amyotrophic lateral sclerosis). *Life Sci* **2000**, *67*, 53–59.

22. J. Chen, P.R. Sanberg, Y. Li, L. Wang, M. Lu, A.E. Willing, J. Sanchez-Ramos, M. Chopp, Intravenous administration of human umbilical cord blood reduces behavioral deficits after stroke in rats. *Stroke* **2001**, *32*, 2682–2688.

23. E.D. Zanjani, G. Almeida-Porada, A.W. Flake, The human/sheep xenograft model: a large animal model of human hematopoiesis. *Int J Hematol* **1996**, *63*, 179–192.

24. J.A. Airey, G. Almeida-Porada, E.J. Colletti, C.D. Porada, J. Chamberlain, M. Movsesian, J.L. Sutko, E.D. Zanjani, Human mesenchymal stem cells form Purkinje fibers in fetal sheep heart. *Circulation* **2004**, *109*, 1401–1407.

25. S.A. Wexler, C. Donaldson, P. Denning-Kendall, C. Rice, B. Bradley, J.M. Hows, Adult bone marrow is a rich source of human mesenchymal 'stem' cells but umbilical cord and mobilized adult blood are not. *Br J Haematol* **2003**, *121*, 368–374.

26. K. Mareschi, E. Biasin, W. Piacibello, M. Aglietta, E. Madon, F. Fagioli, Isolation of human mesenchymal stem cells: bone marrow versus umbilical cord blood. *Haematologia* **2001**, *86*, 1099–1100.

27. R. Sarugaser, D. Lickorish, D. Baksh, M.M. Hosseini, J.E. Davies, Human umbilical cord perivascular (HUCPV) cells: a source of mesenchymal progenitors. *Stem Cells* **2005**, *23*, 220–229.

28. T.W. Wharton, *Adenographia*. Translated by S. Freer. Oxford, UK: Oxford University Press, 1996, pp. 242–248.

29. K.D. McElreavey, A.I. Irvine, K.T. Ennis, W.H. McLean, Isolation, culture and characterisation of fibroblast-like cells derived from the Wharton's jelly portion of human umbilical cord. *Biochem Soc Trans* **1991**, *19*, 29S.

30. B.A. Naughton, J. San Roman, K. Liu, Cells isolated from Wharton's jelly of the human umbilical cord develop a cartilage phenotype when treated with TGFβ in vitro (abstract). *FASEB J* **1997**, *11*, A19.

31. A. Kadner, S.P. Hoerstrup, J. Tracy, C. Breymann, C.F. Maurus, S. Melnitchouk, G. Kadner, G. Zund, M. Turina, Human umbilical cord cells: a new cell source for cardiovascular tissue engineering. *Ann Thorac Surg* **2002**, *74*, S1422–1428.

32. Y.A. Romanov, V.A. Svintsitskaya, V.N. Smirnov, Searching for alternative sources of postnatal human mesenchymal stem cells: candidate MSC-like cells from umbilical cord. *Stem Cells* **2003**, *21*, 105–110.

33. A. Kadner, G. Zund, C. Maurus, C. Breymann, S. Yakarisik, G. Kadner, M. Turina, S.P. Hoerstrup, Human umbilical cord cells for cardiovascular tissue engineering: a comparative study. *Eur J Cardiothorac Surg* **2004**, *25*, 635–641.

34. K.E. Mitchell, M.L. Weiss, B.M. Mitchell, P. Martin, D. Davis, L. Morales, B. Helwig, M. Beerenstrauch, K. Abou-Easa, T. Hildreth, D. Troyer, S. Medicetty, Matrix cells from Wharton's jelly form neurons and glia. *Stem Cells* **2003**, *21*, 50–60.

6
Good Manufacturing Practices: Clinical-Scale Production of Mesenchymal Stem Cells

Luc Sensebé, Philippe Bourin, and Luc Douay

6.1
Introduction

The concept of progenitors/stem cells for cells of mesenchymal origin, which led to the description of mesenchymal stem cells (MSCs), arose during the 1970s with Friedenstein and colleagues [1], who found nonhematopoietic progenitors in bone marrow cells which they called colony-forming unit fibroblasts (CFU-Fs) [1]. CFU-Fs were a heterogeneous population of progenitors with plastic adherence properties that can differentiate both *in vitro* and *in vivo* to osteogenic, chondrogenic, and/or stromal cells. Later, during the 1980s, Owen [2, 3] described the "stromal system", whereby bone marrow-resident stem cells maintain a level of self-renewal and give rise to cells that differentiate into various connective tissue lineages. Since the 1990s, increasing experimental data have allowed for the definition of MSCs as multipotent adult stem cells that can differentiate into connective skeletal tissue, bone, cartilage, marrow stroma, and adipocytes [3–5]. Like the CFU-Fs directly isolated from bone marrow, cultured MSCs remain heterogeneous. Although some differentiate into three lineages (adipogenic, osteogenic, and chondrogenic), most display an osteochondrogenic potential leading to a hierarchical model, the early tripotent mesenchymal progenitors losing their adipogenic then chondrogenic differentiation potential with increasing cell doubling [6]. In addition, controversial data suggest that MSCs may give rise to sarcomeric muscle (skeletal and cardiac) [7–9], endothelial cells [10], and even cells of non-mesodermal origin, such as hepatocytes [11] and neural cells [12, 13].

The wide range of differentiation potential of MSCs, the possibility of their engraftment [14, 15], their immunosuppressive effect [16, 17], and their expansion through culture led to increasing clinical interest in the use of MSCs, through either intravenous infusion or site-directed administration, in numerous pathologic situations. Since the first Phase I study that demonstrated the good tolerance and safety of MSC transplantation [18], clinical trials have investigated large bone defects [19], genetic bone disease as osteogenesis imperfecta [14, 20], hematopoie-

Stem Cell Transplantation. Biology, Processing, and Therapy.
Edited by Anthony D. Ho, Ronald Hoffman, and Esmail D. Zanjani
Copyright © 2006 WILEY-VCH Verlag GmbH & Co. KGaA, Weinheim
ISBN: 3-527-31018-5

tic stem-cell transplantation for repair of hematopoietic stroma [21], and treatment or prevention of graft-versus-host disease (GvHD) [22, 23].

In order to conduct clinical trials, billions rather than millions of MSCs are required, either solely or linked to biomaterials. However, standards for the production of MSCs in this environment are lacking. This type of production necessitates adherence to good manufacturing practices (GMP) to ensure the delivery of a "cell drug" that is safe, reproducible, and efficient. All parts of the process must be defined, specifically the starting material (tissue origin, separation or enrichment procedures), cell density in culture, and medium (fetal calf serum [FCS] or human serum, cytokines with serum-free medium for target). In order to reach the GMP goal, cells must be cultured in as close to a closed system as possible, though unfortunately such a system is not yet available. Analytical methods are needed to assay the active compound and impurities. At a minimum, the quality control (QC) of cells must consider the phenotype, functional potential, and microbiological safety. Finally, quality assurance (QA) system procedures specific to the production of MSCs as a cell drug must be determined and implemented.

6.2
Prerequisites for the Clinical-Scale Production of MSCs

Before beginning the clinical-grade production of MSCs, four main parameters must be defined: the starting material; the cell seeding density; the number of passages; and the medium.

6.2.1
Starting Material

Since the first description of CFU-Fs, the main source of MSCs remains bone marrow [19–23]. The CFU-F assay reveals the estimated number of MSCs in bone marrow to be between 1 per 10^4 and 1 per 10^5 mononuclear cells. The age of the donor is important. The bone marrow of children contains more CFU-Fs than that of adults: 29 per 10^6 versus 3.2 per 10^6 mononuclear cells, respectively [24].

MSCs are present in the mononuclear fraction of bone marrow cells [5,18–23]. After being separated by density gradient, most of the MSC populations are isolated because of their physical property of plastic adherence [5,18–23]. However, cells such as macrophages and endothelial cells also have plastic adherence properties, which necessitates several passages before these cells are removed [25].

Although the *in-vivo* phenotype of the MSC is not known, cultured MSCs express a number of markers, none of which is specific individually or in combination [25–27]. MSCs are devoid of hematopoietic and endothelial markers such as CD45 and CD31. Human MSCs have shown CD34$^-$. The absence of hematopoietic antigen expression has been used to select some rare mesodermal stem cells (Multipotent Adult Progenitor Cells [MAPCs]) present in the fraction of CD45$^-$

GlyA$^-$ bone marrow cells [28, 29]. The real relationship or hierarchy between MAPCs and MSCs is not known. Human MSCs express CD73, CD90 and CD105 at various levels [5, 25]. A subset of multipotential bone marrow stromal cells is recognized by the STRO-1 antibody [30–32]. CFU-Fs from fresh bone marrow are present in the STRO-1$^+$ population [30]. Before culture, this STRO-1$^+$ population can be selected for enrichment with use of immunomagnetic devices, which leads to a 10-fold increase. Fluorescence-activated cell sorting (FACS) of the STRO-1^{bright} fraction leads to a 950-fold enrichment of CFU-Fs , and STRO-1$^+$/CD106$^+$ cells represent a highly purified MSC population [33]. Other cell antigens, such as α1 integrin subunit (CD49a), can be used to enrich the bone marrow mononuclear cell fraction in MSCs. CD49a is expressed in MSCs, not CD34$^+$ cells, and all CFU-Fs are present in the CD49a$^+$ fraction [34, 35].

All of the procedures used enrich MSCs contain of a starting material rather than a real selection of MSCs. If these procedures function well in the research laboratory, then two main questions remain:

• Does the selected population contain all the MSCs?
• Do we have, as in CD34 for hematopoietic stem cells (HSC), a "clinical-grade" antibody that could be used?

The bone-marrow MSC population contains a small fraction of rapidly self-renewing cells (RS cells) with the greatest potential for multilineage differentiation [36]. In cultured MSCs, the RS cells can be defined by their light-scattering characteristics, the FS^{lo}/SS^{lo} cells being up to 90 % clonogenic [37], but they are STRO-1$^-$, and only some express CD90 [36]. As a consequence, immunoselection with these antibodies could lead to a loss of the most primitive progenitors/stem cells. Moreover, although we have numerous antibodies for experimental selection procedures, "clinical-grade" antibodies for MSC enrichment are not yet available. Usually, MSCs are cultured from bone marrow aspirates, but they can be obtained from trabecular bone fragments obtained during total hip arthroplasty. The number of CFU-Fs in the cells from these fragments are in the same range as those in the bone marrow aspirate (1 CFU-F/3.2 × 10^4 cells versus 1 or 10 CFU-F/1 × 10^5 cells) [38]. These MSCs retain the normal potential of differentiation [39], but they can present a rapid decrease in their proliferative potential after the first passage (personal data).

Finally, for MSC production, the commonly used population is mononuclear bone marrow cells selected by plastic adherence after 1 to 3 days of culture [19–23].

6.2.1.1 Are there Alternative Sources of MSCs?

Although this topic has been controversial for many years [38, 40], recent data have shown that umbilical *cord blood* (CB) is a potential source [41–43]. MSCs from such blood can be isolated following a technique close to that used for bone marrow MSCs, the critical parameters being time from collection to isolation, volume of CB, and amount of mononuclear cells [43]. Compared to that in bone marrow, the frequency of MSCs in umbilical CB is lower, from 1 per 10^5 to 1 per 10^8 mononuclear cells. These cells seem to be *bona fide* MSCs and are negative for hematopoietic antigens; they express CD73, CD90 and CD105 antigens; they have a high potential for proliferation; and they are capable of osteogenic, chondrogenic, and adipogenic differentiation. Moreover, with umbilical CB mononuclear cells, after a negative immunodepletion step to eliminate $CD3^+$, $CD14^+$, $CD19^+$, $CD38^+$, $CD66b^+$, and $GlyA^+$ cells, multipotent MSCs with the ability to differentiate into cell types of all three germ layers can be obtained [42]. The existence of networks of registered umbilical CB banks and the high potential of differentiation could lead to the use of such blood to produce MSCs on a clinical scale.

Amniotic fluid represents another source of MSCs because of its phenotype and potential for differentiation similar to that of bone marrow MSCs [44].

However, in adults *peripheral blood* does not seem to be a reliable source of MSCs. Although cells with the characteristics of MSCs are detected in some peripheral blood stem cell grafts [45], usually MSCs are absent from these grafts and from the blood of normal donors [38, 45, 46].

Finally, *adipose tissues* may represent an important source of easily available and abundant MSCs [47–49]. The cells of this tissue have already shown very interesting properties of endothelial and cardiac differentiation, both *in vivo* and *in vitro* [9, 50].

6.2.2
Cell Plating Density

Cell plating density is a critical parameter to ensure good expansion rate and maintenance of the differentiation potential of MSCs. Most experimental research and some clinical trials use high cell density, such as 170×10^3 mononuclear cells cm^{-2}, initially plated in cultures for the co-infusion of MSCs and HSCs [21]. For the next passages, plating density is decreased to 6×10^3 cells cm^{-2} [20, 21, 51]. The team of Prockop demonstrated that the development of early progenitors/stem cells as RS cells depended on very low plating density, from 3 to 59 cells cm^{-2} [36, 52]. Increasing the plating density from 10 cells cm^{-2} to 1×10^3 cells cm^{-2} led to a decrease in the amount of expansion, from 500- to 30-fold, and in CFU-F cloning efficiency, from 36 % to 12 % [52]. These results could explain why some data showed a rapid decrease in proliferation rate and multiple differentiation potential with 6×10^3 cells cm^{-2} [51]. In the context of clinical-scale production of MSCs, a plating density of 10 cells cm^{-2} is difficult to achieve because of the need for very wide culture surfaces and nonoptimal cell production. For example, to obtain 10^8 cells requires a culture surface of 4×10^4 cm^2 at 10 cells cm^{-2}, and only 3×10^3 cm^2 at 10^3 cells cm^{-2}. A plating density of 1000 cells cm^{-2} seems a reasonable compromise, because this density allows for a high number of harvested cells [52].

6.2.3
Number of Passages

MSCs are adherent cells that grow on plastic and have normal growth inhibition at confluence. Moreover, after the first adherence step of bone marrow mononuclear cells and the subsequent expansion period, some macrophages, lymphocytes and endothelial cells remain within the layer of MSCs. All these characteristics led to the use of successive passages for obtaining large amounts of pure MSCs devoid of any other cells. If further passages are necessary, these can alter the quality of MSCs. In rats, bone marrow MSCs retain their proliferative and differentiation potential until passage 15 [53]. In humans, after the first 3 weeks of initial culture and 12 to 15 doublings, successive passages slow the proliferation rate, and cells progressively show loss of multipotentiality, retaining only their ability to differentiate in osteoblasts [6, 51, 54]. These alterations are more pronounced with material from adults than from children [24]. The aging of MSCs by expansion may be related to the constant decrease in telomere length [24]. Factors include MSC purity, the number of cells needed, and increasing age with number of passages. Use of the classic medium containing FCS can result in MSC expansion with a minimal number of passages.

6.2.4
Medium

The medium is the final main parameter to consider in clinically up-grading MSC culture. The basic composition does not seem a real problem. DMEM or αMEM [20, 21, 23, 54] is often used, but other compositions work well [19, 55]. The two pivotal compounds in medium composition are the serum (FCS versus human serum or plasma) and growth factors. Classically, the optimal conditions for MSC expansion require FCS-supplemented media, the standard being 10%. This FCS needs to be carefully tested to ensure the best expansion rate [5, 20, 36, 37]. Although FCS carefully tested for viruses and ensuring good traceability exists, the risk of transmission of infectious disease, even the new variant of Creutzfeld–Jakob disease, cannot be excluded. Moreover, in such a culture medium, MSCs retain in their cytoplasm some FCS proteins, and a standard preparation of 100×10^6 MSCs contains 7 to 30 mg of FCS proteins [56]. Such amounts of proteins in MSC grafts may elicit immunologic response *in vivo*, and humoral responses could explain some of the failure in cell therapy with MSCs [20, 56]. Recently, some very early progenitor cells, with longer telomeres and expressing genes found in embryonic cells, have been selected in serum-deprived medium [57].

All of these data led to an effort to decrease or eliminate the use of FCS in culturing MSCs; an alternative to reducing the FCS amount may be effective [56]. Some teams attempted to replace FCS with human autologous or AB serum. MSC expansion in AB serum alone or in autologous serum is less efficient, but supplementing the medium with human AB serum by FGF2 overcomes this deficit [58, 59]. If MSCs cultured with autologous serum retain their osteogenic potential [60], the bone formed is less extensive than that formed with FCS culture [59].

Although MSCs can be expanded in 10 % FCS medium without supplementation of exogenous growth factors, these factors can increase the expansion rate and are mandatory in serum-free media. Some serum-free media have been developed in the research process [31, 61], but they are not yet available for clinical-scale production that follows GMP. The growth factor requirement of MSCs is not completely known: it includes at least platelet-derived growth factor (PDGF), epidermal growth factor (EGF), transforming growth factor β (TGFβ), insulin-like growth factor (IGF), and fibroblast growth factor 2 (FGF2) [31, 61]. Of these, FGF2 appears to be the predominant cytokine involved in MSC biology. FGF2 stimulates the proliferation of all mesoderm-derived cells [62]. MSCs in culture express receptors for FGF1 and FGF2 [63]. Under all culture conditions, FGF2 promotes the proliferation of MSCs with loss of contact inhibition and growth in multilayered sheets [55, 58, 64, 65], beginning with a low concentration (0.2 ng mL^{-1}) to 20 ng mL^{-1} [66]. FGF2 does not have a detectable effect on colony formation, but increases proliferative potential [55, 67]; moreover, it seems to decrease the number of CFU-Fs by about 30 % [68]. FGF2 maintains the multipotentiality and osteogenicity of MSCs and may increase the frequency of tripotential progenitors [6, 68]. Recent data involving 1 ng mL^{-1} FGF2 showed an increase of MSC lifespan to more than 70 doublings, with maintenance of differentiation potential to 50 doublings. In these cultures, the telomere length increased early and then steadily decreased with doubling. Telomerase activity was never detected; thus, immature progenitors could be selected by FGF2 [68]. In conclusion, in order to produce large amounts of MSCs, the use of FGF2 allows for better expansion efficiency, preserved differentiation potential, and a decreased time of culture for the same number of cells.

6.2.5
Culture with Biomaterials

Musculoskeletal lesions can be repaired by using biomaterials combined with MSCs. This solution has the advantage of allowing for better integration of the biomaterial, which in the long term offers an adapted cicatrization and increases the solidity of repair. Certain authors have proposed that the MSCs are cultivated independently, and then combined with the biomaterial immediately before grafting [19, 69]. To implement this solution, particular techniques for the culture of the cells are not needed. In other cases, cells can be cultured directly on or in the biomaterial [70, 71]. This strategy thus imposes a more sophisticated culture system than the usual ones. Currently, no bioreactor is adapted for clinical use.

6.3
Clinical-Scale Production: the French Experience

In 2001, the French society for HSC transplantation and cellular therapy, Société Française de Greffe de Moelle et Thérapie Cellulaire (SFGM-TC), started a program to co-transplant MSCs and HSCs in allografts and autografts. To produce clinical-grade MSCs, the first task was to define precisely the conditions of MSC culture and the process to reach GMP.

6.3.1
Culture Conditions

The first step consists of the choice of medium. The medium must be as simple as possible to avoid nontraceable products or molecules, except for FCS. After several tests, αMEM without desoxyribonucleotides and ribonucleotides and supplemented with 10 % FCS gave the best results. The starting material was bone marrow harvested from the iliac crest or sternum. A concentration step was not needed because of sufficient number of CFU-Fs: 32 CFU-F/10^6 bone marrow nucleated cells (SFGM-TC data). Bone marrow nucleated cells were seeded without density gradient separation or use of immunomagnetic devices. The medium was changed twice weekly until the cells were confluent. At culture inception, three different cell concentrations were tested (Table 6.1). After 21 days, the number of cells obtained was identical among concentrations, but the proportion of recovery was better for the lowest concentration. In additional passages, two cell-plating densities were compared, and again, the lowest concentration gave the best results. At 1×10^3 cells cm^{-2}, the increase in cell number was 17.8-fold at the

Table 6.1 Percentage of recovered cells at the end of first period of culture of mesenchymal stem cells.

Cell plating density cm^{-2} day 0	0.5×10^5 (n = 17)	1×10^5 (n = 17)	2×10^5 (n = 22)
No. of cells cm^{-2} day 21	0.19×10^5	0.19×10^5	0.22×10^5
Δ	0.38^a	0.19^b	0.11

At culture inception, whole bone marrow nucleated cells were seeded at three different concentrations: 0.5×10^5, 1×10^5, and 2×10^5 cells cm^{-2}. Cells were grown in medium supplemented with 10 % FCS without FGF2. The medium was renewed twice each week until cells were confluent (obtained in 3 weeks). At day 21, the adherent cells were harvested with use of trypsin-EDTA and counted. The proportion of cell recovery (Δ) was calculated as:
No. of adherent cells at day 21/No. of seeded cells at day 0
[a]0.5×10^5 versus 1×10^5, p <0.0002.
[b]1×10^5 versus 2×10^5, p <0.001.

first passage and 7.3-fold at the second. Finally, the addition of 1 ng mL^{-1} FGF2 allowed for the highest increase of cell number (51-fold at the first passage and 47-fold at the second) for the lowest concentration. At this concentration, FGF2 increased not only the cell expansion but also the maintenance of CFU-F efficiency within the cells.

Thus, it was decided that culture with whole nucleated bone marrow cells at a concentration of 50×10^4 cells cm^{-2} in 10% FCS medium supplemented with 1 ng mL^{-1} FGF2 gave the best results. For further passages, the cell plating density was decreased to 1×10^3 cells cm^{-2}.

6.3.2
Devices for MSC Culture

Today, totally closed systems for MSC culture are not yet available, but GMP conditions can be reached with use of laminar air flow cabinets for culture inception, medium changes, trypsinization, and cell removal. Wide surface containers, manufactured under GMP conditions by different companies (e.g., CellSTACK from Corning, USA or CellFactory from Nunc, Denmark), are available in different

Table 6.2 Mesenchymal stem cells obtained in clinical-scale culture.

	P0 (day 21)	P1 (day 40)
Total no. of cells	122×10^6	153×10^6
Amount of increase	1.67-fold	123-fold
CD45 (%)	0	0
CD73 (%)	99.4 (241)	99.9 (91)
CD90 (%)	99.5 (273)	99.9 (174)
CD49a (%)	54 (3.8)	47 (2.3)
HLA-DR	0	0

Results of a representative experiment. Whole bone marrow nucleated cells (63.6×10^6 cells, corresponding to 4 mL bone marrow aspirate) were plated at 0.5×10^5 cells cm^{-2} in a two-level CELL-STACK (Corning, USA) (1272 cm^2). The medium consisted of αMEM with 10% FCS, supplemented with 1 ng mL^{-1} FGF2, and was replaced twice weekly until day 21 (passage 0 [P0]). Confluence occurred between days 16 and 18. At day 21, the cells were harvested using trypsin-EDTA and plated at 1×10^3 cells cm^{-2} in a second two-level CELLSTACK (1.24×10^6 cells in 300 mL medium). Culture was performed under previous conditions until day 18 (first passage [P1]). The amount of increase was calculated by dividing the number of cells at the end of each culture period by the number of starting cells (see Table 6.1). At the end of each passage and after trypsin was applied, cells were analyzed by flow cytometry (FACSCalibur; Becton Dickinson, USA). All antibodies used were from Becton Dickinson. During culture, all interventions were performed using specific devices developed by Macopharma (Roubaix, France).

sizes, with 1 to 40 levels and surfaces ranging from 636 cm^2 to 25 440 cm^2. To perform MSC culture in a near-closed system, specific connecting systems with bags containing medium were developed by Macopharma (Tourcoing, France). With previous defined culture conditions and use of these devices, a specific non-time-consuming process is implemented. Starting with 5 to 10 mL bone marrow, 1 or 2 billion MSCs can be obtained in 5 weeks with just one passage. Results of one experiment are shown in Table 6.2 (L. Sensebé, personal data).

6.4
QA and QC

The production of cells for therapeutic use and the production of a drug differ according to the "active ingredient". On one hand, the ingredient is eminently variable because of its complexity, whereas on the other, it is defined perfectly by its chemical formula. In the pharmaceutical industry, the initial control of the "active ingredient" allows for insuring the reproducibility of the pharmacological effect. However, in cellular therapy, reproducibility of the therapeutic effect is reached mainly by strictly framing the cell production process, which will maximally limit intrinsic variability. This control is ensured by the system of QA and quality control (QC) carried out at all stages of the transformation process. This section describes a French software experiment that allows for thorough management of the QA and QC specific to the production of MSCs.

6.4.1
French Experimental System GESAQ

GESAQ is an acronym for "Groupe d'Étude du Système d'Assurance de la Qualité". This group, composed of five teams, defined the schedule of conditions to manage all components of QA of a cell-therapy unit. The system ensures the traceability of the grafts used from harvest to use. A data-processing service company developed the software, which is the property of the Etablissement Français du Sang (EFS).

The activity of the cell-therapy unit is managed in real time, on the basis of a convivial single system. All components are taken into account:
- the building (rooms with controlled dust contamination, storage);
- matter (initial cell product, ancillary products);
- materials (laminar flow hood, centrifugal machine, control freezing apparatus);
- personnel; and
- working methods (implemented in the step-by-step protocols).

For example, the software manages the program of preventive maintenance of the materials and the buildings. It controls the number and delayed use of stocks of devices and ancillary products. Thus, at the time of transformation of a cellular product, only one competent technician, with data on validated materials and products, can perform the transformation process.

The program operates under PC Windows with a Microsoft Internet Explorer interface, and all functions are accessible from a principal menu with 12 sub-menus. This software is used by a growing number of cell-therapy units in France, and greatly facilitates the QA management.

6.4.2
Quality Controls

Quality control must be carried out at each of the three stages of the process to produce MSCs (i.e., the starting materials, culture and final product). The goal and the nature of QC differs according to each stage. Sometimes, QC must follow contradictory objectives such as being discriminating and fast enough to allow passage to the next stage. In published clinical trials, the initial control of the harvested cells is never mentioned [18–21,23,72].

QC during culture is simple (visual examination, bacterial contamination), whereas that before injection of the MSCs is disparate and includes, in general, a more or less complete phenotypical analysis and test of bacterial contamination. In certain cases, a functional test [20] or search for contaminant cancerous cells [21] is performed.

6.4.2.1   **Control of the Harvested Graft**
These controls ensure that the graft contains quality cells and a sufficient number to reach the necessary number of produced cells, and that it is not likely to transmit infectious disease. Thus, the controls include serologic tests of the donor before and/or when cells are taken, mainly for infections due to immunodeficiency viruses, hepatitis B and C and human T-cell leukemia virus. Currently, no test equivalent to the evaluation of CD34+ exists to determine the quantity of MSCs in a bone marrow graft. The only quantitative test is the evaluation of CFU-Fs, and the length of the test prevents its use in determining, *a priori*, the quantity of MSCs present initially in the graft. Thus, only the number and viability of the nucleate cells can be used to validate the cellular starting suspension.

Figure 6.1   Flow cytometry analysis for quality control of produced MSCs at first passage. Phenotypic analysis of a typical clinical-scale culture of MSCs. The cultured MSCs at first passage (P1) were analyzed by flow cytometry with use of a FACSCalibur (Becton Dickinson) and antibodies against CD34, CD45, HLA-DR, CD49a, CD73, and CD90. For each antibody, 10^5 cells were incubated with the relevant antibody, and isotypic control was performed with use of a corresponding Ig isotype and fluorochrome. For analysis, 10 000 events were acquired. Cultivated MSCs were CD45$^-$, CD34$^-$, HLA-DR$^-$, CD90$^+$, CD73$^+$, and CD49a$^+$. ▶

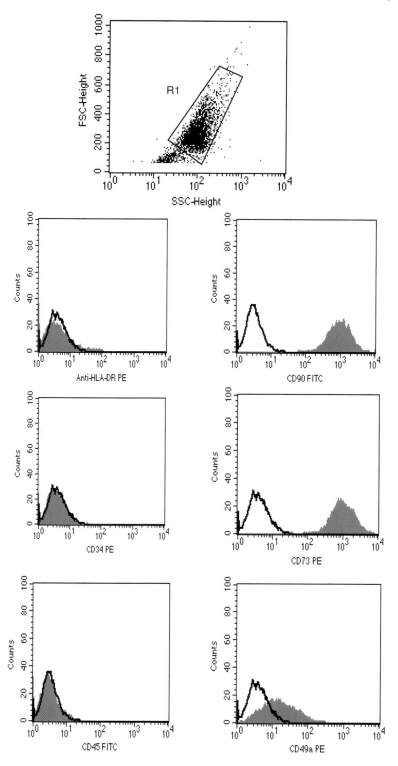

6.4.2.2 **Controls During Culture**

Visually monitoring the cells during culture remains a significant element of follow-up. It is thus imperative that devices allow it. Objective QC is carried out when cells are detached, and aims to ensure the quality of the cells to continue the culture and to check for bacterial contamination. These controls include numeration, evaluation of cellular viability, phenotypical analysis and determination of the CFU-Fs. The phenotypical analysis, in particular, allows for judging contamination in cells of hematopoietic origin (CD45+). A functional test of the CFU-Fs is useful here also, only for *a posteriori* taking into account the time of realization.

6.4.2.3 **Controls During Release of the Graft**

These controls are essential, because they determine whether the graft is used, or not. They must be fast (less than a few hours) and precise. They are adapted to the required therapeutic goal. For example, one would check the absence of expression of CMH II for treatment of GvHD. Such controls include numeration, evaluation of cellular viability, and phenotypical analysis. The markers tested include at least the antigens CD45, CD90, and CD73 [5] (Fig. 6.1). The evaluation of infection by mycoplasma is also possible through detection of an ADN fluorescent marker such as DAPI.

6.5
Future Prospects

The clinical production of MSCs according to GMP is one of the main steps in realizing the promise of MSCs in therapy. The final goal is standardized "cell drug" production. For this purpose, we need to improve QA and QC of not only the processes but above all the media and devices. There is a growing consensus to avoid animal serum in cell therapy production because of its immunological [20, 56] and, more importantly, infectious risk. To produce serum-free media, five main categories of molecules must be tested: transporters; metabolites; enzyme inhibitors; antioxidants; general hormones; and growth factors. This testing will allow for defining either serum-free media for MSC expansion or differentiation in a specific cell type. As pointed out earlier, MSCs are adherent cells that require a suitable surface on which to grow and expand. Moreover, the prospects for their use in therapy force us as much as possible to handle them under GMP conditions. Thus, the new devices must offer surface qualities allowing for their adhesion and culture under completely closed conditions. This system will allow for the creation of an automated culture process. Finally, for local use, specific scaffolds and matrices must be developed to ensure the local development and suitable differentiation of MSCs. All of these developments require close cooperation between research laboratories, cell-therapy units, and private companies.

In conclusion, the clinical-scale production of MSCs according to GMP is now possible. However, before starting such a production, it is important to define precisely all of the parameters and different steps of the process, including the controls, and to use traceable reagents and devices. Moreover, all of these prerequisites must be linked to a QA system in order to obtain approval for human use from regulatory authorities. Taking into account the previously described security requirements, it is both feasible and reasonable to begin production of MSCs for large Phase II clinical studies. The main future developments will be the implementation of serum-free cultures in closed systems or in bioreactors.

Abbreviations/Acronyms

CB	cord blood
CFU-Fs	colony-forming unit fibroblasts
EGF	epidermal growth factor
FCS	fetal calf serum
FGF2	fibroblast growth factor 2
GMP	Good Manufacturing Practice
HSC	hematopoietic stem cells
IGF	insulin-like growth factor
MAPCs	multipotent adult progenitor cells
MSCs	mesenchymal stem cells
PDGF	platelet-derived growth factor
QA	quality assurance
QC	quality control
RS (cells)	rapidly self-renewing (cells)
TGFβ	transforming growth factor β

References

1. A.J. Friedenstein, J.F. Gorskaja, N.N. Kulagina, *Exp Hematol* **1976**, *4*, 267–274.
2. M. Owen, in: W.A. Peck (Ed.), *Bone and Mineral Research*, Volume 3. Elsevier, New York, **1985**, pp. 1–25.
3. A.I. Caplan, *J Orthop Res* **1991**, *9*, 641–650.
4. D.J. Prockop, *Science* **1997**, *276*, 71–74.
5. M.F. Pittenger, A.M. Mackay, S.C. Beck, R.K. Jaiswal, R. Douglas, J.D. Mosca, M.A. Moorman, D.W. Simonetti, S. Craig, D.R. Marshak, *Science* **1999**, *284*, 143–147.
6. A. Muraglia, R. Cancedda, R. Quarto, *J Cell Sci* **2000**, *113*, 1161–1166.
7. S. Wakitani, T. Saito, A.I. Caplan, *Muscle Nerve* **1995**, *18*, 1417–1426.
8. S. Makino, K. Fukuda, S. Miyoshi, F. Konishi, H. Kodama, J. Pan, M. Sano, T. Takahashi, S. Hori, H. Abe, J. Ihata, A. Umezawa, S. Ogawa, *J Clin Invest* **1999**, *103*, 697–705.
9. V. Planat-Bénard, C. Menard, M. André, M. Puceat, A. Perez, J.-M. Garcia-Verdugo, L. Pénicaud, L. Casteilla, *Circ Res* **2004**, *94*, 223–229.
10. J. Oswald, S. Boxberger, B. Jorgensen, S. Feldmann, G. Ehninger, M. Bornhäuser, C. Werner, *Stem Cells* **2004**, *22*, 377–384.

11. J. Chagraoui, A. Lepage-Noll, A.A. Anjo, G. Uzan, P. Charbord, *Blood* **2003**, *101*, 2973–2982.

12. D. Woodbury, E.J. Schwarz, D.J. Prockop, I.B. Black, *J Neurosci Res* **2000**, *61*, 364–370.

13. A. Hermann, R. Gastl, S. Liebau, M. Oana Popa, J. Fiedler, B.O. Boehm, M. Maisel, H. Lerche, J. Schwarz, R. Brenner, A. Storch, *J Cell Sci* **2004**, *117*, 4411–4422.

14. E.M. Horwitz, D.J. Prockop, L.A. Fitzpatrick, W.W. Koo, P.L. Gordon, M. Neel, M. Sussman, P. Orchard, J.C. Marx, R.E. Pyeritz, M.K. Brenner, *Nat Med* **1999**, *5*, 309–313.

15. S.M. Devine, A.M. Bartholomew, N. Mahmud, M. Nelson, S. Patil, W. Hardy, C. Sturgeon, T. Hewett, T. Chung, W. Stock, D. Sher, S. Weissman, K. Ferrer, J. Mosca, R. Dean, A. Moseley, R. Hoffmann, *Exp Hematol* **2001**, *29*, 244–255.

16. M. Di Nicola, C. Carlo-Stella, M. Magni, M. Milanesi, P.D. Longoni, P. Matteucci, S. Grisanti, A.M. Gianni, *Blood* **2002**, *99*, 3838–3843.

17. W.T. Tse, J.D. Pendleton, W.M. Beyer, M.C. Egalka, E.C. Guinan, *Transplantation* **2003**, *75*, 389–397.

18. H.M. Lazarus, S.E. Haynesworth, S.L. Gerson, N.S. Rosenthal, A.I. Caplan, *Bone Marrow Transplant*, **1995**, *16*, 557–564.

19. R. Quarto, M. Mastrogiacomo, R. Cancedda, S.M. Kupetov, V. Mukhachev, A. Lavroukov, E. Kon, M. Marcacci, *N Engl J Med* **2001**, *344*, 385–386.

20. E.M. Horwitz, P.L. Gordon, W.W. Koo, J.C. Marx, M. Neel, R.Y. McNall, L. Muul, T. Hofmann, *Proc Natl Acad Sci USA* **2002**, *99*, 8932–8937.

21. O.N. Koç, S.L. Gerson, B.W. Cooper, S.M. Dyhouse, S.E. Haynesworth, A.I. Caplan, H.M. Lazarus, *J Clin Oncol* **2000**, *18*, 307–316.

22. F. Frassoni, A. Bacigalupo, E. Gluckman, V. Rocha, B. Bruno, H. Lazarus, S. Devine, K. Holland, P. McCarthy, P. Curtin, R. Maziaz, E. Shpall, A.M. Moseley (abstract) *Bone Marrow Transplant* **2002**, *29*, 75.

23. K. Le Blanc, I. Rasmusson, B. Sundberg, C. Götherström, M. Hassan , M. Uzunel, O. Ringdén, *Lancet* **2004**, *363*, 1439–1441.

24. M.A. Baxter, R.F. Wynn, S.N. Jowitt, J. Ed Wraith, L.J. Fairbairn, I. Bellantuono, *Stem Cells* **2004**, *22*, 675–682.

25. R.J. Deans, A.B. Moseley, *Exp Hematol* **2000**, *28*, 875–884.

26. P. Charbord, E. Tamayo, F. Deschaseaux, J.-P. Remy-Martin, L. Pelletier, L. Sensebé, M. Deschaseaux, B. Peault, P. Hervé, *Hematology* **1999**, *4*, 257–282.

27. E.H. Javazon, K.J. Beggs, A.W. Flake, *Exp Hematol* **2004**, *32*, 414–425.

28. M. Reyes, T. Lund, T. Lenvik, D. Aguiar, L. Koodie, C.M. Verfaillie, *Blood* **2001**, *98*, 2615– 2625.

29. Y. Jiang, B.N. Jahagirdar, R.L. Reinhardt, R.E. Schwartz, C.D. Keene, X.R. Ortiz-Gonzalez, M. Reyes, T. Lenvik, L. Lund, M. Blackstad, J. Du, S. Aldrich, A. Lisberg, W.C. Low, D.A. Largaespada, C.M. Verfaillie, *Nature* **2002**, *418*, 41–49.

30. P.J. Simmons, B. Torok-Storb, *Blood* **1991**, *78*, 55–62.

31. S. Gronthos, P.J. Simmons, *Blood* **1995**, *85*, 929–940.

32. J.E. Dennis, J.-P. Carbillet, A.I. Caplan, P. Charbord, *Cells Tissues Organs* **2002**, *170*, 73–82.

33. S. Gronthos, A.C.W. Zannettino, S.J. Hay, S. Shi, S.E. Graves, A. Kortesidis, P.J. Simmons, *J Cell Sci* **2003**, *116*, 1827–1835.

34. F. Deschaseaux, P. Charbord, *J Cell Physiol* **2000**, *184*, 319–325.

35. F. Deschaseaux, F. Gindraux, R. Saadi, L. Obert, D. Chalmers, P. Hervé, *Br J Haematol* **2003**, *122*, 506–517.

36. D.C. Colter, I. Sekiya, D.J. Prockop, *Proc Natl Acad Sci USA* **2001**, *98*, 7841–7845.

37. J.R. Smith, R. Pochampally, A. Perry, S.-C. Hsu, D.J. Prockop, *Stem Cells* **2004**, *22*, 823–831.

38. S.A. Wexler, G. Donaldson, P. Denning-Kendall, C. Rice, B. Bradley, J.M. Hows, *Br J Haematol* **2003**, *121*, 368–374.

39. R. Tuli, S. Tuli, S. Nandi, M.L. Wang, P.G. Alexander, H. Haleem-Smith, W.J. Hozack, P.A. Manner, K.G. Danielson, R.S. Tuan, *Stem Cells* **2003**, *21*, 681–693.

40. K. Mareschi, E. Biasin, W. Piacibello, M. Aglietta, E. Madon, F. Fagioli, *Haematologia* **2001**, *86*, 1099–1100.

41. A. Erices, P. Conget, J. Minguell, *Br J Haematol* **2000**, *109*, 235–242.

42. O.K. Lee, T.K. Kuo, W.-M. Chen, K.-D. Lee, S.-L. Hsieh, T.-H. Chen, *Blood* **2004**, *103*, 1669–1675.

43. K. Bieback, S. Kern, H. Klüter, H. Eichler, *Stem Cells* **2004**, *22*, 625–634.

44. P.S. in't Anker, S.A. Scherjon, C. Kleijburg-van der Keur, W.A. Noort, F.H.J. Claas, R. Willemze, W.E. Fibbe, H.H.H. Kanhai, *Blood* **2003**, *102*, 1548–1549.

45. M. Fernandez, V. Simon, G. Herrera, C. Cao, H. Del Favero, J.J. Minguell, *Bone Marrow Transplant* **1997**, *20*, 265–271.

46. H.M. Lazarus, S.E. Haynesworth, S.L. Gerson, A.I. Caplan, *J Hematother* **1997**, *6*, 447–455.

47. S. Gronthos, D.M. Franklin, H.A. Leddy, P.G. Robey, R.W. Storms, J.M. Gimble, *J Cell Physiol* **2001**, *189*, 54–63.

48. P.A. Zuk, M. Zhu, P. Ashjian, D.A. De Ugarte, J. Huang, H. Mizuno, Z.C. Alfonso, J.K. Frazer, P. Benhaim, M.H. Hedrick, *Mol Biol Cell* **2002**, *13*, 4279–4295.

49. G.R. Erickson, J.M. Gimble, D.M. Franklin, H.E. Rice, H. Awad, F. Guilak, *Biochem Biophys Res Commun* **2002**, *294*, 371–379.

50. V. Planat-Benard, J.S. Silvestre, B. Cousin, M. Andre, M. Nibbelink, R. Tamarat, M. Clergue, C. Manneville, C. Saillan-Barreau, M. Duriez, A. Tedgui, B. Levy, L. Penicaud, L. Casteilla *Circulation* **2004**, *109*, 656–663.

51. A. Banfi, A. Muraglia, B. Dozin, M. Mastrogiacomo, R. Cancedda, R. Quarto, *Exp Hematol* **2000**, *28*, 707–715.

52. I. Sekiya, B.J. Larson, J.R. Smith, R. Pochampally, J.-G. Cui, D.J. Prockop, *Stem Cells* **2002**, *20*, 530–541.

53. S. Wislet-Gendebien, P. Leprince, G. Moonen, B. Rogister, *J Cell Sci* **2003**, *116*, 3295–3302.

54. C.M. DiGirolamo, D. Stokes, D. Colter, D.G. Phinney, R. Class, D.J. Prockop, *Br J Haematol* **1999**, *107*, 275–281.

55. I. Martin, A. Muraglia, G. Campanile, R. Cancedda, R. Quarto, *Endocrinology* **1997**, *138*, 4456–4462.

56. J.L. Spees, C.A. Gregory, H. Singh, H.A. Tucker, A. Peister, P.J. Lynch, S.-C. Hsu, J. Smith, D.J. Prockop, *Mol Ther* **2004**, *9*, 747–756.

57. R.R. Pochampally, J.R. Smith, J. Ylostalo, D.J. Prockop, *Blood* **2004**, *103*, 1647–1652.

58. M. Yamaguchi, F. Hirayama, S. Wakamoto, M. Fujihara, H. Murahashi, N. Sato, K. Ikebuchi, K. Sawada, T. Koike, M. Kuwabara, H. Azuma, H. Ikeda, *Transfusion* **2002**, *42*, 921–927.

59. S.A. Kuznetsov, M.H. Mankani, P.G. Robey, *Transplantation* **2000**, *70*, 1780–1787.

60. N. Yamamoto, M. Isobe, A. Negishi, H. Yoshimasu, H. Shimokawa, K. Ohya, T. Amagasa, S. Kasugai, *J Med Dent Sci* **2003**, *50*, 63–69.

61. D.P. Lennon, S.E. Haynesworth, R.G. Young, J.E. Dennis, A.I. Caplan, *Exp Cell Res* **1995**, *219*, 211–222.

62. D. Gospodarowicz, *Ann N Y Acad Sci* **1991**, *638*, 1–8.

63. C. van den Bos, J.D. Mosca, J. Winkles, L. Kerrigan, W.H. Burgess, D.R. Mar, *Hum Cell* **1997**, *10*, 45–50.

64. L.J. Oliver, D.B. Rijkin, J. Gabrilove, M.J. Hannocks, E.L. Wilson, *Growth Factors* **1990**, *3*, 231–236.

65. S. Pitaru, S. Kotev-Emeth, D. Noff, S. Kaffuler, N. Savion, *J Bone Miner Res* **1993**, *8*, 919–929.

66. L. Sensebé, J. Li, M. Lilly, C. Crittenden, P. Hervé, P. Charbord, J.W. Singer, *Exp Hematol* **1995**, *23*, 507–513.

67. S. Walsh, C. Jefferiss, K. Stewart, G.R. Jordan, J. Screen, J.N. Beresford, *Bone* **2000**, *27*, 185–195.

68. G. Bianchi, A. Banfi, M. Mastrogiacomo, R. Notaro, L. Luzzatto, R. Cancedda, R. Quarto, *Exp Cell Res* **2003**, *287*, 98–105.

69. C.A. Vacanti, L.J. Bonassar, M.P. Vacanti, J. Shufflebarger, *N Engl J Med* **2001**, *344*, 1511–1514.

70. W. Bensaïd, J.T. Triffit, C. Blanchat, K. Oudina, L. Sedel, H. Petite, *Biomaterials* **2003**, *24*, 2497–2502.

71. Y. Açil, H. Terheyden, A. Dunsche, B. Fleiner, S. Jepsen *J Biomed Mater Res* **2000**, *51*, 703–710.

72. O.N. Koç, J. Day, M. Nieder, S.L. Gerson, H.M. Lazarus, W. Krivit, *Bone Marrow Transplant* **2002**, *30*, 215–222.

7

The Clonal Activity of Marked Hematopoietic Stem Cells

Jingqiong Hu, Manfred Schmidt, Annette Deichmann, Hanno Glimm, and Christof von Kalle

7.1
Introduction

Hematopoietic stem cells (HSCs) have been defined as multipotent cells with self-renewal capability. Despite great progress having been achieved during the past decade towards characterizing HSCs, many aspects regarding their *in-vivo* behavior, such as clonal activity, lineage contribution and the life-span of individual stem cell clones, remain to be answered. Genetic marking of HSCs, in particular by using integrating vectors, provides a powerful method to follow an individual cell and its clonal progeny in the physiological context. Highly sensitive and specific retroviral integration site analysis has been successfully exploited in animal models as well as in human gene therapy clinical trials. While the first successful human gene therapy clinical study (SCID-X1) has demonstrated a polyclonal re-population pattern of T lymphopoiesis [1], and brought renewed hope to the feasibility of efficient gene therapy, the occurrence of a T-cell leukemia-like disease in three patients of the very same trial [2] showed that the clonal activity of gene-modified stem or progenitor stem cells might help or harm, in being therapeutic, or might lead to malignancy. Thus, a thorough understanding of the clonal activity of HSCs is indispensable for improving cellular and gene therapy for hematological malignancies and other disorders.

7.2
Characterization of *In-vivo* Clonal Activity of HSCs by Genetic Marking

The first gene marking studies aimed at tracking the fate of transduced cells in humans without direct therapeutic intent [3]. In principle, gene marking can be used to track the *in-vivo* behavior of any cell type. Among these approaches, the genetic marking of HSCs has particular significance because there is a lack of any *in-vitro* surrogate assays which could define a "true" HSC, and the *in-vivo* behavior of these cells is otherwise difficult to follow. The *in-vivo* clonal activity of

Stem Cell Transplantation. Biology, Processing, and Therapy.
Edited by Anthony D. Ho, Ronald Hoffman, and Esmail D. Zanjani
Copyright © 2006 WILEY-VCH Verlag GmbH & Co. KGaA, Weinheim
ISBN: 3-527-31018-5

HSCs represents a vital feature of functional HSCs. It refers to the ability of HSCs to undergo repeated mitosis without differentiating, contributing thereby to multiple lineages without loss of self-renewal and multipotency capabilities. This field has become increasingly attractive during recent years because the elucidation of this intriguing process holds the key to the limitations of current cell, *ex-vivo* molecular and gene therapy clinical trials aimed at treating various hematologic disorders and malignancies.

7.3
Retroviral Integration Site Analysis

Analyses based on X-linked polymorphic alleles such as the glucose-6-phosphate dehydrogenase (G6PD) isoenzyme have pioneered molecular approaches to assess the clonal composition of HSCs *in-vivo* [4]. However, this type of analysis could only provide an approximate estimation of the degree of polyclonality present in the human hematopoietic system, and cannot possibly follow the fate of individual stem cell clones. Furthermore, its application is limited to females heterozygous for this allele or, alternatively, to female donor stem cells after transplantation into male recipients [5]. Currently, retroviral integration site analysis is much more widely used for the same purpose. Retroviral vectors insert into the host genome and thereby introduce a unique molecular marker for each transduced cell and its clonal progeny in the form of a genomic-proviral fusion sequence, which can be seen as a clonal signature to the founder cell and all of its progenies. The existence of identical fusion sequences in different progenitor cells is therefore proof of *in-vivo* clonal activity of the originally marked founder cell. Furthermore, the insertion locus sequence can be used to design clone-specific PCR primers to follow individual clones over time and across different lineages.

As a prerequisite to any meaningful clonality study, successful identification of the retroviral-genomic integration sites calls for a highly sensitive and specific method, as the *in-vivo* situation of marked clones is highly complex at the molecular level. Various methods have been used for identifying proviral-genomic junction sequences, most of which are based on PCR technology. These methods include two-step-PCR, anchored-PCR, capture-PCR, panhandle-PCR, inverse PCR, and ligation-mediated PCR [6–10]. All of these methods have to process the DNA to be analyzed through multiple steps, and therefore are limited in terms of specificity and sensitivity due to one or several inefficient reaction steps. This results in a considerable loss of the sample material, thereby usually requiring 100 to 1000 copies of a particular flanking sequence [11, 12]. These methods can therefore be helpful for analyzing clonal samples such as cultured cell clones or clonal tumor cell populations, but they are not sensitive and/or specific enough for *in-vivo* clonality analysis of the hematopoietic system, especially when characterizing a polyclonal composition. For this purpose, we have developed a highly sensitive and specific linear amplification-mediated PCR (LAM-PCR), which combines a unique succession of biotinylated primer extension, hexanucleotide priming and restriction digestion on solid phase, to allow the simultaneous identification of multiple individual retrovirally transduced cells in highly polyclonal samples – if necessary, down to the single cell level [13]. Important insights into the *in-vivo* clonality of marked HSCs have been obtained using this method.

7.4
Clonality Analysis in Animal Model and Human Gene Therapy Trials

7.4.1
Clonal Activity of Marked HSCs in Mouse Models

The characterization of *in-vivo* hematopoiesis by using retroviral integration site analysis [14] has been carried out in mouse transplantation models as early as 1985. Although insensitive methods (Southern blot) were used at that time, these studies manifested two theories: one is the clonal succession theory which suggests that hematopoiesis is carried out by a periodically changing subset of active stem cells [4, 15, 16]; another alternative theory hypothesizes that hematopoiesis is carried by HSC indefinitely capable of self-renewal and multipotency [17]. In mouse studies, evidence for both theories has been found. It has been shown that some clones contributed only transiently for a certain period post-transplantation, while others began contributing later, suggesting that different clones are responsible for the short-term and long-term repopulation, which is more consistent with the clonal succession theory. Others observed an oligoclonal hematopoietic repopulation with a small number of clones contributing stably and to all lineages, which is more consistent with the alternative theory. Marked changes in clonal contributions during the first 4–6 months after transplantation were observed, which could be best interpreted with a stochastic model. It has also been shown that a single transduced HSC can reconstitute both myeloid and lymphoid lineages for an entire life-span. This evidence seems to point at an unlimited clonal activity of marked HSCs. However, in one gene marking study, leukemia has developed in secondary and tertiary transplant recipients, which shows that in some models, gene marking can be of influence on the clonal activity, in its most extreme form, leading directly to malignancy [18].

A limitation in using mouse models to characterize human *in-vivo* hematopoiesis is seen in the differences regarding hematopoietic demand, stem cell turnover and retroviral receptor expression levels. The stem cell engraftment, proliferation and differentiation in these models may also differ greatly from that of human hematopoiesis; furthermore, the life-span of the NOD/SCID mice is significantly shortened, and therefore long-term follow-up studies in xenografts have been impossible until very recently [19]. These limitations have restricted extrapolation of the model to human hematopoiesis, thus far. Recent findings have shown that, through selective removal of the genetic origin of thymoma in new

NOD/SCID mouse models, long-term follow-up studies of steady-state human hematopoiesis in xenografts might be available in the near future [20].

7.4.2
Clonal Activity of Marked HSCs in Non-Human Primate Models

To date, non-human primate models have been found far superior to mouse models with relevance to human *in-vivo* hematopoiesis in several aspects:
- they are genetically more closely related to humans;
- they have HSCs that respond to human cytokines, and anti-human antibodies to HSCs surface markers such as CD34 often cross-react with their primate analogue; and
- their-life span allows longer follow-up.

To date, by using retroviral integration site analysis, a highly polyclonal repopulation of transduced autologous HSCs has been observed in this model [21]. Molecular evidence was also accumulated for the existence of short-term and long-term repopulating cell clones. Recently, Kiem and colleagues reported long-term follow-up results in 42 rhesus macaques, 23 baboons, and 17 dogs with significant levels of gene transfer for an average of 3.5 years after infusion of marked HSCs. Integration site analysis confirmed stable, polyclonal retrovirally marked hematopoiesis, without progression towards mono- or oligoclonality over time [22]. Many of these animals continue to be maintained in long-term follow-up studies, and this will allow a long-term risk assessment for various forms of genetic modification involving the hematopoietic system.

7.4.3
Clonal Activity of Marked HSCs in Human Gene Therapy Clinical Trials

7.4.3.1 Clonality Analysis in ADA-SCID Gene Therapy Clinical Trial
A human gene therapy trial for adenosine deaminase deficiency (ADA-SCID) was initiated by Kohn and colleagues in 1993. Significant gene marking, though without evidence of immune reconstitution, was achieved [23]. Several interesting findings have been obtained by means of retroviral integration site analysis. Overall, a monoclonal to oligoclonal hematopoietic repopulation profile was found in all patients following the transplantation of autologous umbilical cord blood CD34+ cells. In one patient, a single insertion site has been identified for as long as 9 years in blood mononuclear cells and sorted T lymphocytes following transplantation. This indicates that a single transduced stem or progenitor cell clone can serve as the primary source of transgene marked cells for an extended period of time, without any manifestation of malignant proliferation [24].

7.4.3.2 Clonality Analysis in SCID-X1 Gene Therapy Clinical Trial

The SCID-X1 trial initiated by Fischer, Cavazzana-Calvo and colleagues has been the first therapeutically successful human gene therapy clinical trial [1]. Hence, the very high efficiency of gene transfer resulted in a therapy-related serious adverse event in the two youngest patients in the study, with both developing a T-cell acute leukemia-like (T-ALL) disease after 3 years of gene therapy. Recently, a third case of lymphoproliferation has been reported for this trial, for which molecular analysis is not yet available at the time of writing. In this study – as in its British counterpart – a murine leukemia virus (MLV)-based retroviral vector was used to carry the transgene interleukin (IL)-2 receptor gamma common chain (γc) into autologous bone marrow CD34+ cells. Immune function was restored in a total of 15 SCID-X1 patients worldwide. A LAM-PCR analysis was performed on all patients to monitor the clonal activity of marked HSCs. Successive retroviral integration site analyses revealed a polyclonal profile after transplantation (>50 clones

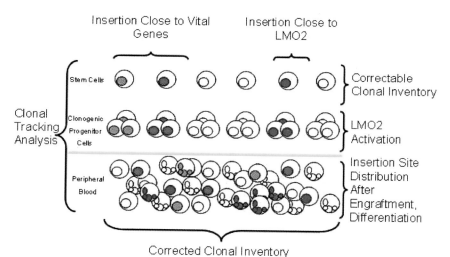

Figure 7.1 Analytical power of insertion site detection. Stem and progenitor cell compartments can be measured and determined by insertion site analysis as described in this figure using the example of the X-SCID trial. The color labeling is indicative of shared insertion site information. The ability to quantify the activity of stem and progenitor cell clones in different lineages for the first time allows the unlimited tracing of clonal contributions by single cells over time simultaneously in different clones with minimally invasive sampling in a live individual. The search for identified insertion sequence information by regular PCR with primers specific for a particular clone can be used to track the presence of clones over time or in different highly purified cell lineages preparations. By determining the number of clones contributing, a clonal repertoire of genetic correction can be contrived and related to the time course of correction. Bioinformatics mining of the sequence catalogue obtained can be used to address questions of vector or target cell-related changes in insertion site distribution, and may eventually help to clarify which characteristics of an insertion locus exactly determine how highly an inserted retrovirus is expressed. (Figure reproduced with permission from *Semin. Hematol.* **2004**;*41*(4):303–318; © Elsevier Inc.)

per patient), indicating multiple distinctive transduced clones contributing to post-transplantation hematopoiesis [2]. Except for the two patients who developed a T-ALL-like disease, the polyclonal profile of hematopoietic repopulation in the patients remained rather constant until 3 years after gene therapy, which was in contrast to the results of a monoclonal to oligoclonal profile observed in the first ADA-SCID trial [24] (Fig. 7.1).

A possible explanation for this significant difference in the two SCID gene therapy trials is the fact that, in SCID-X1, a much enlarged pool of common lymphocytes progenitors [25], due to the complete block in T-cell differentiation and maturation, may be present at the time of treatment. Furthermore, in the ADA-SCID trial, the patients were given an enzyme replacement therapy (PEG-ADA) at the same time as a substitute treatment, and therefore no selective pressure was imposed on the transduced cells [26–28]. In contrast, there is a much greater selective advantage of transduced common lymphocytes progenitors over nontransduced cells in SCID-X1 [29]. Meanwhile, further gene therapeutic treatment of ADA-SCID could be achieved by facilitating engraftment of transduced cells with partially myeloablative chemotherapy conditioning [30]. A separate SCID-X1 trial [31] might successfully correct the genetic disorder in further patients, so far without signs of side effects. These two examples show that the clonal activity of marked HSCs can be affected by the nature of the disease itself and by pharmaceutical interventions.

7.4.3.3 LMO2 Insertion Leads to Malignant Expansion of Marked HSCs

Of note, the two first T-ALL-like disease cases which resulted from retroviral induced insertional mutagenesis demonstrate vividly that insertion into a proto-oncogene region can lead to malignancy [32, 33]. Both cases represent the example of a worst-case scenario of retroviral integration to marked HSCs – namely, the monoclonal malignant expansion of gene-corrected HSC clones. Interestingly, in another context of ADA-SCID gene therapy, it could be demonstrated unequivocally that monoclonal expansion of a single transduced clone for longer than 9 years can still be benign. These two examples suggested that the potential malignant clonal expansion should be carefully considered in each different setting.

The mechanisms of malignant expansion of certain transduced HSCs is unclear. The clonal activity of marked HSCs in the particular setting of SCID-X1 is complicated by several issues. First, although the selective advantage conferred to the transduced cells is crucial for the success of gene therapy, it also likely enables the proliferation possibility of every transduced clone that successfully expresses the transgene. A particular clone now already possesses the ability to proliferate and may take advantage of this to expand in an uncontrollable manner by a slight growth advantage compared to its "normal" competitors. This possibility is a likely scenario to have occurred in the two leukemia cases, because the clonal proliferation profile detected developed slowly after reinfusion of the gene-modified cells, and a steady expansion was observed, just as would be predicted from observations in the LMO2 transgenic mouse model [34]. Second, other factors –

including a familiar predisposition to cancer, additional genetic alterations (chromosomal translocation in patient #4 and trisomy in patient #5), infection of varicella zoster virus (VZV), patient age, and an expanded progenitor cell pool and immune deficit of the patient – might also be involved. Furthermore, the transgene product itself might also have played a role, promoting the affected clones initially to engraft and survive *in vivo*. Taken together, a synergetic interaction of several factors may have contributed to the leukemogenesis in this setting [32, 33]. Therefore, this seemingly greater risk of malignancy may not necessarily extrapolate directly to other gene therapy trials outside of SCID-X1, though the capacity of activating proto-oncogenes is not restricted to this particular vector or setting. As a consequence, stringent scrutiny needs to be brought to the close interaction of the retroviral integration site with the clonal activity of the respective transduced HSC clone, in order to understand which vector elements activate the surrounding genetic environment, why they do, and whether their function is really indispensable for vector applications.

7.5
Interaction of Retroviral Integration Site and Transgene Expression with Clonal Activity of the Respective HSC

Despite the widespread and successful exploitation of retroviral integration site analysis, several core issues of retroviral gene transfer still require careful clarification.

The interaction of a retroviral integration site with the clonal activity of the respective transduced clone could have several outcomes. Depending on each different scenario, the result of the interaction could be categorized into the following working hypotheses:

- Insertion of a vector into a specific gene locus: This could result in the expansion of respective transduced clones, in the worst-case scenario, malignant expansion of certain transduced clones – insertional mutagenesis, as shown in recent gene therapy clinical trials as well as in animal studies. The possibility of insertional mutagenesis was previously assumed to be extremely low, but recent findings in both human gene therapy clinical trials and animal models strongly challenge this notion [33, 35]. Genome-wide insertion site distribution analysis has suggested gene coding regions as being preferential targets in retroviral integration [36–39]. It may be hypothesized that more genes, including some proto-oncogenes, are actively transcribed in the HSCs, since these cells are responsible for an unlimited self-renewal and multipotent differentiation. Therefore, the HSCs, albeit desirable targets for clinical gene therapy of hematologic diseases, might at the same time be vulnerable targets for retroviral integration, since actively transcribed genes may favor the integration into such loci due to easier accessibility for the viral pre-integration complex [40], rendering their unwanted proliferation more likely.

- Extinction of transduced clones: In certain unfavorable genome loci, vector integration might produce a growth disadvantage and subsequent elimination from the hematopoietic compartment due to triggering apoptosis, impaired engraftment or proliferation, thus leading to the elimination of the transduced clone. Although an issue of effectiveness, this scenario is clearly less of a safety concern.
- Maintenance of the respective clones: In most cases, insertions are likely to take place at unimportant gene loci or to insert in a manner that does not have much impact on the clonal activity of the respective transduced clone, resulting in a normal growth of these clones.

In general, retroviral integration will result in a heterogeneous HSC repertoire with the above-noted scenarios. Of note is that the malignant expansion of certain transduced clones seems to be self-limiting and exceedingly rare, considering the high number of transduced cells (10^7–10^9) transferred per patient in the affected retroviral gene therapy trial [35]. Indeed, this fact may offer additional access to safer modes of application, considering how low the number of corrected clones probably really is.

7.5.1
Impact of Transgene Expression on Clonal Activity of Marked HSCs

The nonphysiologic overexpression of the transgene could render the respective transduced clones either at a selective advantage or disadvantage, thereby affecting the clonal activity of those marked HSCs. In general, those genes involved in a signaling pathway, for example a cytokine receptor subunit such as IL-2RG that is involved in regulating proliferation, differentiation and/or apoptosis of the affected cell clone are considered to tend towards transformation. This mechanism has been suggested in the SCID-X1 gene therapy trial.

7.6
Clinical Interventions Affect the Clonal Activity of Marked HSCs

The clonal activity of marked HSCs can further be impacted by clinically relevant manipulations such radiation therapy, chemotherapy, and the administration of nonmyeloablative agents. Kuramoto et al. reported that low-dose busulfan (a nonmyeloablative agent) might cause significant changes in the clonal dynamics of HSCs [41]. Irradiation studies have shown that even low-dose irradiation can affect the clonal activity of HSCs, causing clonal fluctuation. Now, similar questions are being asked of chemotherapy. For example, the question was raised as to whether, after chemotherapy, hematopoiesis is carried out by newly activated clones or, alternatively, by individual transduced clones that were contributed before the chemotherapy and could remain active after such treatment. We have attempted to address this question by tracking clones identified before chemotherapy in post-chemotherapy samples. It was possible to show that some clones were only partially suppressed by chemotherapy; these clones could be reactivated after chemotherapy. However, at the present time it remains unclear as to whether the impact of chemotherapy only suppresses – rather than abolishes – the activity of some stem/progenitor cell clones. The tracking of more clones will determine whether this finding is a common phenomenon, or is restricted to only some stem/progenitor clones with greater surviving capability.

7.7
Perspectives

The clonal capabilities and behavior of stem cells, with their ability to build the entire blood-forming and immune system of an individual from a single cell, continue to fascinate and amaze us. Much has been learned from the molecular marking of stem cell activity afforded to us by early gene therapy studies and their preclinical development. While we still do not understand this activity in its entirety, we have learned that individual stem cells can contribute to blood formation for continuous periods of time, and that – at least in the post- transplantation setting, and also in large-animal models – the number of repopulating stem cells is rather finite. It will be extremely interesting to learn from future experiments how the blood-forming process is then built through the various levels of differentiation, what is the long-term fate of such clones in the steady state, and how such activities are regulated at the cellular level. New findings in these areas are imminent, and will be of great importance for further progress in molecular, cellular and gene therapy involving stem cells.

Acknowledgments

These studies were supported by grants from the NIH, NHLBI, Dr. Mildred Scheel Stiftung fuer Krebsforschung, Bonn, Germany, and the Verein zur Foerderung der Leukaemie und Tumor Forschung, Freiburg, Germany to H.G.; and grants Ka 976/4-1 from the Deutsche Forschungsgemeinschaft and KV9527/7 from the German Minister for Education and Research to C.v.K.

Abbreviations/Acronyms

ADA	adenosine deaminase
G6PD	glucose-6-phosphate dehydrogenase
HSC	hematopoietic stem cell
LAM-PCR	linear amplification-mediated PCR
NOD/SCID	non-obese diabetic/severely compromised immunodeficient
PEG	polyethylene glycol
T-ALL	T-cell acute leukemia-like
VZV	varicella zoster virus

References

1. Cavazzana-Calvo M, Hacein-Bey S, de Saint BG, et al. Gene therapy of human severe combined immunodeficiency (SCID)-X1 disease. *Science* **2000**;*288*:669–672.
2. Hacein-Bey-Abina S, Von KC, Schmidt M, et al. LMO2-associated clonal T cell proliferation in two patients after gene therapy for SCID-X1. *Science* **2003**;*302*:415–419.
3. Shi PA, De AM, Donahue RE, et al. In vivo gene marking of rhesus macaque long-term repopulating hematopoietic cells using a VSV-G pseudotyped versus amphotropic oncoretroviral vector. *J Gene Med* **2004**;*6*:367–373.
4. Abkowitz JL, Catlin SN, Guttorp P. Evidence that hematopoiesis may be a stochastic process in vivo. *Nat Med* **1996**;*2*:190–197.
5. Okamoto T, Okada M, Wada H, et al. Clonal analysis of hematopoietic cells using a novel polymorphic site of the X chromosome. *Am J Hematol* **1998**;*58*:263–266.
6. Jones DH, Winistorfer SC. Amplification of 4-9-kb human genomic DNA flanking a known site using a panhandle PCR variant. *Biotechniques* **1997**;*23*:132–138.
7. Lagerstrom M, Parik J, Malmgren H, et al. Capture PCR: efficient amplification of DNA fragments adjacent to a known sequence in human and YAC DNA. *PCR Methods Appl* **1991**;*1*:111–119.
8. Nolta JA, Dao MA, Wells S, Smogorzewska EM, Kohn DB. Transduction of pluripotent human hematopoietic stem cells demonstrated by clonal analysis after engraftment in immune-deficient mice. *Proc Natl Acad Sci USA* **1996**;*93*:2414–2419.
9. Pfeifer GP, Steigerwald SD, Mueller PR, Wold B, Riggs AD. Genomic sequencing and methylation analysis by ligation mediated PCR. *Science* **1989**;*246*:810–813.
10. Mueller PR, Wold B. In vivo footprinting of a muscle specific enhancer by ligation mediated PCR. *Science* **1989**;*246*:780–786.
11. Schmidt M, Glimm H, Lemke N, et al. A model for the detection of clonality in marked hematopoietic stem cells. *Ann N Y Acad Sci* **2001**;*938*:146–155.
12. Schmidt M, Hoffmann G, Wissler M, et al. Detection and direct genomic sequencing of multiple rare unknown flanking DNA in highly complex samples. *Hum Gene Ther* **2001**;*12*:743–749.
13. Schmidt M, Glimm H, Wissler M et al. Efficient characterization of retro-, lenti-, and foamy vector-transduced cell populations by high-accuracy insertion site sequencing. *Ann N Y Acad Sci* **2003**;*996*:112–121.
14. Dick JE, Magli MC, Huszar D, Phillips RA, Bernstein A. Introduction of a selectable gene into primitive stem cells capable of long-term reconstitution of the hemopoietic system of W/Wv mice. *Cell* **1985**;*42*:71–79.
15. Abkowitz JL, Linenberger ML, Newton MA, et al. Evidence for the maintenance of hematopoiesis in a large animal by the

sequential activation of stem-cell clones. *Proc Natl Acad Sci USA* **1990**;*87*:9062–9066.

16. Kay HE. The control of leucopoiesis. *Proc R Soc Med* **1967**;*60*:1025–1027.

17. Jordan CT, Lemischka IR. Clonal and systemic analysis of long-term hematopoiesis in the mouse. *Genes Dev* **1990**;*4*:220–232.

18. Li Z, Dullmann J, Schiedlmeier B, et al. Murine leukemia induced by retroviral gene marking. *Science* **2002**;*296*:497.

19. Plum J, De SM, Verhasselt B, et al. Human T lymphopoiesis. In vitro and in vivo study models. *Ann N Y Acad Sci* **2000**;*917*:724–731.

20. Glimm H, Schmidt M, Fischer M, et al. Evidence of similar effects of short-term culture on the initial repopulating activity of mobilized peripheral blood transplants assessed in NOD/SCID-beta2microglobulin(null) mice and in autografted patients. *Exp Hematol* **2005**;*33*:21–25.

21. Schmidt M, Zickler P, Hoffmann G, et al. Polyclonal long-term repopulating stem cell clones in a primate model. *Blood* **2002**;*100*:2737–2743.

22. Kiem HP, Sellers S, Thomasson B, et al. Long-term clinical and molecular follow-up of large animals receiving retrovirally transduced stem and progenitor cells: no progression to clonal hematopoiesis or leukemia. *Mol Ther* **2004**;*9*:389–395.

23. Kohn DB, Hershfield MS, Carbonaro D, et al. T lymphocytes with a normal ADA gene accumulate after transplantation of transduced autologous umbilical cord blood CD34+ cells in ADA-deficient SCID neonates. *Nat Med* **1998**;*4*:775–780.

24. Schmidt M, Carbonaro DA, Speckmann C, et al. Clonality analysis after retroviral-mediated gene transfer to CD34+ cells from the cord blood of ADA-deficient SCID neonates. *Nat Med* **2003**;*9*:463–468.

25. Kohn DB, Sadelain M, Glorioso JC. Occurrence of leukaemia following gene therapy of X-linked SCID. *Nat Rev Cancer* **2003**;*3*:477–488.

26. Cavazzana-Calvo M, Hacein-Bey S, Yates F, et al. Gene therapy of severe combined immunodeficiencies. *J Gene Med* **2001**;*3*:201–206.

27. Kohn DB. Gene therapy for hematopoietic and immune disorders. *Bone Marrow Transplant* **1996**;*18*(Suppl.3):S55–S58.

28. Kohn DB, Weinberg KI, Nolta JA, et al. Engraftment of gene-modified umbilical cord blood cells in neonates with adenosine deaminase deficiency. *Nat Med* **1995**, Oct;*1(10)*:1017–23.

29. Hacein-Bey-Abina S, Fischer A, Cavazzana-Calvo M. Gene therapy of X-linked severe combined immunodeficiency. *Int J Hematol* **2002**;*76*:295–298.

30. Aiuti A, Vai S, Mortellaro A, et al. Immune reconstitution in ADA-SCID after PBL gene therapy and discontinuation of enzyme replacement. *Nat Med* **2002**;*8*:423–425.

31. Gaspar HB, Parsley KL, Howe S, et al. Gene therapy of X-linked severe combined immunodeficiency by use of a pseudotyped gammaretroviral vector. *Lancet* **2004**;*364*:2181–2187.

32. Baum C, Fehse B. Mutagenesis by retroviral transgene insertion: risk assessment and potential alternatives. *Curr Opin Mol Ther* **2003**;*5*:458–462.

33. Baum C, von Kalle C, Staal FJ, et al. Chance or necessity? Insertional mutagenesis in gene therapy and its consequences. *Mol Ther* **2004**;*9*:5–13.

34. Drynan LF, Hamilton TL, Rabbitts TH. T cell tumorigenesis in Lmo2 transgenic mice is independent of V-D-J recombinase activity. *Oncogene* **2001**;*20*:4412–4415.

35. Baum C, Dullmann J, Li Z, et al. Side effects of retroviral gene transfer into hematopoietic stem cells. *Blood* **2003**;*101*:2099–2114.

36. Laufs S, Nagy KZ, Giordano FA, et al. Insertion of retroviral vectors in NOD/SCID repopulating human peripheral blood progenitor cells occurs preferentially in the vicinity of transcription start regions and in introns. *Mol Ther* **2004**;*10*:874–881.

37. Mitchell RS, Beitzel BF, Schroder AR, et al. Retroviral DNA integration: ASLV, HIV, and MLV show distinct target site preferences. *PLoS Biol* **2004**;*2*:E234.

38. Schroder AR, Shinn P, Chen H, et al. HIV-1 integration in the human genome favors active genes and local hotspots. *Cell* **2002**;*110*:521–529.

39. Wu X, Li Y, Crise B, Burgess SM. Transcription start regions in the human genome are favored targets for MLV integration. *Science* **2003**;*300*:1749–1751.

40. Bushman F. Targeting retroviral integration? *Mol Ther* **2002**;*6*:570–571.

41. Kuramoto K, Follman D, Hematti P, et al. The impact of low-dose busulfan on clonal dynamics in nonhuman primates. *Blood* **2004**;*104*:1273–1280.

Part III
On the Threshold to Clinical Applications

Stem Cell Transplantation. Biology, Processing, and Therapy.
Edited by Anthony D. Ho, Ronald Hoffman, and Esmail D. Zanjani
Copyright © 2006 WILEY-VCH Verlag GmbH & Co. KGaA, Weinheim
ISBN: 3-527-31018-5

8

A Large Animal Non-Injury Model for Study of Human Stem Cell Plasticity

Graça Almeida-Porada, Christopher D. Porada, and Esmail D. Zanjani

Abstract

An experimental model system that allows assessment of the full differentiative potential of human stem cells (HSCs) under normal physiological conditions, in the absence of genetic or injury-induced dysfunction, could reveal the true capabilities of HSCs as well as providing a valuable tool to elucidate the mechanism(s) underlying stem cell differentiation. Because studies have shown that, in the absence of selective pressure, the differentiation of stem cells into cells of a different germinal layer is highly inefficient, it is very unlikely that a healthy adult animal can fulfill these requirements. The naturally occurring stem cell migratory patterns, the availability of expanding homing and engraftment sites, and the presence of tissue/organ-specific signals in the developing mammalian fetus provide the ideal setting for stem cells to display their full biological potential. These characteristics, combined with the relative immunological naiveté of the early gestational age fetus, that permits the engraftment and long-term persistence of allogeneic and xenogeneic donor stem cells, make it possible to use the developing fetus to assess the *in-vivo* potential of a variety of stem cells. In this chapter, we describe the advantages of the permissive aspects of the developing early gestational age preimmune fetus that led us to develop a large animal model of HSC plasticity in sheep.

8.1
Introduction

Although our understanding of the processes that regulate events controlling the formation of functional tissues during ontogeny is still in its infancy, conventional teaching has held that multipotent stem cells are gradually lost during the process of embryogenesis in which the blastula gave rise to the three germ layers. It was thought that each one of these germ layers was responsible for the formation of specific mature tissues such that brain and skin were of ectodermal derivation,

Stem Cell Transplantation. Biology, Processing, and Therapy.
Edited by Anthony D. Ho, Ronald Hoffman, and Esmail D. Zanjani
Copyright © 2006 WILEY-VCH Verlag GmbH & Co. KGaA, Weinheim
ISBN: 3-527-31018-5

blood and muscle were mesodermal, and intestine and hepatocytes were formed from endoderm. This dogma led investigators to assume, reasonably, that further development of these tissues – and their repair when necessary – stemmed from precursor elements with established commitment to the tissue type involved. In support of this presumption are recent studies showing that adult tissues such as skin [1], mammary gland [2], muscle [3], and even brain [4–6] all contain regenerative stem cells. However, the recent discovery that cells within the bone marrow have the ability to give rise to nonhematopoietic cellular elements in various tissues has suggested that alternative mechanisms might also be contributing to tissue maintenance and repair throughout life. This, in turn, has raised the exciting possibility that it may be possible to use stem cells from the hematopoietic system for the functional repair of damaged tissues and organs in clinical settings. This possibility is strongly supported by the results of several studies in animal models that have shown convincingly that significant numbers of functional donor-derived hepatocytes are formed within the liver of transplant recipients and can mediate the correction of a liver defect [7–14]. Even in humans, indirect observations have shown that donor-derived hepatocytes, neural cells, and cardiomyocytes can arise after bone marrow transplantation [15–18]. Despite the exciting nature of these demonstrations of cellular and molecular plasticity, an in-depth understanding of the processes controlling this apparent plasticity and ways of efficiently steering the cells along a specific lineage will be needed before the full therapeutic potential of these various stem cell populations can be realized. For this to happen, an experimental model that would allow donor stem cells to participate in the generation of cells from other unrelated tissues under normal physiological conditions, in the absence of genetic or injury-induced dysfunction within a specific organ will be required. Ideally, such a model would also permit the robust formation of various donor-derived tissue-specific cells. Because studies have now provided evidence that, in the absence of selective pressure, the differentiation of bone marrow cells into other cells types such as mature hepatocytes is highly inefficient [12, 13], it is very unlikely that a healthy adult animal can fulfill these requirements.

Arguably, the ideal way to assess the complex issue of stem cell plasticity would be *in vitro*, as this would allow delineation of the pathways by which the stem cells being tested adopt alternate cellular fates. In addition, definitive evidence of plasticity needs to be answered by demonstrating the multipotentiality of a single cell or of a clonally derived cell population. Thus, performing *in-vitro* studies with highly purified populations of stem cells from various organs under defined culture conditions would appear to offer distinct advantages over transplantation experiments performed *in vivo* in which the experimenter has little or no control over the conditions within the recipient. Indeed, *in-vitro* cultures have been pivotal in establishing the concept that adult stem cells from various tissues possess plasticity. This is especially true in the case of marrow-derived mesenchymal stem cells (MSCs), likely due to the ease with which these adherent cells can be cultured and the fact that the conditions for growing many of the mesodermal cells to which MSCs give rise have already been delineated. Unfortunately, this is not the case with other sources of stem cells that do not readily propagate in culture. *In-vitro* studies are further complicated by the fact that the mediators required for stem cells to undergo the dramatic changes in cellular fate that have thus far been observed *in vivo* are largely unknown, making it difficult to reproduce *in vitro* the conditions present within an organ microenvironment *in vivo*. It is also important to note that *in-vitro* approaches are unable to duplicate conditions that affect the migratory patterns and homing of stem cells to different tissues/organs *in vivo*. Thus, at the present time, the best way in which to assess the differentiative potential of stem cells is by performing *in-vivo* transplantation studies. However, since ethical and practical considerations prevent limiting dilution studies of highly defined populations of stem cells in humans, investigators have either employed animal stem cells or have been forced to test the ability of human stem cells (HSCs) to engraft/differentiate within a xenogeneic setting.

To date, most *in-vivo* model systems that have been employed to demonstrate the versatility of stem cells have used either an external stress (e.g., radiation- or chemical-induced injury) or an experimentally created shortage of a specific cell type in the recipient to induce the transplanted cells to differentiate into the specific missing or injured organ cells [8, 15, 19, 20]. In other studies, mdx mice possessing a specific gene defect were used as recipients [21, 22]. The results of these studies proved that the presence of surrounding activated cells and/or signaling from the organ-specific microenvironment were essential for the transplanted cells to be induced to divide and to differentiate into cells of the injured/deficient organ. These studies have however, by nature of their design, restricted the fate of the transplanted cells to one particular organ/system, thus preventing evaluation of the full potential of the transplanted cell populations. In more recent studies, murine neural stem cells (NSCs) were microinjected into chick and mouse embryos and shown to give rise to differentiated cells of all three germ layers. This confirmed the findings of earlier studies in which investigators had driven transplanted stem cells to differentiate along a single lineage but had greatly underestimated the full potential of these cells [4]. At this point it is important to note that most studies providing evidence for stem cell plasticity have used murine models. Unfortunately, data acquired from mice cannot always be extrapolated to reflect the situation in higher animals and humans, and consequently a clinically relevant model is required in which the full potential of HSC plasticity can be evaluated.

8.2
The Uniqueness of the Fetal Sheep Model

During fetal life, a series of naturally occurring biological events – for example, the migratory processes of stem cells via the circulatory system – assure not only that the appropriate stem/progenitor cells migrate to the target tissues/organs when needed, but also that an interchange of cells is allowed between the bone marrow and other organs during development. In the fetal sheep model, the homing of circulating stem cells to the various target organs is possible due to both the vast array of adhesion molecules that are up- or down-regulated at specific points of gestation, and the expression of related tissue-specific chemokines that provide the correct signaling for the attraction and lodging of the circulating stem cells. Then, under the permissive milieu of the target organ – and because of the continuous need for new cells due to the vast cell proliferation that occurs at this time – the lodged stem cells are able to produce the required type(s) of cell(s). Thus, the transplantation of HSCs into fetal sheep recipients permits, through environmental conditioning, the re-encoding or modification of stem cell differentiative patterns, assuming that the transplanted cells possess such potential. During the fetal period, however, organs do not all proliferate at the same time; moreover, within the same organ differentiation into different cell types occurs at different time points. Furthermore, in contrast to other model systems used for to study stem cell transplantation, whereas fetal sheep recipients have a normal functioning immune system, this immune system is still largely immuno-naïve, allowing not only the engraftment/differentiation of HSCs but also the creation of donor-specific tolerance [23–29]. Thus, the transplantation of HSCs early in gestation, during the "preimmune" stage of development, will condition the fetal sheep recipients to receive additional stem cell transplants from the same donor later during the fetal period. Consequently, multiple HSC transplants can be performed at precise times during gestation, into a tolerized animal, to target the differentiation of human cells into specific cell types within a particular organ or system. For instance, transplantation of human hematopoietic cells in the fetal sheep model with the intent of producing human hepatocytes will likely be best at a time when the liver is still permissive to the homing of hematopoietic cells and before the metabolic functions of the fetal liver are fully developed (Fig. 8.1). Likewise, transplantation of HSCs for the successful production of human pneumocytes cannot be performed until the correct time point during development, when pneumocyte production has commenced; otherwise, few human pneumocytes would be generated in the fetal sheep model (Fig. 8.2). Assuming that not all HSCs transplanted will have the same ability to differentiate into a particular cell type [30], the number of HSCs reaching an organ may be important with regard to both the type and numbers of cells generated within a model [31].

In addition to the high rate of proliferation and the presence of permissive microenvironments within numerous tissues, there is another feature of the developing fetus that cannot be disregarded. Namely, it is suspended in a fluid-filled sac during development, rendering it essentially weightless, at least during the early stages of development [32–35]. Much evidence has shown that conditions of weightlessness can alter not only the repertoire of genes expressed by numerous cell types [36–41], but also the ability of *in-vitro*-grown cells to accurately reiterate their *in-vivo* functionality [42, 43]. Thus, the relative weightless state of the early gestational fetus may not only play a role in the normal process of embryonic/fetal development, but also make the fetal recipient a unique model system in which transplanted stem cell populations can adopt alternate cellular fates that have, as yet, not been observed *in vitro* or in other *in-vivo* adult model systems.

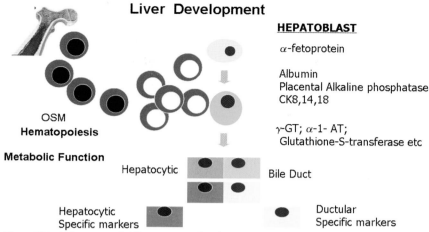

Liver Development

HEPATOBLAST

α-fetoprotein

Albumin
Placental Alkaline phosphatase
CK8,14,18

γ-GT; α-1- AT;
Glutathione-S-transferase etc

OSM
Hematopoiesis

Metabolic Function

Hepatocytic

Bile Duct

Hepatocytic
Specific markers

Ductular
Specific markers

Figure 8.1 Induction and differentiation of organ-specific stem cells at different fetal ages. Transplantation of human hematopoietic cells in the fetal sheep model with the intent of producing human hepatocytes will likely be best at a time when the liver is still permissive to the homing of hematopoietic cells and before the metabolic functions of the fetal liver are fully developed. OSM: Oncostatin M.

Lung Development

	IIGlandular	Canalicular	Saccular	Alveolar
Human	42-112	112-196	196-252	252-childhood
Sheep	-95	95-120	120-134	134-

Main airway system

↑ Injection of stem cells

Development and maturation acinar epithelium

From smooth-walled cylindrical
structures to thinning of interstitium

Appearance Alveoli

Figure 8.2 Induction and differentiation of specific cell types within the lung at different fetal ages in human and sheep. Transplantation of human stem cells (HSCs) to successfully produce specific human cell types within the lung of fetal sheep cannot be performed until the proper time point during development when differentiation of the desired cell type has commenced. Furthermore, the amount of donor stem cells reaching an organ within the appropriate temporal window may be important to both the type and numbers of cells generated.

Likewise, because of the large size of the sheep and its relatively long life span, this model can provide experimental opportunities not easily possible with murine models. For example, a large body size allows evaluation of donor cell activity in the same animal for years after transplantation, and also allows the investigator easily to obtain sufficient human cells from the primary recipients to perform serial transplantation studies. Indeed, successful engraftment and multilineage differentiation has now been observed in primary, secondary, and tertiary recipients using this model system [24–28]. In addition, because of the absence of any myeloablative conditioning, transplanted HSCs compete with endogenous ovine stem cells for available niches within the bone marrow or other organs. This suggests that events occurring after the transplantation of HSCs into sheep fetuses may be biologically relevant to events in humans.

8.3
Differentiative Potential of Human Cells in the Fetal Sheep Model

The versatility of this non-injury fetal model of HSC plasticity was demonstrated by revealing the differentiative potential of different populations of both HSCs and human MSCs upon transplantation into the fetal sheep model. In one study, highly purified populations of HSCs and mobilized peripheral blood, when transplanted into the fetal sheep model, gave rise to human bone marrow-derived mesenchymal cells. This provided evidence that the model was able to support both hematopoietic engraftment and the switch of engrafted cells from one mesodermal fate to another [44]. More recently, we reported that enriched populations of human adult bone marrow, mobilized peripheral blood (mPB), and umbilical cord blood (CB) HSCs all have the ability to generate functional human hepatocytes within the developing fetal liver [45]. On a cell-per-cell basis, HSCs derived from adult bone marrow had the highest hepatocytic potential of the three HSC sources tested at early time points after transplantation. By contrast, in sheep analyzed at later time points, CB HSCs seemed to have a higher hepatocytic potential. Since at earlier time points this effect was not evident, it is possible that CB HSCs need more time to give rise to hepatocytes than adult HSCs. Mobilized peripheral blood HSCs were also able to generate hepatocytes, raising the possibility that HSC mobilization could be used to mediate liver repair in clinical patients in whom liver function has been compromised [45].

We also analyzed the differences between phenotypically distinct HSCs with respect to their plasticity, and addressed possible differentiative pathways that allow for the conversion of HSCs to hepatocytes. We found that for the same source of HSCs, a direct correlation between hematopoietic engraftment and the generation of hepatocytes was established, and that the number of human hepatocytes generated in the fetal sheep model increased with time [46] (Fig. 8.3). The functionality of the human hepatocytes produced was demonstrated by the presence of human albumin in the serum of the transplanted animals (Fig. 8.4). Importantly, this differentiation from mesodermal HSCs to endodermal hepatocytes occurred in the absence of any injury or stimulus, demonstrating the importance of the microenvironmental influence of a fetal milieu for evaluating the full range of plasticity of a stem cell. Recently, controversy has arisen regarding the origin of donor liver cells post-transplantation, and several investigators have shown that cell fusion is at least one of the possible mechanisms responsible for the apparent differentiation of hematopoietic cells into liver cells [46]. The sheep model allows the robust formation of relatively large numbers of human-derived hepatic cells under normal physiologic conditions, and thus provides a valuable tool for the study of the mechanisms underlying the involvement of cells of human hematopoietic compartments in hepatopoiesis. Furthermore, a xenograft model such as the sheep is ideally suited for studying the issue of fusion, since it will be possible to determine whether chromatin from both species exists within the same nucleus. Thus far, all human hepatocytes examined in our studies have shown exclusively human genetic material, demonstrating that fusion does not likely play a major role in the generation of hepatocytes from transplanted human cells in

A

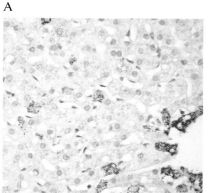

B

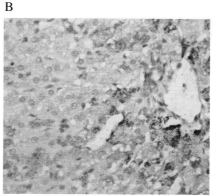

Figure 8.3 Human stem cells (HSCs) generate human hepatocytes in a sheep fetal model. Liver sections obtained at (a) 3 weeks and (b) 4 months post-transplant from sheep transplanted with human BM CD34+Lin– cells, stained with an anti-human hepatocyte antibody (Clone OCH1E5). It was found that for the same source of HSCs, a direct correlation existed between hematopoietic engraftment and hepatocyte generation, and that the number of human hepatocytes generated in the fetal sheep model increased with time (human hepatocytes identified by dark brown staining).

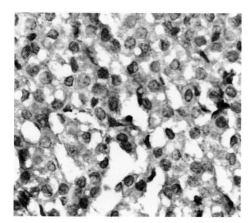

Figure 8.4 Human hepatocytes generated in the sheep model were functional, as shown by human albumin production in the livers of transplanted animals. The figure shows a section from the liver of a sheep transplanted with BM CD34+Lin– cells; human albumin production is indicated by light brown staining.

this model system. However, because these sections were analyzed at a time during gestation when liver cells did not display a high degree of polyploidy, it is possible that later in life, if circulating human cells lodge within the livers of these animals, the mechanism to generate new human hepatocytes would involve cell fusion.

Recently, we also examined the ability of human fetal brain-derived stem cells (neural stem cells, NSCs) to contribute to blood cell production after transplantation in the fetal sheep model. Human NSCs produced not only differentiated cells of multiple hematopoietic lineages, but also CD34+ cells that were able to repopulate secondary transplant recipients. However, further studies showed that significant qualitative differences existed between the hematopoietic cells generated upon NSC transplantation and those derived from marrow HSCs [47]. Nevertheless, our results showed that human fetal brain-derived NSCs possessed, to some extent, an ability to respond to the microenvironmental cues present within the intact fetal hematopoietic system and to reprogram their differentiative agenda, generating a range of differentiated hematopoietic cells types.

One of the most promising stem cell sources for tissue replacement therapy are the so-called bone marrow MSCs. These cells are able to differentiate in several different models into a number of other cells such as chondrocytes, adipocytes, myocytes and endothelium, as well as cells of alternate germinal derivation such as neural cells, skin, and liver [48–55]. We evaluated the ability of these human cells to give rise to other cell types *in vivo* using the fetal sheep model. To this end, several clonal MSC populations were isolated from adult bone marrow and their ability to give rise to donor (human)-derived blood and other organ-specific cell types was evaluated. All of the transplanted MSC clones generated multilineage hematopoietic cells, including CD34$^+$ cells. An analysis of the livers from these animals confirmed that some clones were capable of giving rise to significant numbers of human hepatocytes, detected by immunohistochemistry using a monoclonal antibody specific for human hepatocytes and by *in-situ* hybridization using a human Alu-specific probe. An examination of the skin from the same sheep showed the presence of human keratinocytes expressing human-specific cytokeratin. Since the nature of the model does not allow the use of single cell transplantation, the ability of the sheep model to distinguish differentiative potential and functionality between several identical clones was of the utmost importance. The results of these MSC studies showed that the sheep model also supports the differentiation of clonal populations of bone marrow-derived stem cells into cells of all three germinal derivations [56], and allowed us to discern disparities in the differentiative potential of the various clones. We also investigated the ability of the sheep model to support the differentiation of MSCs derived from tissues other than bone marrow. We reasoned that the fetal kidney might represent a rich source of pluripotent MSCs, based on both the mesenchymal origin of the kidney and the role played by the developing mesonephros in primitive hematopoiesis [57]. Therefore, we transplanted fetal kidney MSCs into preimmune fetal sheep, and then evaluated the recipients for the presence/engraftment of human cells, beginning at 2 months post-transplant and continuing at intervals thereafter until 9 months post-transplant. Human hematopoietic cells were readily detectable in the hematopoietic system of the transplanted sheep at 2 months post-transplant, and the sheep maintained their chimeric status for at least 9 months. We also examined the ability of fetal kidney MSCs to give rise to human hepatocytes in sheep. In a similar approach to that used in the studies described above, human hepatocyte-like cells could readily be detected within the sections from sheep transplanted *in utero* with the human MSCs. Furthermore, human hepatocyte-like cells present within the liver sections stained positively with an antibody specific for human albumin, showing that the donor-derived hepatocytes generated from the metanephric MSCs were functional. In all of the sections which stained positive for human hepatocyte-like cells, the human origin of the hepatocyte-like cells was confirmed by *in-situ* hybridization with a human-specific Alu probe.

In conclusion, studies using the fetal sheep model of human stem cell transplantation showed that, by taking advantage of the naturally occurring phenomena during fetal development, this randomly bred large animal non-injury, xenogeneic model system could be used to study the *in-vivo* behavior and plasticity of HSCs from a variety of tissues following transplantation. This success was achieved by virtue of the model's ability to permit the expression of such a degree of HSC plasticity in the absence of injury or induced regeneration within a specific tissue.

Acknowledgments

These studies were supported by grants HL70566, HL73737, NAG9-1340, HL49042, HL66058, HL52955 from the National Institutes of Health, and NASA.

Abbreviations

CB   cord blood
HSC   human stem cell
mPB   mobilized peripheral blood
MSC   mesenchymal stem cell
NSC   neural stem cell

References

1. Watt, F.M., Epidermal stem cells: markers, patterning and the control of stem cell fate. *Philos. Trans. R. Soc. Lond. B. Biol. Sci.* 1998; *353*: 831–837.
2. Ormerod, E.J., Rudland P.S., Regeneration of mammary glands in vivo from isolated mammary ducts. *J. Embryol. Exp. Morphol.* 1986; *96*: 229–243.
3. Seale, P., Rudnicki M.A., A new look at the origin, function, and "stem-cell" status of muscle satellite cells. *Dev. Biol.* 2000; *218*: 115–124.

4. Clarke, D.L., Johansson, C.B., Wilbertz, J., Veress, B., Nilsson, E., Karlström H., Lendahl, U., Frisén, J., Generalized potential of adult neural stem cells. *Science* **2000**; *288*: 1660–1663.

5. Gage, F.H., Ray, J., Fisher, L.J., Isolation, characterization, and use of stem cells from the CNS. *Annu. Rev. Neurosci.* **1995**; *18*: 159–192.

6. Eriksson, P.S., Perfilieva, E., Bjork-Eriksson, T., Alborn, A.M., Nordborg, C., Peterson, D.A., Gage, F.H., Neurogenesis in the adult human hippocampus. *Nat. Med.* **1998**; *4*: 1313–1317.

7. Lagasse, E., Connors, H., Al-Dhalimy, M., Reitsma, M., Dohse, M., Osborne, L., Wang, X., Finegold, M., Weissman, I.L., Grompe, M., Purified hematopoietic stem cells can differentiate into hepatocytes in vivo. *Nat. Med.* **2000**; *6*: 1229–1234.

8. Petersen, B.E., Bowen, W.C., Patrene, K.D., Mars, W.M., Sullivan, A.K., Murase, N., Boggs, S.S., Greenberger, J.S., Goff, J.P., Bone marrow as a potential source of hepatic oval cells. *Science* **1999**; *284*: 1168–1170.

9. Theise, N.D., Badve, S., Saxena, R., Henegariu, O., Sell, S., Crawford, J.M., Krause, D.S., Derivation of hepatocytes from bone marrow cells of mice after radiation-induced myeloablation. *Hepatology* **2000**; *31*: 235–240.

10. Krause, D.S., Theise, N.D., Collector, M.I., Henegariu, O., Hwang, S., Gardner, R., Neutzel, S., Sharkis, S.J., Multi-organ, multi-lineage engraftment by a single bone marrow-derived stem cell. *Cell* **2001**; *105*: 369.

11. Wagers, A.J., Sherwood, R.I., Christensen, J.L., Weissman, I.L., Little evidence for developmental plasticity of adult hematopoietic stem cells. *Science* **2002**; *297*: 2256.

12. Wang, X., Montini, E., Al-Dhalimy, M., Lagasse, E., Finegold, M., Grompe, M., Kinetics of liver repopulation after bone marrow transplantation. *Am. J. Pathol.* **2002**; *161*: 565.

13. ;Mallet, V.O., Mitchell, C., Mezey, E., Fabre, M., Guidotti, J.E., Renia, L., Coulombel, L., Kahn, A., Gilgenkrantz, H., Bone marrow transplantation in mice leads to a minor population of hepatocytes that can be selectively amplified in vivo. *Hepatology* **2002**; *35*: 799.

14. Wang, X., Ge, S., McNamara, G., Hao, Q.L., Crooks, G.M., Nolta, J.A., Albumin expressing hepatocyte-like cells develop in the livers of immune-deficient mice transmitted with highly purified human hematopoietic stem cells. *Blood* **2003**; *101*: 4201–4208.

15. Theise, N.D., Nimmakayalu, M., Gardner, R., Illei, P.B., Morgan, G., Teperman, L., Henegariu, O., Krause, D.S., Liver from bone marrow in humans. *Hepatology* **2000**; *32*: 11–16.

16. Alison, M.R., Poulsom, R., Jeffery, R., Dhillon, A.P., Quaglia, A., Jacob, J., Novelli, M., Prentice, G., Williamson, J., Wright, N.A., Cell differentiation: hepatocytes from non-hepatic adult stem cells. *Nature* **2000**; *406*: 257.

17. Korbling, M., Katz, R.L., Khanna, A., Ruifrok, A.C., Rondon, G., Albitar, M., Champlin, R.E., Estrov, Z., Hepatocytes and epithelial cells of donor origin in recipients of peripheral-blood stem cells. *N. Engl. J. Med.* **2002**; *346*: 738.

18. Danet, G.H., Luongo, J.L., Butler, G., Lu, M.M., Tenner, A.J, Simon, M.C., Bonnett, D.A., $C1qR_p$ defines a new human stem cell population with hematopoietic and hepatic potential. *Proc. Natl. Acad. Sci. USA* **2002**; *99*: 10441–10445.

19. Ferrari, G., Cusella-De Angelis, G., Coletta, M., Paolucci, E., Stornaiuolo, A., Cossu, G., Mavilio, F., Muscle regeneration by bone marrow-derived myogenic progenitors. *Science* **1998**; *279*: 1528–1530.

20. Bjornson, C.R.R., Rietze, R.L., Reynolds, B.A., Magli, M.C., Vescovi, A.L., Turning brain into blood: a hematopoietic fate adopted by adult neural stem cells in vivo. *Science* **1999**; *283*: 534–537.

21. Bittner, R.E., Popoff, I., Shorny, S., Hoger, H., Wachtler, F., Dystrophin expression in heterozygous mdx/+ mice indicates imprinting of X chromosome inactivation by parent-of-origin-, tissue-, strain- and position-dependent factors. *Anat Embryol [Berl].* **1997**; *195*: 175–182.

22. Gussoni, E., Soneoka, Y., Strickland, C.D., Buzney, E.A., Khan, M.K., Flint.

A.F., Kunkel, L.M., Mulligan, R.C., Dystrophin expression in the mdx mouse restored by stem cell transplantation. *Nature* 1999; *401*: 390–394.

23. Zanjani, E.D., Pallavicini, M.G., Ascensao, J.L., Flake, A.W., Langlois, R.G., Reitsma, M., MacKintosh, F.R., Stutes, D., Harrison, M.R., Tavassoli, M., Engraftment and long-term expression of human fetal hematopoietic stem cells in sheep following transplantation in utero. *J. Clin. Invest.* 1992; *89*: 1178–1788.

24. Civin, C.I., Almeida-Porada, G., Lee, M.-J., Olweus, J., Terstappen, L.W.M.M., Zanjani, E.D., Sustained, retransplantable, multilineage engraftment of highly purified adult human bone marrow stem cells in vivo. *Blood* 1996; *88*: 4102–4109.

25. Yin, A.H., Miragi, A.S., Zanjani, E.D., Ogawa, M., Olweus, J., Kearney, J., Buck, D.W., AC133, A novel marker for human hematopoietic stem and progenitor cells. *Blood* 1997; *90*: 5002–5012.

26. Shimizu, Y., Ogawa, M., Kobayashi, M., Almeida-Porada, G., Zanjani, E.D., Engraftment of cultured human hematopoietic cells in sheep. *Blood* 1998; *91*: 3688–3692.

27. Zanjani, E.D., Almeida-Porada, G., Livingston, A.G., Flake, A.W. Ogawa, M., Human bone marrow CD34⁻ cells engraft in vivo and undergo multilineage expression including giving rise to CD34⁺ cells. *Exp. Hematol.* 1998; *26*: 353–360.

28. Giesert, C., Almeida-Porada, G., Scheffold, A., Kanz, L., Zanjani, E.D., Buhring, H.J., The monoclonal antibody W7C5 defines a novel surface antigen on hematopoietic stem cells. *Ann. N. Y. Acad. Sci.* 2001; *938*: 175–183.

29. Verfaillie, C.M., Almeida-Porada, G., Wissink, S., Zanjani, E.D., Kinetics of engraftment of CD34(-) and CD34(+) cells from mobilized blood differs from that of CD34(-) and CD34(+) cells from bone marrow. *Exp. Hematol.* 2000; *28*: 1071–1079.

30. Kucia, M., Reca, R., Jala, V.R., Dawn, B., Ratajczak, J., Ratajczak, M.Z., Bone marrow as a home of heterogeneous populations of nonhematopoietic stem cells. *Leukemia* 2005; *19*: 1118–1127.

31. Chamberlain, J., Frias, A., Porada, C., Yamagami, T., Zanjani, E.D., Almeida-Porada, G., Tissue availability of donor mesenchymal stem cells determines the degree of plasticity following transplantation in an non-injury model. *Blood* 2004; *104*: 676a.

32. Reid, G.M., Tervit H.M., Sudden Infant Death Syndrome: near weightlessness and delayed neural transformation. *Med. Hypotheses* 1996; *46*: 383–387.

33. Serova, L.V., The maternal-fetal system as an object for the study of mechanisms of the physiologic effect of weightlessness. *Kosm. Biol. Aviakosm.* 1987; *21*: 63–66.

34. Wood, C., Weightlessness: its implications for the human fetus. *J. Obstet. Gynaecol. Br. Commonw.* 1970; *77*: 333–336.

35. Susse, H.J., Wurterle, A., The shifting center of gravity in the growing fetus. *Zentralbl. Gynakol.* 1967; *89*: 980–984.

36. Hammond, T.G., Benes, E., O'Reilly, K.C. Wolf D.A., Linnehan, R.M., Taher, A., Kaysen, J.H., Allen, P.L., Goodwin, T.J., Mechanical culture conditions effect gene expression: gravity-induced changes on the space shuttle. *Physiol. Genomics* 2000; *3*: 163–173.

37. de Groot, R.P., Rijken P.J., den Hertog, J., Boonstra, J, Verkleij, A.J., de Laat, S.W., Kruijer, W., Microgravity decreases c-fos induction and serum response element activity. *J. Cell Sci.* 1990; *97*: 33–38.

38. de Groot, R.P., Rijken P.J., den Hertog, J., Boonstra, J., Verkleij, A.J., de Laat, S.W., Kruijer, W., Nuclear responses to protein kinase C signal transduction are sensitive to gravity. *Exp. Cell Res.* 1991; *197*: 87–90.

39. Walther, I., Pippia, P., Meloni, M.A., Turrini, F., Mannu, F., Cogoli, A., Simulated microgravity inhibits the genetic expression of interleukin-2 and its receptor in mitogen-activated T lymphocytes. *FEBS Lett.* 1998; *436*: 115–118.

40. Hammond, T.G., Lewis, F.C., Goodwin, T.J., Linnehan, R.M., Wolf, D.A., Hire, K.P., Campbell, W.C., Benes, E., O'Reilly, K.C., Globus, R.K., Kaysen, J.H., Gene expression in space. *Nat. Med.* 1999; *355*: 359.

41. Stein, G.S., van Wijnen A.J., Stein J.L., Lian, J.B., Pockwinse, S.H., McNeil, S.,

Implications for interrelationships between nuclear architecture and control of gene expression under microgravity conditions. *FASEB J.* **1999**; *13*: 157–166.

42. Margolis, L., Hatfill, S., Chuaqui, R., Vocke, C., Emmet-Buck, M., Linehan, W.M., Duray, P.H., Long term organ culture of human prostate tissue in a NASA-designed rotating wall bioreactor. *J. Urol.* **1999**; *161*: 290–297.

43. Akins, R.E., Schroedl, N.A., Gonda, S.R., Hartzell, C.R., Neonatal rat heart cells cultured in simulated microgravity. *In Vitro Cell Dev. Biol. Anim.* **1997**; *33*: 337–343.

44. Almeida-Porada, G., Hoffman, R., Manalo, P., Gianni, A.M., Zanjani, E.D., Detection of human cells in human/sheep chimeric lambs with in vitro human stroma-forming potential. *Exp. Hematol.* **1996**; *24*: 482–487.

45. Almeida-Porada, G., Porada C.D., Chamberlain. J., Torabi, A., Zanjani, E.D., Formation of human hepatocytes by human hematopoietic stem cells in sheep. *Blood* **2004**; *104*: 2582–2590.

46. Wang, X., Willenbring, H., Akkari, Y., Torimaru, Y., Foster, M., Al-Dhalimy, M., Lagasse, E., Finegold, M., Olson, S., Grompe, M., Cell fusion is the principal source of bone-marrow-derived hepatocytes. *Nature* **2003**; *422*: 897.

47. Almeida-Porada, G., Crapnell, K., Porada, C., Benoit, B., Nakauchi, H., Quesenberry, P., Zanjani, E.D., In vivo haematopoietic potential of human neural stem cells. *Br. J. Haematol.* **2005**; *130*: 276–283.

48. Reyes, M., Lund, T., Lenvik, T., Aguiar, D., Koodie, L., Verfaillie, C.M., Purification and ex vivo expansion of postnatal human marrow mesodermal progenitor cells. *Blood* **2001**; *98*: 2615.

49. Liechty, K.W., MacKenzie, T.C., Shaaban, A.F., Radu, A., Moseley, A.B., Deans, R., Marshak, D.R., Flake, A.W., Human mesenchymal stem cells engraft and demonstrate site-specific differentiation after in utero transplantation in sheep. *Nat. Med.* **2000**; *6*: 1282.

50. Kopen, G.C., Prockop, D.J., Phinney, D.G., Marrow stromal cells migrate throughout forebrain and cerebellum, and they differentiate into astrocytes after injection into neonatal mouse brains. *Proc. Natl. Acad. Sci. USA* **1999**; *96*: 10711–10716.

51. Sanchez-Ramos, J., Song, S., Cardozo-Pelaez, F., Hazzi, C., Stedeford, T., Willing, A., Freeman, T.B., Saporta, S., Janssen, W., Patel, N., Cooper, D.R., Sanberg, P.R., Adult bone marrow stromal cells differentiate into neural cells in vitro. *Exp. Neurol.* **2000**; *164*: 247–256.

52. Woodbury, D., Schwarz, E.J., Prockop, D.J., Black, I.B., Adult rat and human bone marrow stromal cells differentiate into neurons. *J. Neurosci. Res.* **2000**; *61*: 364–370.

53. Pittenger, M.F., Mackay, A.M., Beck, S.C., Jaiswal, R.K., Douglas, R., Mosca, J.D., Moorman, M.A., Simonetti D.W., Craig, S., Marshak, D.R., Multilineage potential of adult human mesenchymal stem cells. *Science* **1999**; *284*: 143.

54. Mackay, A.M., Beck, S.C., Murphy, J.M., Barry, F.P., Chichester, C.O., Pittenger, M.F., Chondrogenic differentiation of cultured human mesenchymal stem cells from marrow. *Tissue Eng.* **1998**; *4*: 415–418.

55. Almeida-Porada, G., Porada, C., El Shabrawy, D., Simmons, P.J., Zanjani, E.D., Human marrow stromal cells represent a latent pool of stem cells capable of generating long-term hematopoietic cells. *Blood* **2001**; *99*: 713a.

56. Almeida-Porada, G., El Shabrawy, D., Porada, C., Ascensao, J.L., Zanjani, E.D., Clonally derived marrow stromal cells [MSC] populations are able to differentiate into blood, liver and skin cells. *Blood* **2001**; *99*: 791a.

57. Almeida-Porada, G., El Shabrawy, D., Porada, C., Zanjani, E.D., Differentiative potential of human metanephric mesenchymal cells. *Exp. Hematol.* **2002**; *30*: 1454–1462.

9

Developmental Potential of Somatic Stem Cells Following Injection into Murine Blastocysts

Michael Dürr, Friedrich Harder, and Albrecht M. Müller

9.1
Introduction

The analysis of the comprehensive developmental potential of somatic stem cells requires an assay system which permits the development of all cell lineages that assemble into a multicellular organism. One such *in-vivo* assay system which allows the development of all cell lineages is the injection of cells into blastocysts, followed by an analysis of the cellular identity of engrafting donor cells in chimeric animals. This experimental strategy exposes the injected stem cells to stimuli and microenvironmental influences of the developing embryo. By using this blastocyst injection assay, it could be shown that embryonic stem (ES) cells are truly pluripotent, as the progeny of injected ES cells gave rise to all cell types present in the developing embryo [1, 2].

The magnitude of the developmental potential of somatic stem cells remains a matter of debate. Various reports published during the past few years have challenged the traditional concept that progenitor or stem cells isolated from differentiated tissues are restricted in the cell types that they can produce. For example, neural stem cells (NSCs) that grow as neurospheres were reported to generate hematopoietic cells after intravenous injection into irradiated recipient mice. Moreover, following injection into blastocysts, NSCs gave rise to differentiated cells of all three germ layers [3, 4]. The results of other studies revealed that isolated hematopoietic stem cells (HSCs) can generate hepatocytes, and that even single HSCs can yield multi-organ engraftment in adult recipient mice [5, 6].

The general plasticity of somatic stem and progenitor cells has also been challenged, as the hematopoietic competence of NSCs was found to be rare and was suggested to depend on epigenetic alterations [7]. Other studies failed to find any evidence that HSCs can differentiate into the neural lineage, and singly transplanted HSCs did not generate non-hematopoietic tissues [8, 9]. In addition, it was shown that stem cells can adopt the phenotype of unrelated cell types by cell fusion, so that the assumed plasticity of somatic stem cells could originate from cell fusion with unrelated cell types [10–13].

Stem Cell Transplantation. Biology, Processing, and Therapy.
Edited by Anthony D. Ho, Ronald Hoffman, and Esmail D. Zanjani
Copyright © 2006 WILEY-VCH Verlag GmbH & Co. KGaA, Weinheim
ISBN: 3-527-31018-5

It seems that not all reported transdifferentiation events are fusion products, however. In a recent report, Jang et al. found that HSCs became liver cells without cell fusion when co-cultured with injured liver separated by a barrier [14]. A further report demonstrated that bone marrow cells can generate epithelial cells in lung, liver and skin without fusion, albeit at low frequencies [15]. Finally, by co-culturing neurosphere cells with endothelial cells, the neurosphere cells were converted to cells that did not express neuronal or glial markers, but showed stable expression of multiple endothelial markers and had the capacity to form capillary networks [16]. The generation of epithelial cells from HSCs or endothelial cells from NSCs is especially surprising as the donor cells and their progeny belonged to different germ layers.

Although some of the above-mentioned reports provide arguments in favor of broader developmental potentials of somatic stem cells, other analyses have failed to find any evidence of transdifferentiation. This indicates that somatic stem cells do not give rise to unrelated cell types, or that transdifferentiation is a rare event, and consequently the full range of developmental potentials of somatic stem cells is still inadequately defined.

In this chapter, we show that blastocyst injection is a suitable experimental tool for analyzing the developmental potential of different somatic progenitor/stem cell types. We also summarize the outcome of the injection of different somatic progenitor/stem cell types into blastocysts, with particular discussion of results from neurosphere cells and HSC injections into blastocysts. Additionally, we show that injected human AML tumor cells expressed differentiation markers when exposed to a murine embryonic microenvironment. These data are consistent with a model in which somatic stem cells are mainly restricted in their developmental potential to the tissue of their origin. However, after *in-vitro* culture, and when in contact with embryonic microenvironment, somatic stem cells can acquire hallmarks of unrelated cell types.

9.2
Neurosphere Cells Generate Erythroid-Like Cells Following Injection into Early Embryos

In order to assess the developmental potential of NSCs, we established nonadherent neurosphere cultures and transplanted donor cells into developing blastocysts. Bulk and clonal neurosphere cultures were isolated from murine embryonic day (E) 14.5 animals by selective growth under serum-free conditions in the presence of endothelial growth factor (EGF) and basic fibroblast growth factor 2 (bFGF2) [3,17–19]. Neurosphere cultures were established from mice ubiquitously expressing enhanced green fluorescent protein (eGFP) under the chicken β-actin promoter and the cytomegalovirus (CMV) enhancer [20]. Some neurosphere cultures were generated from human bcl2-transgenic animals that express the transgene under control of the mouse H-2k^b promoter to counteract apoptosis-inducing challenges [21]. All neurospheres differentiated into GFAP$^+$ glial and β3-tubulin$^+$

neuronal cell types *in-vitro*. Using the procedure of micromanipulation, five to eight *in vitro*-expanded eGFP transgenic neurospheres consisting of two to eight cells that developed after overnight culture of dissociated neurospheres were injected. In order to follow the fate of the cells shortly after injection, injected blastocysts after 15 h of *in-vitro* culture were analyzed for the presence of eGFP⁺ donor cells. eGFP⁺ donor cells were detected in the blastocoel immediately after injection, whereas at 15 h after injection, eGFP⁺ cells were visible in the inner cell mass and in the trophectoderm (Fig. 9.1). These results indicate that donor cells survive and seed, developing blastocysts. Unfortunately, eGFP⁺ donor cells could not be detected in post-implantation embryos, but this was most likely due to low donor chimerism.

The transfer of blastocysts injected with wild-type neurosphere cells into fosters resulted in the development of 38 % of blastocysts into E12.5 embryos. To screen entire embryos with a sensitive method for donor cells, we injected male neurosphere cells and used a Y-chromosome-specific (YMT/2B) PCR on genomic DNA of developing female embryos. Analysis of the developing embryos at E12.5 by YMT/2B PCR showed that all embryos (9/9) carried donor cells, with about 0.2 % chimerism (Fig. 9.2A). The donor cells were most often detected in hematopoietic tissues such as yolk sac and fetal blood (6/9 and 8/9, respectively). Fewer

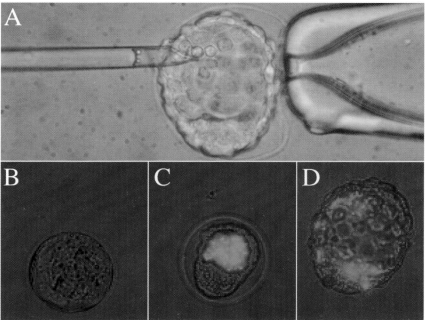

Figure 9.1 Injection of neurosphere cells into murine blastocysts. (A) Neurosphere cells were injected into blastocysts as described [19]. Shown are fluorescence images of a non-injected blastocyst (B), of a blastocyst immediately following injection of eGFP transgenic neurosphere cells (C), and of a blastocyst 15 h after injection of eGFP neurosphere cells (D).

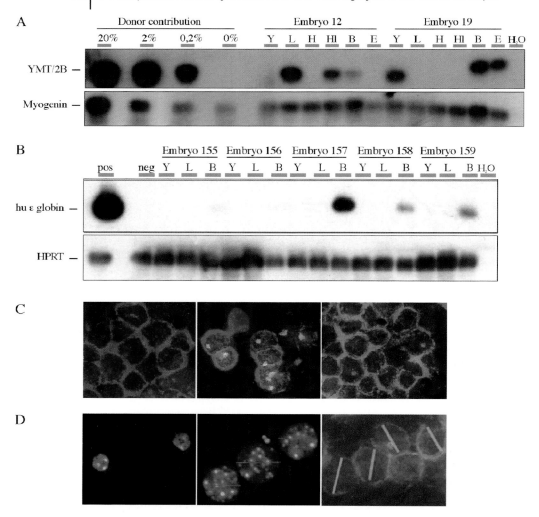

animals carried donor signals in the fetal head or hind limbs (2/9 and 3/9, respectively). Wild-type and bcl2 transgenic neurospheres showed similar seeding pattern in embryos, but the level of chimerism was increased following injection of bcl2 transgenic neurospheres [19].

In order to determine whether donor cells in chimeric embryos express hematopoiesis-specific transcripts, neurosphere cells from a human β-globin locus transgenic mouse line showing erythroid- and developmental stage-specific globin transgene expression were established [22]. β-globin locus transgenic neurosphere cells were injected into blastocysts, and E14.5 embryos were analyzed for the expression of the erythroid-specific ε-globin transgene. RT-PCR analysis showed that the transgenic neurospheres did not express human globin trans-

◀ **Figure 9.2** Erythroid-type donor contribution in chimeric embryos following blastocyst injection of neurosphere cells. (A) Bulk neurosphere cells isolated from male embryos were injected into murine blastocysts. Developing E12.5 female embryos were isolated, the individual tissues dissected, and genomic DNA was prepared. Neurosphere-derived male donor cells were detected in embryos by Y-chromosome-specific PCR (YMT2/B). Male donor contribution was quantified by diluting male into female genomic DNA. Myogenin-specific PCRs were performed as normalization controls. (B) Neurosphere cells from human (hu) β-globin locus transgenic embryos were injected and human ε-globin gene-specific RT-PCR was performed on RNA samples isolated from E14.5 embryos that developed from blastocyst injections. Positive control: RT-PCR on E12.5 fetal liver RNA of a human β-globin locus transgenic animal; negative control: RT-PCR on wild-type liver RNA. Cultured neurosphere cells did not express human globin transgenes. HPRT-specific RT-PCRs were performed for normalization. (C) Clonally derived male, bcl-2 transgenic neurosphere cells were injected into blastocysts and at E14.5 the fetal liver of a chimeric animal was subjected to a combined Y-chromosome-specific *in-situ* hybridization (green)/TER119-specific immuno- histochemistry (red, right panel). Also shown are control female (left) and male (middle) fetal liver cells analyzed by Y-chromosome-specific *in-situ* hybridization/TER119-specific immunohistochemistry. Nuclei were counterstained with DAPI (blue). (D) Analysis of nuclear diameters of 2n and 4n adult hepatocytes and of diameters of donor and recipient cells in a chimeric fetal liver. Hepatocytes isolated from adult donors were stained with DAPI, followed by flow cytometric separation for 2n (left panel) and 4n (middle panel) DNA content and cytocentrifugation. Also shown is the analysis of fetal liver cells of a chimeric E14.5 female embryo that developed after blastocyst injection of cloned male, bcl-2 transgenic neurosphere cells by TER119-specific immunohistochemistry (red)/Y-chromosome-specific *in-situ* hybridization (green, right panel). Nuclear diameters were determined using Adobe Photoshop software. Injected bulk neurosphere cells were of passages 10–25; cloned neurosphere cells were established from bulk cultures after 20 passages and further expanded. (A) and (B) are autoradiograms of Southern blots. (C,D) Original magnifications, ×400. Abbreviations: Y, yolk sac; L, fetal liver; H, head; Hl, hind limbs; B, fetal blood; E, rest of embryo.

genes. However, after injection and development of chimeric embryos, expression of the embryonic human ε-globin transgene was detected in the hematopoietic tissues of 14/16 embryos (Fig. 9.2B). To collect additional evidence that the progeny of injected neurosphere cells gain erythroid features, we used a combined male donor-specific *in-situ* hybridization/erythroid-specific immunohistochemical analysis on chimeric female embryos. The TER-119 monoclonal antibody recognizes an association partner of glycophorin A and marks the late differentiation stages of the murine erythroid lineage [23]. As shown in Figure 9.2C, TER119-positive male donor cells were present in the fetal liver of chimeric E14.5 embryos. About 1–3 % of cells in the fetal liver of engrafted female embryos were TER119 and Y-chromosome double-positive, showing that neurosphere cells after injection into early embryos generate progeny with hallmarks of erythroid cells.

As fused cells have enlarged nuclei [10, 11], we measured the nuclear diameters of donor and recipient cells to determine whether or not TER119$^+$ donor cells are products of cell fusion (Fig. 9.2D). The nuclear diameter of TER119$^+$ donor cells in fetal liver cells was 32.1 ± 4.4 pixel, which was similar to the diameters of recipient-type fetal liver cells (33.8 ± 4.8 pixel). Analysis of hepatocytes sorted for 2n and 4n DNA content had a diameter of 20.7 ± 2.4 and 42.1 ± 4.0 pixel, respectively. These results show that nuclear diameters of donor and host cells are similar. Although this strategy does not analyze cell fusion directly, the results argue against preceding cell fusion of TER119+ donor cells. In summary, the progeny of the injected neurosphere cells were detected in the hematopoietic tissues of developing embryos. In addition, donor cells initiate the expression of an erythroid-specific globin transgene, and neurosphere-derived cells are recognized by the erythroid-specific lineage antibody TER119. Collectively, these data indicate that cultured neurosphere cells acquire features of hematopoietic cells in a murine embryonic microenvironment.

9.3
Hematopoietic Chimerism by Human Cord Blood-Derived HSCs

To determine whether the murine embryo can support the engraftment and differentiation of injected human somatic stem cells, cord blood (CB)-derived CD34+/CD38− HSCs were isolated, and 70–100 human cells injected into each murine blastocyst [24]. By employing human-specific 17αmod PCR, human donor cells were detected preferentially in hematopoietic tissues (Fig. 9.3A). The yolk sac of 2/14 E8.5 embryos showed donor cells, while the peripheral blood of eight out of 20 (8/20) and the head of 1/20 E11.5 embryos were chimeric. In E16.5 embryos, the peripheral blood (5/23) and thymus (2/23), but not the embryonic head (0/23), was seeded with human cells.

To further analyze the cellular nature of donor cells in chimeric embryos, we used a multiplex RT-PCR to detect human globin transcripts. Injected CD34+/CD38− cells did not express globin genes, whereas ε- and β-globin transcripts could be detected in chimeric embryonic blood, suggesting erythroid differentiation of the injected human donor HSCs (Fig. 9.3B). It is worth noting that, upon injection into the early embryonic microenvironment, human CB-derived HSCs preferentially activate embryonic-type globin transcription. In summary, a human donor contribution was observed in yolk sac and fetal liver, but most often in the embryonic blood of developing murine embryos. Thus, injected human HSCs or their differentiated progeny seed – appropriately – the active sites of murine hematopoiesis, lending further support to the notion that donor cells are of hematopoietic nature.

9.4
Injection of Leukemic Cells into Blastocysts

In order to determine whether the murine embryonic microenvironment is also able to host and control leukemic cells, five to 30 human KG-1 myelogenous leukemia cells were injected into each murine blastocyst [25]. The progeny of human KG-1 cells were most frequently detected at E12.5 in yolk sac and the peripheral

A

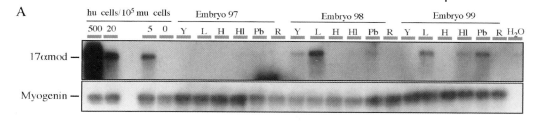

B

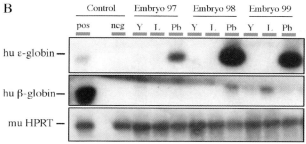

Figure 9.3 Human (hu) donor cells in murine embryos after injection of human cord blood-derived CD34$^+$/CD38$^-$ hematopoietic stem cells (HSCs) into blastocysts. (A) HSCs were isolated by flow cytometry, injected into blastocysts, and donor contribution was determined by a human chromosome 17-specific (17α-mod) PCR on genomic DNA extracted from different tissues of developing E11.5 embryos [42]. Human donor contribution was quantified by diluting human cells into murine cells. (B) Human ε- and β-globin-specific RT-PCRs on RNA isolated from E12.5 embryonic tissues are shown. As positive controls, RT-PCR was performed on RNA from unseparated human cord blood samples; a negative control was performed on RNA from murine embryonic blood of a wild-type embryo. Autoradiograms of Southern blots are shown. Abbreviations: Y, yolk sac; L, fetal liver; Pb, peripheral embryonic blood; R, rest of embryo. Other abbreviations as Figure 9.2.

blood of developing embryos (Fig. 9.4A). The level of engraftment was between five and 20 human cells per 100 000 murine cells. Based on the total number of peripheral blood cells of E12.5 embryos, this corresponds to 300 to 1200 human cells in total. As the hematopoietic tissues were preferentially engrafted, we analyzed the cellular nature of human cells in these tissues. First, RT-PCRs specific for human globin genes were performed (Fig. 9.4B). The transcripts of erythroid-specific human ε- and β-globin genes could be detected in yolk sac, fetal liver and the peripheral blood of E12.5 embryos. Additionally, an immuno-histochemical analysis was carried out on the peripheral blood from transplanted embryos, the results revealing that nucleated cells expressing the erythroid marker glycophorin A (Fig. 9.4C). The frequency of glycophorin A-positive cells in the peripheral blood of embryos was comparable to the number of human donor cells estimated from PCR analysis.

This finding illustrates that, following the injection of human AML cells into blastocysts, human leukemic cells can engraft developing murine embryos. Furthermore, erythroid-specific genes were activated, indicating that differentia-

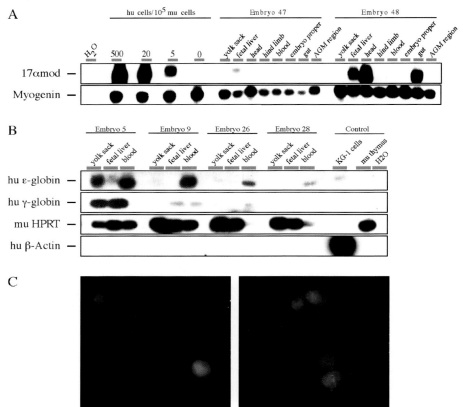

Figure 9.4 Human (hu) donor contribution to developing murine E12.5 embryos after injection of human KG-1 AML cells into blastocysts. KG-1 cells were injected into blastocysts and donor chimerism, erythroid-specific transcripts and expression of the erythroid-specific differentiation marker glycophorin A (CD235a) expression was analyzed. Progeny of injected human cells in tissues of murine embryos were detected by: (A) 17α-mod PCR on genomic DNA; (B) RT-PCR with human globin-specific primers; and (C) CD235a immunostaining (green) of cytospun embryonic blood cells of chimeric embryos. Nuclei were stained with DAPI (blue). (Original magnification, ×400.)

tion of malignant cells originally isolated from an erythroleukemic patient was towards the erythroid lineage.

9.5
Discussion

The injection of somatic cells into preimplantation murine blastocysts represents a powerful tool for the analysis of developmental potentials of diverse cell types. Since Gardner reported the first successful transfer of cells from the inner cell

Table 9.1 Developmental potential of somatic progenitor/
stem cells following injection into murine blastocysts.

Cell type	Results	Reference
Hematopoietic stem cells (HSCs)/ progenitor cells	Transplantation of bone marrow cells induces graft tolerance	43
	Donor cells seed hematopoietic tissues and adult-type HSCs reactivate embryonal globin transcripts	44
	Human HSCs contribute to murine embryos and activate erythroid gene expression	24
	FDCPmix cells preferentially engraft hematopoietic tissues and mainly express hematopoietic immuno-phenotypes	26
Neurosphere cells	NSCs give rise to differentiated cells of all three germ layers	4
	HSCs and neurosphere cells show distinct engraftment patterns in adult chimeras	18
	ES cells but not NSCs engraft developing embryos	30
	NSCs seed blastocysts but are not detected in mid-gestational embryos	31
	Neurosphere cells differentiate to erythroid-like cells in embryos, but not in adults	19
Mesenchymal stem cells	Multipotent adult progenitor cells isolated from adult bone marrow contribute to all three germ layers	29
Epidermal stem cells	Epidermal stem cells engraft ectodermal, mesenchymal and neural-crest tissues and activate appropriate differentiation markers	45
Leukemia cells	Human AML cells engraft hematopoietic tissues and express erythroid differentiation markers	25

mass of a murine blastocyst into the blastocoelic cavity of a synchronous blastocyst in 1968, various murine cell types from embryonic and adult tissues and xenogenic cells have been successfully transplanted into murine blastocysts (Table 9.1) [1]. This indicates that developing embryos provide a suitable environment for the survival of even heterochronically injected cells. The generation of chimeras by blastocyst injection of fetal/adult cell types is intriguing, as many cell generations and differentiation stages separate the injected cells and the receiving blastocysts. Apparently, developing murine embryos are able to provide appropriate environments for the survival (and control) of injected cells.

In principle, the injected cells can suffer three distinct fates. One possibility is that they die in the blastocyst environment due to inappropriate growth conditions and, as a result, they are lost during further development of the embryo. Alternatively, it is conceivable that the injected cells proliferate and by doing so destroy the developing embryo. In fact, if certain tumor cell lines are injected, then no embryos develop (S. Petrovic and A.M. Müller, unpublished results). None of these two scenarios leads to the generation of chimeras. This occurs only, if cells survive and a limited proliferation of donor cells takes place.

Our results show that murine neurosphere cells, human CB $CD34^+/CD38^-$ HSCs and human KG-1 AML cells engraft developing embryos after blastocyst injection. All three injected cell types mainly seed embryonic hematopoietic tissues and generate progeny with erythroid-specific marker expression [18, 19, 24, 25]. The PCR signals in non-hematopoietic tissues could arise from contaminating blood cells. This hematopoietic seeding pattern resembles the distribution of FDCPmix cells, a hematopoietic progenitor cell line, after blastocyst injection. Although some progeny of injected FDCPmix cells were found to be $CD45^-$, most FDCPmix donor cells in chimeric tissues maintained CD45 expression, indicating that they are hematopoietic cells [26].

The results from both the RT-PCR and the immunophenotypic analyses provide evidence that donor signals in PCR analyses of genomic DNA neither originated from unspecific DNA uptake by host cells, nor from donor cells resting inertly in chimeric tissues. Instead, injected cells may selectively migrate to hematopoietic tissues due to the existence of specific chemoattractants, chemokines, or growth factors. Alternatively, the special microenvironment may favor the survival and differentiation of injected cells. The high frequency of donor cells with embryonic-type globin gene expression may be due to the fact that injected cells are exposed to the blood-island-inducing microenvironment of the embryonic yolk sac.

Interestingly, the progeny of injected neurosphere cells also acquire features of hematopoietic cell types. Whether neurosphere cells transdifferentiate into the hematopoietic lineage in developing chimeric embryos remains to be investigated further, however. Of note for the interpretation of this study is that we employed cultured neurosphere cells. It has been reported that the potentials of oligodendrocyte precursor cells and of primordial germ cells were increased by specific *in-vitro* culture conditions [27, 28]. Interestingly, cultured neurosphere cells and cultured mesenchymal stem cells (multipotent adult progenitor cells, MAPCs) also contributed to unrelated cell types following blastocyst injection [4, 29]. It is therefore possible that freshly isolated stem cells have no, or very limited, potential to generate unrelated cell types, and that they acquire this activity only after an extended time in culture. The apparent differences in the ability to engraft developing embryos and adult animals between this and other studies may be due to differences in culture time of the injected neurosphere cells [30, 31]. Thus, whilst our experiments provide some insights into the developmental potentials of cultured neurospheres, we did not investigate whether noncultured neurospheres could contribute to unrelated cell lineages after blastocyst injection.

An alternative explanation for the neurosphere-derived cells with erythroid markers in chimeric embryos is that of cell fusion [10–13]. In order to investigate whether donor cells in peripheral blood and fetal liver are fusion products, we compared the nuclear diameters of both donor and recipient cells. It was observed that donor and recipient cells showed similar nuclear diameters, which was consistent with the view that donor cells are not products of preceding cell fusion events. However, since these analyses study cell fusion only indirectly (due to the low frequencies of donor cells in chimeric animals), the possibility that neurosphere-derived cells in chimeric tissues are fusion products cannot be completely ruled out.

Following injection into murine blastocysts, human leukemic cells were also able to engraft hematopoietic tissues of developing embryos. Moreover, human cells activated globin and glycophorin A expression, indicating differentiation into the erythroid lineage. In contrast to embryonal carcinoma (EC) cells, where the malignant phenotype is caused in part by epigenetic changes [32, 33], most leukemias adopt their malignant phenotype after genetic mutations. To reverse the malignancy of acute myeloid leukemia (AML) cells, it would therefore be necessary to overcome the mutation. In fact, the reversibility of bcr-abl-induced leukemias has been shown by gene deletion and through kinase inhibitors [34, 35], and the genome of cancer cells could be reprogrammed by nuclear transplantation into a pluripotent embryonic state [36]. Originally, KG-1 cells were isolated from a patient suffering from erythroleukemia. The embryonic microenvironment of the blastocysts may have supported a silencing of genetic abnormalities and thereby have enabled the differentiation into the erythroid lineage. Our observation complements earlier reports that the proliferation and differentiation capacity of carcinoma cells can be controlled in certain cellular contexts [37–39].

At this point, an important question is whether it is ethically acceptable to transplant human cells into murine embryos. Karpowicz et al. outlined criteria for an ethically compliant transplantation of human cells into murine embryos [40], and these were followed in the present studies. Consequently, low numbers of donor cells were injected and single cell suspensions transplanted that generated low degrees of "humanization", rather than entire tissues.

The results also revealed that, in contrast to ES cells, somatic stem cells showed no pluripotency but rather a restricted developmental potential with a low level of chimerism after blastocyst injection. One approach to increase chimerism is to provide injected cells with a selective advantage over endogenous cells. Engraftment of the hematopoietic system following the injection of HSCs into a blastocyst could be augmented by using blastocysts from W-mice (which are characterized by an underlying stem cell defect due to a c-kit mutation [41]), or by the injection of donor cells that are protected against apoptosis-inducing challenges. Indeed, the level of chimerism could be raised by injecting neurosphere cells from bcl-2 transgenic mice [19].

Taken together, we show that the engraftment pattern of developing embryos after blastocyst injection of somatic progenitor/stem cells is cell type-specific. While human HSCs and human leukemic cells preferentially seed hematopoietic tissues of their origin, murine neurosphere cells could be detected in heterologous hematopoietic organs. Furthermore, the injection of neurosphere cells into blastocysts led to a seeding of embryonic hematopoietic tissues, and donor cells were observed in the hematopoietic tissues of embryos that transcribed the embryonic-type globin gene. Importantly, after injecting cloned neurosphere cells with *in-vitro* neural differentiation potential, donor cells expressing an erythroid-specific cell surface marker were detected in chimeric fetal liver and peripheral blood. The present studies show that stem cells are not irrevocably fixed in their fate, but rather their character can be modified by contact with developing embryonic microenvironments.

Abbreviations/Acronyms

AML	acute myeloid leukemia
bFGF2	basic fibroblast growth factor
CMV	cytomegalovirus
EC	embryonal carcinoma
EGF	endothelial growth factor
eGFP	enhanced green fluorescent protein
GFAP	glial fibrillary acidic protein
HSCs	hematopoietic stem cells
MAPCs	multipotent adult progenitor cells
NSCs	neural stem cells

Acknowledgments

The authors thank Tri Nguyen for comments on the manuscript.

References

1. Gardner, R.L. Mouse chimaeras obtained by the injection of cells into the blastocyst. *Nature* **1968**, *220*, 596–597.
2. Beddington, R.S., Robertson, E.J. An assessment of the developmental potential of embryonic stem cells in the midgestation mouse embryo. *Development* **1989**, *105*, 733–737.
3. Bjornson, C.R.R., Rietze, R.L., Reynolds, B.A., Magli, M.C., Vescovi, A.L. Turning brain into blood: a hematopoietic fate adopted by adult neural stem cells in vivo. *Science* **1999**, *283*, 534–537.
4. Clarke, D.L., Johansson, C.B., Wilbertz, J., Veress, B., Nilsson, E., Karlstrom, H., Lendahl, U., Frisen, J. Generalized potential of adult neural stem cells. *Science* **2000**, *288*, 1660–1663.
5. Lagasse, E., Connors, H., Al-Dhalimy, M., Reitsma, M., Dohse, M., Osborne, L., Wang, X., Finegold, M., Weissman, I.L., Grompe, M. Purified hematopoietic stem cells can differentiate into hepatocytes in vivo. *Nat Med* **2000**, *6*, 1229–1234.
6. Krause, D.S., Theise, N.D., Collector, M.I., Henegariu, O., Hwang, S., Gardner, R., Neutzel, S., Sharkis, S.J. Multi-organ, multi-lineage engraftment by a single bone marrow-derived stem cell. *Cell* **2001**, *105*, 369–377.
7. Morshead, C.M., Benveniste, P., Iscove, N.N., van der Kooy, D. Hematopoietic competence is a rare property of neural stem cells that may depend on genetic and epigenetic alterations. *Nat Med* **2002**, *8*, 268–273.

8. Castro, R.F., Jackson, K.A., Goodell, M.A., Robertson, C.S., Liu, H., Shine, H.D. Failure of bone marrow cells to transdifferentiate into neural cells in vivo. *Science* **2002**, *297*, 1299.

9. Wagers, A.J., Sherwood, R.I., Christensen, J.L., Weissman, I.L. Little evidence for developmental plasticity of adult hematopoietic stem cells. *Science* **2002**, *297*, 2256–2259.

10. Terada, N., Hamazaki, T., Oka, M., Hoki, M., Mastalerz, D.M., Nakano, Y., Meyer, E.M., Morel, L., Petersen, B.E., Scott, E.W. Bone marrow cells adopt the phenotype of other cells by spontaneous cell fusion. *Nature* **2002**, *416*, 542–545.

11. Ying, Q.L., Nichols, J., Evans, E.P., Smith, A.G. Changing potency by spontaneous fusion. *Nature* **2002**, *416*, 545–548.

12. Wang, X., Willenbring, H., Akkari, Y., Torimaru, Y., Foster, M., Al-Dhalimy, M., Lagasse, E., Finegold, M., Olson, S., Grompe, M. Cell fusion is the principal source of bone-marrow-derived hepatocytes. *Nature* **2003**, *422*, 897–901.

13. Vassilopoulos, G., Wang, P.R., Russell, D.W. Transplanted bone marrow regenerates liver by cell fusion. *Nature* **2003**, *422*, 901–904.

14. Jang, Y.Y., Collector, M.I., Baylin, S.B., Diehl, A.M., Sharkis, S.J. Hematopoietic stem cells convert into liver cells within days without fusion. *Nat Cell Biol* **2004**, *6*, 532–539.

15. Harris, R.G., Herzog, E.L., Bruscia, E.M., Grove, J.E., Van Arnam, J.S., Krause, D.S. Lack of a fusion requirement for development of bone marrow-derived epithelia. *Science* **2004**, *305*, 90–93.

16. Wurmser, A.E., Nakashima, K., Summers, R.G., Toni, N., D'Amour, K.A., Lie, D.C., Gage, F.H. Cell fusion-independent differentiation of neural stem cells to the endothelial lineage. *Nature* **2004**, *430*, 350–356.

17. Reynolds, B.A., Weiss, S. Clonal and population analyses demonstrate that an EGF-responsive mammalian embryonic CNS precursor is a stem cell. *Dev Biol* **1996**, *175*, 1–13.

18. Kirchhof, N., Harder, F., Petrovic, S., Kreutzfeldt, S., Schmittwolf, C., Dürr, M., Mühl, B., Merkel, A., Müller, A.M.

19. Harder, F., Kirchhof, N., Petrovic, S., Wiese, S., Müller, A.M. Erythroid-like cells from neural stem cells injected into blastocysts. *Exp Hematol* **2004**, *32*, 673–682.

20. Okabe, M., Ikawa, M., Kominami, K., Nakanishi, T., Nishimune, Y. 'Green mice' as a source of ubiquitous green cells. *FEBS Lett* **1997**, *407*, 313–319.

21. Domen, J., Gandy, K.L., Weissman, I.L. Systemic overexpression of BCL-2 in the hematopoietic system protects transgenic mice from the consequences of lethal irradiation. *Blood* **1998**, *91*, 2272–2282.

22. Strouboulis, J., Dillon, N., Grosveld, F. Developmental regulation of a complete 70-kb human beta-globin locus in transgenic mice. *Genes Dev* **1992**, *6*, 1857–1864.

23. Kina, T., Ikuta, K., Takayama, E., Wada, K., Majumdar, A.S., Weissman, I.L., Katsura, Y. The monoclonal antibody TER-119 recognizes a molecule associated with glycophorin A and specifically marks the late stages of murine erythroid lineage. *Br J Haematol* **2000**, *109*, 280–287.

24. Harder, F., Henschler, R., Junghahn, I., Lamers, M.C., Müller, A.M. Human hematopoiesis in murine embryos after injecting human cord blood-derived hematopoietic stem cells into murine blastocysts. *Blood* **2002**, *99*, 719–721.

25. Dürr, M., Harder, F., Merkel, A., Bug, G., Henschler, R., Müller, A.M. Chimaerism and erythroid marker expression after microinjection of human acute myeloid leukaemia cells into murine blastocysts. *Oncogene* **2003**, *22*, 9185–9191.

26. Petrovic, S., Cross, M., Müller, A.M. Differentiation potential of FDCPmix cells following injection into blastocysts. *Cells Tissues Organs* **2004**, *178*, 78–86.

27. Kondo, M., Scherer, D.C., Miyamoto, T., King, A.G., Akashi, K., Sugamura, K., Weissman, I.L. Cell-fate conversion of lymphoid-committed progenitors by instructive actions of cytokines. *Nature* **2000**, *407*, 383–386.

Developmental potential of hematopoietic and neural stem cells: unique or all the same? *Cells Tissues Organs* **2002**, *171*, 77–89.

28. Matsui, Y., Zsebo, K., Hogan, B.L. Derivation of pluripotential embryonic stem cells from murine primordial germ cells in culture. *Cell* **1992**, *70*, 841–847.

29. Jiang, Y., Jahagirdar, B.N., Reinhardt, R.L., Schwartz, R.E., Keene, C.D., Ortiz-Gonzalez, X.R., Reyes, M., Lenvik, T., Lund, T., Blackstad, M., Du, J., Aldrich, S., Lisberg, A., Low, W.C., Largaespada, D.A., Verfaillie, C.M. Pluripotency of mesenchymal stem cells derived from adult marrow. *Nature* **2002**, *418*, 41–49.

30. D'Amour, K.A., Gage, F.H. Genetic and functional differences between multipotent neural and pluripotent embryonic stem cells. *Proc Natl Acad Sci USA* **2003**, *100*, 11866–11872.

31. Greco, B., Low, H.P., Johnson, E.C., Salmonsen, R.A., Gallant, J., Jones, S.N., Ross, A.H., Recht, L.D. Differentiation prevents assessment of neural stem cell pluripotency after blastocyst injection. *Stem Cells* **2004**, *22*, 600–608.

32. Mintz, B., Illmensee, K., Gearhart, J.D. Developmental and experimental potentialities of mouse teratocarcinoma cells from embryoid body cores, in: Sherman MI, Solter D (Eds.), *Teratomas and Differentiation*. Academic Press, New York, **1975**, pp. 59–82.

33. Blelloch, R., Hochedlinger, K., Yamada, Y., Brennan, C., Kim, M., Mintz, B., Chin, L., Jaenisch, R. Nuclear cloning of embryonal carcinoma cells. *Proc Natl Acad Sci USA* **2004**, *101*, 13985–13990. Erratum in: *Proc Natl Acad Sci USA* **2004**, *101*, 13985–14305.

34. Huettner, C.S., Zhang, P., Van Etten, R.A., Tenen, D.G. Reversibility of acute B-cell leukaemia induced by BCR-ABL1. *Nat Genet* **2000**, *24*, 57–60.

35. Sawyers, C.L. Rational therapeutic intervention in cancer: kinases as drug targets. *Curr Opin Genet Dev* **2002**, *12*, 111–115.

36. Hochedlinger, K., Blelloch, R., Brennan, C., Yamada, Y., Kim, M., Chin, L., Jaenisch, R. Reprogramming of a melanoma genome by nuclear transplantation. *Genes Dev* **2004**, *18*, 1875–1885.

37. Illmensee, K., Mintz, B. Totipotency and normal differentiation of single teratocarcinoma cells cloned by injection into blastocysts. *Proc Natl Acad Sci USA* **1976**, *73*, 549–553.

38. Mintz, B., Illmensee, K. Normal genetically mosaic mice produced from malignant teratocarcinoma cells. *Proc Natl Acad Sci USA* **1975**, *72*, 3585–3589.

39. Rossant, J., Papaioannou, V.E. Outgrowth of embryonal carcinoma cells from injected blastocysts in vitro correlates with abnormal chimera development in vivo. *Exp Cell Res* **1985**, *156*, 213–220.

40. Karpowicz, P., Cohen, C.B., van der Kooy, D. It is ethical to transplant human stem cells into nonhuman embryos? *Nat Med* **2001**, *10*, 331–335.

41. Fleischman, R.A., Custer, R.P., Mintz, B. Totipotent hematopoietic stem cells: normal self-renewal and differentiation after transplantation between mouse fetuses. *Cell* **1982**, *30*, 351–359.

42. Warburton, P.E., Greig, G.M., Haaf, T., Willard, H.F. PCR amplification of chromosome-specific alpha satellite DNA: definition of centromeric STS markers and polymorphic analysis. *Genomics* **1991**, *11*, 324–333.

43. Brinster, R.L. The effect of cells transferred into the mouse blastocyst on subsequent development. *J Exp Med* **1974**, *140*, 1049–1056.

44. Geiger, H., Sick, S., Bonifer, C., Müller, A.M. Globin gene expression is reprogrammed in chimeras generated by injecting adult hematopoietic stem cells into mouse blastocysts. *Cell* **1998**, *93*, 1055–1065.

45. Liang, L., Bickenbach, J.R. Somatic epidermal stem cells can produce multiple cell lineages during development. *Stem Cells* **2002**, *20*, 21–31.

10

Testing the Limits: The Potential of MAPC in Animal Models

Felipe Prósper and Catherine M. Verfaillie

10.1
Introduction

Although an unlimited self-renewal and pluripotent differentiation potential makes embryonic stem (ES) cells the most attractive source of stem cells for tissue regeneration, the potential for inducing tumor formation and the histoincompatibility between donor and recipient posses important limitations to their clinical use [1, 2]. Stem cells have also been identified in many postnatal tissues. Compared with ES cells, tissue-specific stem cells have less self-renewal ability and are not pluripotent, although they differentiate into multiple lineages. Many recent studies have suggested that tissue-specific stem cells may have the ability to generate cells of tissues from unrelated organs [3]. This area of research is however, not devoid of intense controversy as many of the studies supporting the existence of adult stem cell plasticity have not been confirmed or, in some cases, the initial observations have been explained by a different mechanism such as cell fusion [4, 5]. The aim of this chapter is not to discuss the mechanism of plasticity; rather, the reader is referred to recent excellent reviews addressing these issues [3, 6]

Among the different populations of stem cells isolated, a population of primitive cells, called multipotent adult progenitor cells (MAPCs), that have multipotent differentiation and extensive proliferation potential at the single-cell level, have attracted particular interest. Initially described by the group of Catherine Verfaille, MAPCs have been isolated from a number of mammalian normal bone marrow (BM) including human, rodent, and other mammalian postnatal tissues [7–12]. Beside the initial description, other groups more recently have also been able to isolate and characterize this population of stem cells [13, 14]. However, the lack of standardization of the MAPC culture – as well the lack of reliable markers that help to define this population of cells – has raised some doubts about whether MAPCs exist as such *in vivo* or are artifacts of specific culture conditions. In this chapter, we review the published – and some unpublished – new informa-

Stem Cell Transplantation. Biology, Processing, and Therapy.
Edited by Anthony D. Ho, Ronald Hoffman, and Esmail D. Zanjani
Copyright © 2006 WILEY-VCH Verlag GmbH & Co. KGaA, Weinheim
ISBN: 3-527-31018-5

tion regarding the nature, isolation, and *in-vitro* and *in-vivo* differentiation ability of MAPCs.

10.2
Characterization of MAPCs

10.2.1
Phenotype of MAPCs

It has been reported that MAPCs can be isolated from human, mouse, rat and nonhuman primate BM [8, 10, 12, 15]. MAPCs show a cell-surface phenotype that distinguishes them from other types of stem cell. Unlike mesenchymal stem cells (MSCs) [16], MAPCs do not express the major histocompatibility (MHC) class I antigens, CD105 (endoglin), CD106 (VCAM) or CD73 with expression of very low or absent levels of the CD44 antigen [8, 10, 12, 17]. MAPCs do not express the CD45, CD34, AC133 or CD36 antigens, expressed by hematopoietic stem cells (HSCs) and endothelial progenitor cells (EPC) [18, 19]. MAPCs express cell-surface markers as well as transcription factors associated with pluripotent embryonic stem cells such as stage-specific embryonic antigen (SSEA)-1, low levels of the transcription factors Oct4 (oct3a in human and monkey) and Rex1, or Nanog [8, 12, 17], which are known to be important for maintaining ES cells in an undifferentiated state and to be down-regulated when these cells undergo somatic cell commitment and differentiation [20]. MAPCs may also be isolated from other tissues and other species, as they have been cultured from mouse brain and mouse muscle [21]. Like ES cells, highly sensitive RT-PCR shows that MAPCs can express mRNA for a number of genes associated to different lineages such as nestin, Flk, Flt, or HNF3β.

10.2.2
Proliferative Capacity of MAPCs and Culture Conditions

Unlike most adult somatic stem cells, MAPCs have active telomerase and proliferate without obvious signs of senescence. In humans, MAPC telomeres are 3–5 kb longer than in neutrophils and lymphocytes, and telomere length is conserved when MAPCs are derived from young or old donors [10]. These observations suggest that MAPCs are derived from a population of cells that either has active telomerase *in vivo* or the cell population is highly quiescent *in vivo*, and therefore has not yet incurred telomere shortening *in vivo*.

An important issue is whether prolonged culture of the cells is associated with cytogenetic abnormalities, as has been the case for ES cells [22]. In human MAPC cultures, cytogenetic abnormalities have not been observed to date, even though cells have been maintained in culture for over 150 cell duplications. However, that is not the case for murine-derived MAPCs, mainly mouse, which have become aneuploid, tetraploid, or hypodiploid. Cytogenetic abnormalities are seen more frequently once mouse and rat MAPCs have been expanded for >60–70 population doublings and following repeated cryopreservation and thawing episodes. This characteristic of mouse MAPCs is similar to that of other mouse cell populations, including mouse ES cells.

MAPC cultures remain a challenge. Stringent culture conditions are required to maintain MAPCs in their undifferentiated state. Major culture factors that play roles in the successful maintenance of MAPCs include cell density, concentrations of CO_2 and O_2, and the specific lot of fetal calf serum (FCS) that is used in the culture medium. The control of cell density appears to be species-specific: mouse and rat MAPCs must be maintained at densities between 150 and 500 cells cm^{-2}, and swine and human MAPCs between 500 and 2000 cells cm^{-2}. When maintained at higher densities, MAPCs tend to lose some of their differentiation capacity. MAPC differentiation is also affected by the individual lot of serum used in the culture conditions. Because FCS increases the proliferation rate of MAPCs and increases the rate of cytogenetic abnormalities seen in mouse (and to a lesser extent, rat) MAPCs, the cells are maintained with only 2% FCS. Despite such a low concentration of the serum, however, differentiation remains affected by the individual lots of FCS used.

In order to reduce the rate of cytogenetic abnormalities seen in mouse MAPCs, the isolation and maintenance of rodent MAPCs are conducted under low (5%) O_2 conditions. However, these conditions also produce a slightly different phenotype of MAPCs. If mouse BM MAPCs are isolated and immediately cultured at 5% O_2, Oct4 transcript levels approach those in ES cells, and >90% of the MAPC cells express nuclear Oct4 protein. Low-oxygen-derived MAPCs also express c-Kit and express other markers such as Utf1, Rex1, or Sox2 at levels between 1 and 10% of those observed in ES cells. However, levels of Nanog remain 10 000-fold lower than those in ES cells, and cell-surface SSEA1 concentrations remain lower than observed with ES cells. The implications of these changes in phenotype on the pluripotency of MAPCs are currently being examined. Of note, when MAPCs isolated under normal oxygen conditions are subsequently switched to 5% O_2, no changes in the transcriptional or cell-surface phenotypes are seen. These observations suggest that the isolation of MAPCs under low-O_2 conditions may select for a different cell population and that the phenotype is not inducible *in vitro*.

Since the initial description of MAPCs was made, a number of research groups in the United States, Europe, and Japan have been trained in the isolation and maintenance of MAPCs from rodents and larger mammals. Currently, at least six research groups have successfully isolated and/or maintained MAPCs, including those from human, rat, and mouse, in their own laboratories [12, 14, 17]. In addition, researchers in Pamplona, Spain have recently reported the isolation of MAPCs from the BM of cynomolgus monkeys [12, 17]. The cynomolgus monkey-derived MAPCs (termed pMAPCs) are similar to other MAPCs in several key ways. The cells are CD44 and MHC-class I-negative and Oct3a-positive. Additionally, these cells do not appear to senesce (they have been maintained for >150 population doublings), and they differentiate into cells of the three germ-layers (neuronal-like, endothelial-like, and hepatocyte-like cells). The expression of Oct4 by Western blot and SSEA-4 by flow cytometry has also been demonstrated in pMAPCs [12].

10.3
In-Vitro Differentiation Potential of MAPCs

In 2001, Verfaillie and colleagues demonstrated that human, mouse and rat MAPCs can be successfully differentiated *in vitro* into typical mesenchymal lineage cells, including osteoblasts, chondroblasts, adipocytes and skeletal myoblasts [10]. In addition, human, mouse and rat MAPCs can be induced to differentiate into cells with morphological, phenotypic and functional characteristics of endothelial cells [9], hepatocytes [11], and neurons [7].

Since the first description of MAPCs, evidence has suggested that MAPCs can be induced to differentiate to additional phenotypes. Specifically, the application of culture conditions for dopaminergic differentiation of ES cells or neural stem cells to mouse MAPCs also induces differentiation to cells with phenotypic and functional characteristics of dopaminergic neurons [7]. Recent unpublished studies in our laboratory aimed to determine the vascular potential of MAPCs and showed that, in comparison with other sources of stem cells with recognized potential for vascular differentiation such as AC133$^+$ cells, MAPCs can be specified to both arterial as well as venous endothelium [17]. Using a combination of cytokines including VEGF$_{165}$, Sonic Hedgehog and Dll-4, we have shown that MAPCs can be induced to acquire markers of arterial endothelium both at the mRNA and at the protein level such as EphrinB1, EphrinB2, Hey-2, or Dll-4. This potential, along with the capacity to differentiate into smooth muscle cells, may establish MAPCs as an attractive cell source for arteriogenesis. Depending on the culture conditions used, human MAPCs can also be specified to lymphatic endothelial phenotypes which cannot be achieved when human AC133 endothelial progenitor cells are subjected to the same culture conditions [17].

10.4
In-Vivo Differentiation Potential of MAPCs

Non-obese diabetic/severe combined immunodeficient (NOD-SCID) mice are often used in transplant experiments because a genetic defect prevents the normal development of cells in their immune systems. These mice "tolerate" transplanted tissue since their immune cells do not work sufficiently well to attack and destroy "foreign" transplanted tissues. MAPCs engraft following transplantation into NOD-SCID mice, either without prior radiation or following sublethal radiation [8]. Engraftment has been shown in the hematopoietic system and the lung, liver, and gut epithelium [8], while none of the animals transplanted exhibited teratomas or other tumor types from 6 months to 2 years after transplantation.

Of significant interest is the *in-vivo* potential of MAPCs as EPC or vascular progenitors. Initial studies conducted in a lung cancer tumor model in NOD-SCID mice showed that human MAPCs transplanted into the mice contributed up to 35 % of the endothelial cells of the tumor vessels, suggesting the *in-vivo* potential for vasculogenesis [9]. Interestingly, the transplantation of endothelial pre-committed cells in comparison with undifferentiated MAPCs significantly increased the number of human-derived endothelial cells.

Since the first description of MAPCs, further studies have evaluated the *in-vivo* potential of these cells. We have studied the *in-vitro* and *in-vivo* arterial-venous specification of human MAPCs in comparison with other sources of EPCs, and shown that MAPCs differentiate into arterial, venous, and lymphatic endothelium. Furthermore, using an *in-vivo* matrigel plug assay in nude mice, we have shown that the combination of MAPCs with cytokines (VEGF$_{165}$, Sonic Hedgehog and Dll-4) induces the formation of vascular structures with endothelial cells, smooth muscle cells and deposition of collagen and elastin consistent with the formation of arteria-like structures [17]

Human MAPCs can also differentiate *in vivo* to muscle fibers, as demonstrated in a NOD/SCID mouse model [14] where injection of uncommitted MAPCs or MAPCs previously treated with myogenic conditions in the *tibialis anterioris* of mice treated with cardiotoxin resulted in the formation of myotubes derived from the transplanted cells. As described above, the transplantation of predifferentiated cells resulted in a higher degree of cell engraftment and differentiation to muscle fibers [14], suggesting that local cues will be more effective if the cell population transplanted has already been predetermined to a certain cell fate.

The *in-vivo* immune responses to MAPCs have been assessed using mice with various degrees of T-, B-, and natural killer (NK)-cell immune competence [13]. To follow the biodistribution of MAPCs in live animals, researchers labeled the cells with the red fluorescent protein, DsRed2, and luciferase and transplanted the cells intravenously into T- and B-cell deficient and T-, B-, and NK-cell-deficient mice. These studies showed that endogenous NK cells and T cells do attack MAPCs. Additional studies suggest that total body irradiation (TBI) conditioning may overcome both NK- and T-cell-mediated transplant destruction, resulting in a widespread homing/migration of MAPCs [13].

Additional studies have shown significant cross-correction of glycosaminoglycan (GAG) accumulation in blood, urine, and several solid organs when mouse MAPCs are transplanted into newborn mice with mucopolysaccharidosis type I [23]. In these studies, widespread engraftment was noted in skeletal muscle and brain in addition to the liver, gut, and lung.

Ongoing studies are evaluating the ability of undifferentiated MAPCs or MAPCs committed to a given lineage to differentiate to neural cells in the brain of newborn animals and those with induced lesions of dopaminergic neurons; to cardiomyocytes, and smooth muscle in models of acute and chronic ischemia; to endothelium and smooth muscle in models of peripheral vascular disease; to hepatocytes in models of liver disease; and to hematopoietic cells following irradiation.

10.5
Mechanisms Underlying the Phenomenon of MAPCs

Currently, the mechanism(s) underlying the selection of ideal culture conditions for MAPCs are not fully understood. However, recent data indicate that the pluripotency of MAPCs cannot be attributed to the co-culture of multiple stem cells. For instance, retroviral marking studies have shown that a single MAPC cell can differentiate *in vitro* to mesodermal cells (mesenchymal and non-mesenchymal), neuroectoderm, and hepatocyte-like cells from human [10, 11], mouse, and rat MAPCs [8]. Moreover, a single mouse MAPC is sufficient to generate chimeric animals [8]. These observations therefore rule out the possibility that the pluripotent nature of these cells is due to the coexistence in culture of multiple somatic stem cells.

A second possible explanation for the increased differentiation potential of MAPCs is that the cells undergo fusion and acquire greater pluripotency via this mechanism. While fusion has been shown responsible for apparent ES characteristics of BM and neural stem cells [5, 24, 25], the phenomenon does not likely explain the pluripotency of MAPCs. With the exception of the final differentiation step for neuroectoderm, cultivation and differentiation *in vitro* do not require MAPCs to be cocultured with other cells, thereby diminishing the likelihood that the pluripotency of MAPCs results from fusion. In a recent commentary, Smith et al. suggested that MAPCs could arise from the fusion of multiple cell types early in the culture process, leading to reprogramming of the genetic information and pluripotency. However, no evidence has been reported to date that human or rodent MAPCs are tetraploid or aneuploid early in the culture process, making this explanation less likely, albeit possible. However, a number of findings suggest that fusion is not the likely cause for engraftment seen postnatally, or for the chimerism observed in blastocyst injection experiments. The frequency of the fusion event described for ES-BM, ES-NSC, and HSC-hepatocyte fusions is generally very low, occurring in approximately one cell per 100 000 cells.

Expansion of such fused cells can be detected only when drug selection is applied in the *in-vitro* systems, and withdrawal of 2-(2-nitro-4-trifluoro-methylbenzoyl)-1,3-cyclohexanedione (NTBC) in the fumarylacetoacetate hydrolase (FAH)-mouse model has been used to select for cells that express the FAH gene. (Mice lacking FAH have symptoms similar to humans with the disease hereditary tyrosinemia type I; HT1.) The percent engraftment seen in these postnatal transplant models was in the range of 1 to 9%. The frequency of chimerism seen in blastocyst injection studies ranged between 33 and 80% when MAPCs were injected. These frequencies are significantly higher than have been described for fusion events with ES cells *in vitro* and in HSC-hepatocyte fusion studies *in vivo*. Furthermore, in contrast to reports indicating that fusion may be responsible for apparent plasticity, all *in-vivo* studies with MAPCs were carried out without selectable pressure, mainly in non-injured animals. Therefore, it is less likely that the pluripotent behavior of MAPCs *in vivo* is due to fusion between the MAPCs and the tissues with which they contribute to or engraft. To rule out this possibility formally, specific studies using transgenic mice are currently being designed.

Currently, there is no proof that MAPCs exist as such *in vivo*, and until positive selectable markers for MAPCs are identified this issue will be difficult to resolve. If the cells exist *in vivo*, it might be hypothesized that they are derived, for instance, from primordial germ cells that migrated aberrantly to tissues outside the gonads during development. It is, however, also possible that the removal of certain stem cells from their *in-vivo* environments results in "reprogramming" of the cells to acquire greater pluripotency. The studies on human MAPCs suggest that a cell capable of undergoing a degree of reprogramming is likely a protected stem cell *in vivo*, as the telomere length of MAPCs from younger and older donors is similar, yet significantly longer than that found in hematopoietic cells from the same donor. There is also evidence that a small population of SSEA1-positive cells can be selected from mouse BM and that such cells coexpress SSEA1, a phenotype consistent with MAPCs. Although this observation suggests that MAPCs may exist *in vivo*, to prove unequivocally that these cells are resident MAPCs will require the demonstration that the freshly isolated cells have multi-lineage differentiation ability *in vitro* and *in vivo*. The fact that MAPCs can be isolated from multiple tissues may suggest that stem cells from each tissue have the capacity to be reprogrammed. However, as indicated earlier, those studies in which different organs were used as initiating cell populations for the generation of MAPCs did not employ purified tissue-specific cells or stem cells. Therefore, an alternative explanation is that the cells isolated from BM that can give rise to MAPCs in culture might circulate, and thus be collected from other organs. However, researchers have been unsuccessful in isolating MAPCs from blood or from umbilical cord blood, arguing against this phenomenon. Finally, cells selected from different organs could, in theory, represent common cells resident in multiple organs, such as MSCs that are present in different locations or cells associated with tissues present in all organs (e.g., blood vessels). Currently, studies are ongoing to determine which of these many possibilities might be correct.

Recently published evidence in a mouse model has shown that, after transplantation of complete BM from syngeneic eGFP mice into wild-type mice, MAPCs isolated from the transplanted mice were – somewhat unexpectedly – mostly of donor origin [26]. Numerous studies of BM chimeras have shown that BM stromal cells, including MSC, with some exceptions are host-derived following stem cell transplantation [27, 28]. The fact that MAPCs have been originally isolated from the adherent fraction of BM and that they are cocultured with MSCs could suggest the same scenario for MAPCs. Thus, this study is the first demonstration that *ex-vivo*-cultured MAPC are physiologically different from stromal cells and MSC, which suggests that they may be unrelated ontologically, even though they all form part of the BM adherent cell component defined in culture.

10.6
Conclusion

Whether they exist *in vivo* or are created *in vitro*, MAPCs will likely have clinical relevance. However, understanding the nature of the MAPC will impact upon the ways that the cell type may be used. If MAPCs exist *in vivo*, it will be important to pinpoint their location(s) and to determine whether their migration, expansion, and differentiation in a tissue-specific manner can be induced and controlled *in vivo*. If the cells are instead created by culture conditions, then understanding the mechanism underlying the reprogramming event may allow researchers to exploit this phenomenon on a more routine and controlled basis.

Regardless of the origin of MAPCs, however, the use of such cells in clinical trials remains a future issue. Before the cells can be used routinely, robust culture systems that allow automatization must be developed. As with other stem cells – including ES cells – scientists will need to determine in preclinical models whether undifferentiated, lineage-committed, or terminally-differentiated cells should be used to treat a variety of disorders. If lineage-committed or terminally-differentiated cells are to be used, then robust differentiation cultures must be developed at the clinical scale. Furthermore, additional studies will need to be performed to demonstrate whether potentially contaminating undifferentiated MAPCs will interfere with engraftment *in vivo*. Likewise, studies to assess the specific aspects of engraftment, such as the level of HLA-mismatch tolerated in transplantation and whether hematopoietic engraftment from MAPCs will be required for such tolerance, will be needed. As with other extensively cultured cells, investigators must also determine if prolonged expansion leads to genetic abnormalities in cells that may cause malignancies when transplanted *in vivo*. Nonetheless, the discovery of MAPCs has opened the door for scientists to investigate the fundamental nature of adult stem cells and their possible applications in future therapies.

Abbreviations/Acronyms

BM	bone marrow
EPC	endothelial progenitor cell
ES	embryonic stem
FAH	fumarylacetoacetate hydrolase
FCS	fetal calf serum
GAG	glycosaminoglycan
HSC	hematopoietic stem cell
HT1	hereditary tyrosinemia type I
MAPC	multipotent adult progenitor cell
MHC	major histocompatibility
MSC	mesenchymal stem cell
NOD-SCID	non-obese diabetic/severe combined immunodeficient

NTBC	2-(2-nitro-4-trifluoro-methylbenzoyl)-1,3-cyclohexanedione
TBI	total body irradiation
VCAM	vascular cell adhesion molecule
VEGF	vascular endothelial growth factor

References

1. Bishop AE, Buttery LD, Polak JM. Embryonic stem cells. *J Pathol* **2002**;*197*:424–429.

2. Brehm M, Zeus T, Strauer BE. Stem cells – clinical application and perspectives. *Herz* **2002**;*27*:611–620.

3. Wagers AJ, Weissman IL. Plasticity of adult stem cells. *Cell* **2004**;*116*:639–648.

4. Alvarez-Dolado M, Pardal R, Garcia-Verdugo JM, et al. Fusion of bone-marrow-derived cells with Purkinje neurons, cardiomyocytes and hepatocytes. *Nature* **2003**;*425*:968–973.

5. Terada N, Hamazaki T, Oka M, et al. Bone marrow cells adopt the phenotype of other cells by spontaneous cell fusion. *Nature* **2002**;*416*:542–545.

6. Poulsom R, Alison MR, Forbes SJ, Wright NA. Adult stem cell plasticity. *J Pathol* **2002**;*197*:441–456.

7. Jiang Y, Henderson D, Blackstad M, Chen A, Miller RF, Verfaillie CM. Neuroectodermal differentiation from mouse multipotent adult progenitor cells. *Proc Natl Acad Sci USA* **2003**;*100*(Suppl.1):11854–11860.

8. Jiang Y, Jahagirdar BN, Reinhardt RL, et al. Pluripotency of mesenchymal stem cells derived from adult marrow. *Nature* **2002**;*418*:41–49.

9. Reyes M, Dudek A, Jahagirdar B, Koodie L, Marker PH, Verfaillie CM. Origin of endothelial progenitors in human postnatal bone marrow. *J Clin Invest* **2002**;*109*:337–346.

10. Reyes M, Lund T, Lenvik T, Aguiar D, Koodie L, Verfaillie CM. Purification and ex vivo expansion of postnatal human marrow mesodermal progenitor cells. *Blood* **2001**;*98*:2615–2625.

11. Schwartz RE, Reyes M, Koodie L, et al. Multipotent adult progenitor cells from bone marrow differentiate into functional hepatocyte-like cells. *J Clin Invest* **2002**;*109*:1291–1302.

12. Clavel C, Aranguren X, Barajas M, et al. Isolation and characterization of multipotent adult progenitor cells from cynomolgus monkey. *Exp Hematol* **2005**;*33*:120 (abstract).

13. Tolar J, Osborn M, Bell S, et al. Real-time in vivo imaging of stem cells following transgenesis by transposition. *Mol Ther* **2005**;*12*:42–48.

14. Muguruma Y, Reyes M, Nakamura Y, et al. In vivo and in vitro differentiation of myocytes from human bone marrow-derived multipotent progenitor cells. *Exp Hematol* **2003**;*31*:1323–1330.

15. Mazo M, Bresolle C, Agbulut O, et al. Bone marrow-derived multipotent adult progenitor cells (MAPC) in a rat model of myocardial infarction. *Exp Hematol* **2005**;*33*:108 (abstract).

16. Deans RJ, Moseley AB. Mesenchymal stem cells: biology and potential clinical uses. *Exp Hematol* **2000**;*28*:875–884.

17. Aranguren X, Clavel C, Luttun A, et al. The arterial and venous potential of human MAPC and AC133 derived endothelial cells is modulated by activation of notch and patched ligands. *Exp Hematol* **2005**;*33*:75 (abstract).

18. Weissman IL, Anderson DJ, Gage F. Stem and progenitor cells: origins, phenotypes, lineage commitments, and transdifferentiations. *Annu Rev Cell Dev Biol* **2001**;*17*:387–403.

19. Khakoo AY, Finkel T. Endothelial progenitor cells. *Annu Rev Med* **2005**;*56*:79–101.

20. Niwa H, Miyazaki J, Smith AG. Quantitative expression of Oct-3/4 defines differentiation, dedifferentiation or self-renewal of ES cells. *Nat Genet* **2000**;*24*:372–376.

21. Jiang Y, Vaessen B, Lenvik T, Blackstad M, Reyes M, Verfaillie CM. Multipotent progenitor cells can be isolated from postnatal murine bone marrow, muscle, and brain. *Exp Hematol* **2002**;*30*:896–904.

22. Draper JS, Smith K, Gokhale P, et al. Recurrent gain of chromosomes 17q and 12 in cultured human embryonic stem cells. *Nat Biotechnol* **2004**;*22*:53–54.

23. Tolar J, Clarke L, Jiang Y, et al. Multipotent adult progenitor cells (MAPCs) reduce glycosaminoglycan (GAG) accumulation in a murine model of Hurler syndrome (MPS 1H). *Blood* **2003**;*102* (abstract):#3115.

24. Ying QL, Nichols J, Evans EP, Smith AG. Changing potency by spontaneous fusion. *Nature* **2002**;*416*:545–548.

25. Lagasse E, Connors H, Al-Dhalimy M, et al. Purified hematopoietic stem cells can differentiate into hepatocytes in vivo. *Nat Med* **2000**;*6*:1229–1234.

26. Reyes M, Li S, Foraker J, Kimura E, Chamberlain JS. Donor origin of multipotent adult progenitor cells in radiation chimeras. *Blood* **2005**;*106*:3646–3649.

27. Bianco P, Gehron Robey P. Marrow stromal stem cells. *J Clin Invest* **2000**;*105*:1663–1668.

28. Simmons PJ, Przepiorka D, Thomas ED, Torok-Storb B. Host origin of marrow stromal cells following allogeneic bone marrow transplantation. *Nature* **1987**;*328*:429–432.

11

Mesenchymal Stem Cells as Vehicles for Genetic Targeting of Tumors

Frank Marini*, Brett Hall*, Jennifer Dembinski, Matus Studeny, A. Kate Sasser, and Michael Andreeff

Abstract

Recent evidence suggests that mesenchymal stem cells (MSCs) selectively home to tumors and contribute to the formation of tumor-associated stroma. These stromal precursors can be genetically modified to produce anticancer agents *in situ* in tumors. The resulting local, high-level production of these agents blunts tumor growth kinetics and inhibits tumor growth. In this chapter, we will review the ability of MSCs and other bone marrow-derived cell populations to integrate into the tumor microenvironment, and their potential roles in that setting. We will also examine the biological rationale for using MSCs and other bone marrow-derived cell populations as delivery vehicles for antitumor proteins.

11.1
Introduction

Tumor metastases are the principal cause of death in cancer patients. Unfortunately, treatment strategies for advanced-stage cancer are typically associated with high toxicity and only modest success rates. The benefits of escalated-dose chemotherapy and radiation therapy have been exhausted, and targeted, less-toxic treatment strategies are desperately needed. Recent investigations by our group have shown that bone marrow (BM)-derived mesenchymal stem cells (MSCs) contribute to stroma formation in tumors and their metastases, and have provided proof of principle for using MSCs as delivery vehicles for antitumor agents [1–3]. Some studies have begun to evaluate the feasibility, safety, and practicality of the therapeutic use of MSCs in tissue regeneration or cell replacement therapies [4–13]. Although the ontogeny and homeostasis of MSCs *in vivo* re-

---

* F. Marini and B. Hall contributed equally to this chapter.

Stem Cell Transplantation. Biology, Processing, and Therapy.
Edited by Anthony D. Ho, Ronald Hoffman, and Esmail D. Zanjani
Copyright © 2006 WILEY-VCH Verlag GmbH & Co. KGaA, Weinheim
ISBN: 3-527-31018-5

mains to be fully elucidated [14–17], several studies in primates and humans have detailed aspects of MSC isolation, expansion, and transplantation [13, 18–21]. Our understanding of the biological role of MSCs within the tumor microenvironment is limited, however, and needs to be improved in order to ensure high efficacy of MSC-targeted anticancer strategies. This chapter will: (i) outline the current knowledge and understanding of why tumors recruit circulating stem cells such as MSCs into their microenvironment; (ii) detail why MSCs persist as tumor stromal cell elements and fulfill an important role in tumor architecture; and (iii) address what the potential roles of MSCs are within the tumor microenvironment. Finally, attention will be focused on the use of MSCs as cellular delivery vehicles, particularly to deliver anticancer agents directly into the tumor microenvironment where these cells selectively engraft and participate in tumor stroma development. This approach could be termed a "Trojan-Horse" concept, where MSCs are incorporated into the tumor architecture and MSC-based cellular mini-pumps produce potent anticancer agents *in situ*.

11.2
The Tumor Stroma and its Components

Solid tumors are multifaceted tissues that are maintained by interactions between tumor cells and the frequently nonmalignant microenvironment: this is often described as the "stroma" – fibroblasts, vasculature and perivasculture cells, the extracellular matrix (ECM) (Fig. 11.1) [21–25]. Stromal cells provide a physical architecture (matrix), growth factors, cytokines, a blood supply, and also remove metabolic and biological waste [26]. Interestingly, new data are available which suggest that premalignant cells can be inhibited by normal stroma, though any alterations in the tumor microenvironment that result in the initiation of tissue remodeling (e.g., inflammatory insult) may result in cancer progression. For example, the inhibition of transforming growth factor-beta (TGF-β) signaling in fibroblasts resulted in tumor formation in the prostate [27], whereas stromal fibroblasts with intact TGF-β signaling were capable of suppressing the oncogenic development from neighboring tumor cells. These data would suggest that continual pressure from the tumor cell compartment maintains stromal cells in an tumor-promoting "activated state" – a condition that, in turn, can select for tumor-enhancing lesions within the stromal cell compartment.

Even small tumors (>1–2 mm) require new access to blood vessels for survival (angiogenesis: the sprouting of new blood vessels from existing blood vessels) [28]. This process requires complex cellular and molecular cooperation between multiple cell types, including endothelial cells, fibroblasts, and macrophages. Under abnormal conditions, such as wound healing or tumorigenesis, endothelial cells can be activated to divide and form new blood vessels. Frequently detected growth factors that activate endothelial cells include vascular endothelial growth factor (VEGF) and basic fibroblast growth factor (bFGF). Tumor cells, tumor-associated macrophages, and tumor-associated fibroblasts (TAFs) produce VEGF and

Tumor Microenvironment

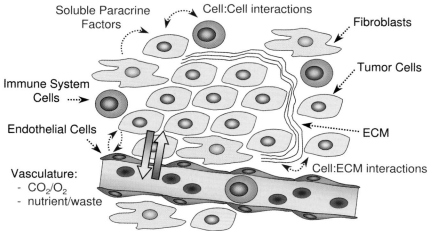

Figure 11.1 The interactions of tumor cells and their nonmalignant microenvironmental partners (stroma) are critical for the survival of the tumor. Stromal cells provide structural support for malignant cells, modulate the tumor microenvironment, and influence phenotypic behavior as well as the aggressiveness of the malignancy. In response, the tumor provides growth factors, cytokines, and cellular signals that continually initiate new stromal reactions and recruit new cells into the microenvironment to further support tumor growth. It is not fully understood how stroma influences the neoplastic cells, but there is evidence for the involvement of soluble paracrine factors, extracellular matrix (ECM) formation, and direct cell-to-cell interaction.

are involved in tumor-induced angiogenesis [29–32]. Both tumor-associated macrophages and TAFs are abundantly detected in tumors and often support tumor growth [33, 34]. The cellular origin of TAFs remains unclear, but accumulating evidence suggests that they originally derive from resident organ fibroblasts [35, 36]. While studies have indicated that organ fibroblasts in the proximity of the developing tumor became TAFs [37, 38], these data do not preclude the possibility that circulating MSCs or mesenchymal progenitor (stem) cells contribute directly to the heterogeneous organ-specific fibroblast population or the TAF cell pool [1–4, 28–30] (Fig. 11.2).

Some TAFs are tissue-resident, matrix-synthesizing or matrix-degrading cells, whereas others are contractile cells (myofibroblasts), circulating precursor cells (fibrocytes), or blood vessel-associated pericytes. TAFs biologically impact the tumor microenvironment through the production of growth factors, cytokines, chemokines, matrix-degrading enzymes, and immunomodulatory mechanisms [31]. The bone marrow (BM) is a unique and accessible source of multiple lineages of cells with therapeutic value, especially progenitor and stem cells. The adherent fraction of BM cells contains differentiated mesenchymal stromal

A

B MSC Migration (72 hours)

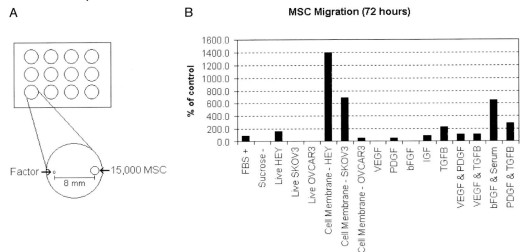

Figure 11.2 Potential factors mediating tropism of mesenchymal stem cells (MSCs) for ovarian tumors. In order to assess what factor(s) could mediate selective migration of MSCs toward a tumor xenograft, an *in-vitro* migration assay was performed. (A) Growth factor-depleted matrigel-coated culture dishes were utilized in an Ochtolony-type fashion, whereby MSCs labeled with GFP were placed into a center well and potential tropic factors were spotted in peripheral wells 8–10 mm away from center. Migration was assessed visually, by counting the number of MSC-GFP cells that migrated from the center well of the plate to the outer rim (8 mm distance), as a percentage of control migrated cells after 72 h (B). Due to the initial observation that 100% FCS spotted into an outside well, stimulated significant MSC migration, this was used as the 100% positive control. A 15% sucrose solution was used as a negative control. As shown in (B), maximal migration of MSCs was observed when the outer wells contained bFGF plus FCS (1 ng per 9 μL) or when the outer wells contained enriched cell membrane fractions from HEY or SKOV3 ovarian carcinomas. Less MSC migration was observed when outer wells contained recombinant human PDGF-BB (10 ng), IGF (0.5 μg), TGF-β (10 ng), or combinations of VEGF + PDGF-BB (10 ng each), VEGF + TGF-β (10 ng each), or PDGF-BB + TGF-β (10 ng each). Little or no migration was observed when VEGF (10 ng) alone or bFGF (1 μg) alone was utilized. These data suggest that secreted factors alone or in combination (e.g., PDGF or TGF-β) may be partially responsible for the selective migration and engraftment of MSCs into these tumors, and that the membrane of these ovarian carcinomas may possess an MSC tropic factor.

cells and pluripotent MSCs, which give rise to differentiated cells belonging to the osteogenic, chondrogenic, adipogenic, myogenic, and fibroblastic lineages [32]. Although MSCs are routinely recovered from BM, they have also been isolated from a number of other tissues, including muscle, synovium, umbilical cord, and adipose tissue.

In addition to the production of numerous growth factors and their support of angiogenesis, TAFs also provide organization to the tumor stroma by producing components of the ECM. Thus, several critical events required for tumor growth are not sufficient unless the tumor cells attract and stimulate fibroblasts. Based on our most recent understanding of tumor stroma development [33], we propose that stromagenesis is a multistep process involving the concomitant recruitment of local tissue fibroblasts and circulating BM-derived stem cells or MSCs from the BM into the tumor, followed by intratumoral proliferation of these cells, and finally, the acquisition of many biological TAF characteristics [33a]. Once in the tumor microenvironment, the MSCs convert into activated myofibroblasts (i.e., TAFs) and may also differentiate into fibrocytes, which produce ECM components, and perivascular or vascular structures. Little is known about the dynamics of fibroblast involvement and the molecules that regulate it. As a side note, the reactive stroma associated with many solid tumors exhibit several of the same biological markers seen in tissue stroma at sites of wound repair [5]. Many biological processes in wound repair, including stromal cell acquisition of the myofibroblast phenotype, deposition of type I collagen, and induction of angiogenesis are observed in reactive stroma during cancer progression.

11.3
The Role of Tumor–Stroma Interactions in Tumor Progression

The development of cancer is a progressive, multistep process during which a cell accumulates multiple and consecutive genetic alterations, eventually resulting in phenotypic changes that lead to malignant tumor growth. This tumor cell-centered view of cancer development has led to numerous useful discoveries that have greatly furthered our understanding of cancer. However, the characteristic changes that enable a cancer cell to grow into a tumor are less well understood, largely because the analysis of those phenotypes has been restricted to the tissue level, which is much more complicated than conventional cell-culture models. During recent years, it has become apparent that the dynamic interplay between different cell types and their microenvironment is crucial for maintaining normal, balanced tissue homeostasis. As a result, the tumor microenvironment also strongly influences many steps during tumor development.

In vivo, a tumor is a complex ecosystem comprising genetically altered neoplastic cells and the tumor stroma, which is best described as a framework of connective nontumor tissue cells associated with the ECM and surrounded vasculature. Although the cell types and matrix composition of the tumor stroma appear to change during tumor development, this aspect of a solid tumor malignancy was considered "inert" – that is, it mainly provides architectural support and nourishment/waste removal for the neighboring cancer cells. However, during the rise in popularity in interest in tumor angiogenesis and recognition of the crucial role of stromas in tumor development, a new more significant view of tumor stroma has emerged. The fact that tumors are composed of many cell types and cannot exist in isolation has been acknowledged, and increasing experimental evidence has emphasized the importance of the microenvironment for tumor development and progression [34]. Recent studies have shown that genetic alterations in the tumor cells themselves are not sufficient to generate a malignant tumor;

rather, it has shown that a "tumor-permissive" stromal environment is also needed [35, 36].

Compared to the connective tissue of normal organs, which is able to maintain homeostasis, tumor stroma has disruptions in both the composition of the ECM and the functional state of local stromal cells [37–39]. These alterations in epithelial–stromal interactions seem to be crucial for tumor growth, invasion, and metastasis [40, 41]. For some time, the stroma of malignant tumors was thought to closely resemble the granular tissue of healing wounds [42], and it is now believed that various features of granular tissue-like stroma may favor or even induce tumor invasion [43–45]. For example, investigators during the 1950s observed enhanced tumor formation when carcinogen-treated "activated" stroma was heterotypically grafted with untreated skin epithelial cells [46]. Activated stromal cells have also been reported to have growth-promoting effects on tumor cells [47–49], indicating persistent functional alterations in these tumor fibroblasts [50, 51]. Furthermore, the importance of tumor–stroma interactions in modulating the carcinogenic potential of epithelial tumors has been demonstrated through dysregulation of matrix-degrading proteases, which are essential to cancer invasion and metastasis (for a review, see [52]). All of these reports stress the importance of the tumor stromal compartment in malignant tumors and strongly indicate that continuous interactions between the carcinoma and stromal cells (resulting in their reciprocal regulation and modulation) are prerequisites for carcinoma development and progression.

11.4
The Similarity of MSC Tumor Tropism to Wound Healing

The homing of MSCs after systemic or local infusion has been studied in animal models in a variety of experimental settings [53–55]. For example, Erices et al. [56] reported the systemic infusion of MSC in irradiated syngeneic mice. After 1 month, 8% of bone cells and 5% of lung cells in the recipient mice were positive for the transplanted cells. In a separate study in baboons administered labeled MSC, the latter were detected in the bone marrow more than 500 days after transplantation [57]. Apparently, the native capacity of MSCs to adhere to matrix components favors their preferential homing to bone, lungs, and cartilage when injected intravenously. However, conditioning regimens prior to cell transplantation (such as irradiation or chemotherapy treatment) may greatly influence the efficiency and sites of MSC homing. In this regard, a growing number of studies of various pathological conditions have demonstrated that MSCs selectively home to sites of injury, irrespective of the tissue or organ [58–62]. This ability of MSCs has been demonstrated in brain injury [63, 64], wound healing, and tissue regeneration [54, 65, 66]. Evidence suggests that tumors can be considered sites of tissue damage, or "wounds that never heal" [67], as well as sites of potential inflammatory cytokine and chemokine production. Thus, these properties may enable MSCs to home and deliver therapeutic agents to tumors.

Although a number of factors have been implicated in the homing of bone marrow cells to sites of injury, it remains unknown what processes and factors underlie MSC homing to tumors (Fig. 11.2). In wound repair, as in cancer, cells that usually divide infrequently are induced to proliferate rapidly, the ECM is invaded, connective tissues are remodeled, epithelial and stromal cells migrate, and new blood vessels are formed. We speculate that many of the same factors involved in wound healing are up-regulated in the tumor microenvironment to initiate the homing process.

11.5
The Rationale for using MSCs as Cellular Delivery Vehicles

The BM is a unique and accessible source of multiple lineages of cells with potential therapeutic value, especially progenitor and stem cells. The adherent fraction of BM cells contain differentiated mesenchymal BM stromal cells and pluripotent MSC, which give rise to differentiated cells belonging to the osteogenic, chondrogenic, adipogenic, myogenic, and fibroblastic lineages [68]. As mentioned above, although MSCs are routinely recovered from BM, they have also been isolated from a number of other tissues, including muscle, synovium, umbilical cord, and adipose tissue [69–72]. There has been increasing interest in recent years in the BM stromal cell system in various fields of cell therapy, such as hematopoietic stem cell transplantation and connective tissue engineering. In these studies [73–76], MSCs have been mainly used to support hematopoietic cell engraftment, repair tissue defects, or to trigger the regeneration of a variety of mesenchymal tissues [77–79].

11.5.1
Recent Studies of MSC as Cellular Vehicles

Many observations support the concept of using MSCs as delivery vehicles for antitumor agents. We, and others, have observed dynamic interactions between BM-derived MSCs and various tumor cell lines *in vitro* (Fig. 11.3) [1–3, 80–83], and we have shown *in vivo* that human MSCs engraft and persist within existing tumor microenvironments [1–3, 84, 85]. In addition, MSCs phenotypically resemble TAFs in the presence of TGF-β1 or in coculture with tumor cells [86, 87]. Finally, two independent studies have shown that BM fibroblasts contribute to the development of stromal cell populations in tumors in mice [88, 89].

Elegant studies conducted by Ischii et al. strongly implicated BM-derived cells in the development of tumors [90]. In their study, severe combined immunodeficient (SCID) mice received BM cells from β-gal+ and RAG-1-deficient mice carrying a unique major histocompatibility complex molecule (H2Kb). Once engraftment was observed in the recipient mice, pancreatic tumors were implanted and allowed to progress. The tumors were subsequently removed, and the contribution of β-gal+ cells to the tumor microenvironment was assessed. The authors

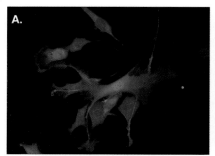

Figure 11.3 Dynamic interactions between murine mesenchymal stem cells (MSCs) and tumor cells *in vitro*. γGFP-labeled murine mesenchymal stem cells were cocultured with RFP-labeled tumor cells. (A) The murine MSCs interact and span the DAOY tumor cells.

(B) The interplay appears differently, with the MSCs preferring to interact only partially with the RH30. Great variability was observed between tumor cells lines in their interaction with MSCs.

found that a small proportion (13 %) of H2Kb+α-SMA+ double-positive myofibroblasts were present around the developing malignancy 2 weeks after tumor xenotransplantation, suggesting that a small number of BM-derived myofibroblasts were incorporated into the cancer stroma during the early stages of tumor growth. By 4 weeks after tumor implantation, approximately 40 % of the myofibroblasts in the cancer-induced stroma were of BM origin. Immunohistochemical staining of serial sections revealed that non-BM-derived myofibroblasts were adjacent to the cancer nest, whereas BM-derived myofibroblasts were outside of the non-BM-derived myofibroblasts, indicating that most BM-derived myofibroblasts incorporate into the cancer-induced stroma mainly in the late stage of tumor development.

Taking these data into consideration, it has been suggested that BM-derived myofibroblasts contribute to the pathogenesis of cancer-induced desmoplastic reaction by altering the tumor microenvironment. The findings of Ischii et al. also suggested that the local tissue reaction starts at the circumference of a neoplasm in the early stage of tumor development. However, it is unclear from these data whether the tumor-resident β-gal+ myofibroblasts are generated from a circulating stem cell population originating from an adherent BM population (such as MSCs or endothelial precursors), or from a circulating hematopoietic population. Nevertheless, the findings demonstrate a specific contribution of BM-derived cells to the developing tumor stroma, which strongly suggests that BM-derived cells target tumors as a result of physiological requests by the expanding tumor [91, 92]. The study by Ischii et al. also supported our initial observations that tumors recruit cells from the circulation and that this population of stromal precursor cells could be used to target novel therapies to growing tumors. A number of other groups have since reported similar findings in ovarian, rhabdomyosarcoma, and breast tumor models [93–95].

The direct targeting of anticancer agents into the tumor microenvironment would increase their efficacy [96, 97]. There are several advantages to using BM-derived MSCs as cellular delivery vehicles:

- Most invasive cancers are likely to induce a desmoplastic reaction to some extent, thereby providing a common target for the treatment of many types of cancers;
- MSCs have a low intrinsic mutation rate and are therefore less likely to acquire a drug-resistant phenotype than the genetically unstable cancer cells;
- MSCs are simple to isolate and culture [98];
- MSCs can engraft after *ex-vivo* expansion [99, 100];
- MSCs have a high metabolic activity and an efficient machinery for the secretion of therapeutic proteins [101];
- MSCs can be expanded for more than 50 population doublings in culture, without any loss of their phenotype or multilineage potential; this is in striking contrast to terminally differentiated cells such as endothelial or muscle cells [102].

Practical attempts to use MSCs as cellular delivery vehicles have focused mainly on the delivery of therapeutic gene products (such as interleukin (IL)-3, growth hormone, and factor IX) via the systemic circulation [101, 103, 104]. Recent studies from our group [1–3] have reported that exogenously administered MSCs migrate and preferentially survive and proliferate within tumor masses (Fig. 11.4). Once in the tumor microenvironment, the MSCs incorporate into the tumor architecture and serve as precursors for stromal elements, predominantly fibroblasts. The preferential distribution of MSCs in lung tumor nodules, but not in lung parenchyma, was demonstrated after systemic administration of MSCs in mice bearing melanoma xenografts. In addition, when MSCs were systemically injected into mice with subcutaneously established tumors, MSC-derived fibroblasts were consistently identified in tumors, but not in healthy organs [105].

We also examined the therapeutic potential of MSCs to deliver interferon-beta (IFN-β) into the tumor microenvironment. Our data showed that IFN-β acts through local paracrine effects after it has been delivered into tumors by cellular *in-vitro*-modified MSC "mini-pumps", thus emphasizing the importance of tumor-targeted delivery of a cytokine. No survival benefits were seen when the same MSCs were made to produce systemic IFN-β at sites distant from tumors, or when recombinant IFN-β protein was delivered systemically without a carrier [2]. MSCs may also have similar capabilities to specifically home to and selectively engraft into established gliomas *in vivo*. Data from our group showed that MSCs injected into the carotid artery of mice can specifically target U87 gliomas, and that the MSCs can migrate from either the contralateral or the ipsilateral carotid artery, suggesting that blood flow or a perfusion effect is not directly responsible for the MSCs entering the tumor [3]. When those MSCs were armed to secrete IFN-β, we observed statistically significant decreases in tumor size, suggesting that MSCs producing IFN-β were able to control tumor growth kinetics. Addi-

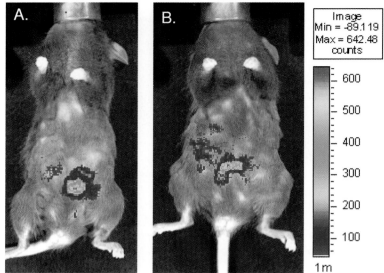

Figure 11.4 Noninvasive detection of migrated and engrafted MSCs into established murine ovarian carcinomas *in vivo*. Renilla luciferase-labeled ID8 ovarian carcinomas were established intraperitoneally (IP) in C56BL6 mice. After 29 days, firefly luciferase-labeled MSC were injected IP, and allowed to home to and engraft into the established tumors for 10 days. (A) Upon renilla-specific substrate injection, the bioluminescent signal emitted by the ID8 tumor was readily visualized, after allowing that substrate signal to die-off. Importantly, upon injection of a firefly-specific substrate the engrafted MSCs could be visualized in the abdomen. (B) The engrafted MSCs that appear to co-localize within the area occupied by the ID8 tumor were visualized.

tional studies by Hamada et al. have shown that genetically modified MSCs expressing IL-2 can also control tumor development when injected intratumorally into established gliomas, again suggesting that intratumoral production of this immunostimulatory cytokine is responsible for controlling tumor kinetics [106, 107]. Thus, the cultured BM-adherent cell population contains cells with extraordinarily high proliferative capacity that can contribute to the maintenance of both tumor stroma and connective tissue in organs remote from the BM [108]. At the same time, the successful engraftment of MSCs into tissues would most likely take place only in a state of increased cell turnover triggered by tissue damage or tumor growth. This property makes MSCs excellent candidates for the cell-based delivery of therapeutics to tumor sites.

11.6
The Challenges in Developing MSC-Based Delivery Strategies

Potential concerns in using MSCs as delivery vehicles stem from how little we understand about the homeostatic maintenance of this cell population *in vivo* [109–111], and the possibility that MSCs themselves might enhance or initiate tumor growth [112, 113]. However, unlike homeostatic MSC niches, tumor microenvironments are pathologically altered tissues that resemble unresolved wounds [67, 114, 115], and the studies that reported tumorigenic properties of MSCs used extensively passaged and/or genetically altered cells [116, 117]. Additional studies have shown that genetic alterations must accumulate in tumor stromal cells, presumably over time, before they can enhance tumor growth [118–121]. We, and others, have shown that lower-passage MSCs do not form tumors *in vivo* [122–126].

Although the use of stem cells for cancer gene therapy usually provides some degree of tumor selectivity, strategies to improve tumor homing may greatly increase the applicability and success of tumor-targeted cell delivery vehicles [127]. One can envision two major approaches for increased targeting. First, other cell populations with specific properties can be identified and tested, such as more primitive stem cells with higher replicative or migratory potential, such as the "RS" phenotype [128]. For instance, isolation methods could be developed to specifically select cell types with innate specific targeting or other phenotypic properties that can be exploited to enhance tumor targeting or incorporation into the tumor microenvironment [129]. Second, cell homing may be manipulated by using specific culture agents or medium treatments to alter the expression of cell-surface receptors [130, 131]. It might be possible to manipulate cell targeting or homing properties by choosing specific pretreatment or isolation protocols. These two approaches represent another area of investigation to enhance the efficacy of cellular vehicle strategies.

11.7
Conclusions

In this chapter, we have described recent progress in the development of BM-derived stem cell populations – particularly MSCs – for use as cellular delivery vehicles with specific tropisms for solid tumors and their microenvironments. This tumor tropism is based primarily on the innate physiological ability of MSCs to home to sites of inflammation and tissue repair. The most obvious advantage offered by MSCs is the local delivery and intratumoral release of therapeutic agents, especially proteins with poor pharmacokinetic profiles. An added benefit of cell-based treatments is the possibility of *ex-vivo* manipulation to maximize cell loading, as shown by our direct comparison of therapeutic effects provided by cell-based or systemic injection of recombinant proteins. In addition to delivering cancer-related protein substances, cellular vehicles could also serve as protective "coatings" for the delivery of payloads such as replicating oncoselective viruses (e.g., Onyx-15 [132] or Delta 24 [133]). Not only would the cells protect these viruses from the immune system, but they would also allow amplification of the initial viral load with the possibility of increasing viral spread through cell-to-cell contact specifically at tumor sites.

Many investigations remain to be performed before the potential therapeutic benefits of MSC-based approaches can be fully exploited. The development of optimally targeted cellular vectors will require advances in cell science and virology, and will certainly benefit from continued innovative application of knowledge gained from basic biology, tissue stem cell engineering, and gene therapy.

Abbreviations/Acronyms

bFGF	basic fibroblast growth factor
BM	bone marrow
ECM	extracellular matrix
GFP	green fluorescent protein
IFN-β	interferon-beta
IGF	insulin-like growth factor
IL	interleukin
MSC	mesenchymal stem cell
PDGF	platelet-derived growth factor
RFP	red fluorescent protein
SCID	severe combined immunodeficient
TAF	tumor-associated fibroblast
TGF-β	transforming growth factor-beta
VEGF	vascular endothelial growth factor

References

1. Studeny M, Marini FC, Champlin RE, Zompetta C, Fidler IJ, Andreeff M. Bone marrow-derived mesenchymal stem cells as vehicles for interferon-beta delivery into tumors. *Cancer Res* 2002;62(13):3603–3608.

2. Studeny M, Marini FC, Dembinski JL, Zompetta C, Cabreira-Hansen M, Bekele BN, Champlin RE, Andreeff M. Mesenchymal stem cells: potential precursors for tumor stroma and targeted-delivery vehicles for anticancer agents. *J Natl Cancer Inst* 2004;96(21):1593–1603.

3. Nakamizo A, Marini F, Amano T, Khan A, Studeny M, Gumin J, Chen J, Hentschel S, Vecil G, Dembinski J, Andreeff M, Lang FF. Human bone marrow-derived mesenchymal stem cells in the treatment of gliomas. *Cancer Res* 2005;65(8):3307–3318.

4. Le Blanc K, Pittenger M. Mesenchymal stem cells: progress toward promise. *Cytotherapy* 2005;7(1):36–45. Review.

5. Garcia-Olmo D, Garcia-Arranz M, Herreros D, Pascual I, Peiro C, Rodriguez-Montes JA. A phase I clinical trial of the treatment of Crohn's fistula by adipose mesenchymal stem cell transplantation. *Dis Colon Rectum* 2005;48(7):1416–1423.

6. Bang OY, Lee JS, Lee PH, Lee G. Autologous mesenchymal stem cell trans-

plantation in stroke patients. *Ann Neurol* 2005;*57*(6):874–882.

7. Lazarus HM, Koc ON, Devine SM, Curtin P, Maziarz RT, Holland HK, Shpall EJ, McCarthy P, Atkinson K, Cooper BW, Gerson SL, Laughlin MJ, Loberiza FR Jr, Moseley AB, Bacigalupo A. Cotransplantation of HLA-identical sibling culture-expanded mesenchymal stem cells and hematopoietic stem cells in hematologic malignancy patients. *Biol Blood Marrow Transplant* 2005;*11*(5):389–398.

8. Kan I, Melamed E, Offen D. Integral therapeutic potential of bone marrow mesenchymal stem cells. *Curr Drug Targets* 2005;*6*(1):31–41. Review.

9. Gojo S, Umezawa A. Plasticity of mesenchymal stem cells – regenerative medicine for diseased hearts. *Hum Cell* 2003;*16*(1):23–30. Review.

10. Chen SL, Fang WW, Qian J, Ye F, Liu YH, Shan SJ, Zhang JJ, Lin S, Liao LM, Zhao RC. Improvement of cardiac function after transplantation of autologous bone marrow mesenchymal stem cells in patients with acute myocardial infarction. *Chin Med J (Engl)* 2004;*117*(10):1443–1448. Erratum in: *Chin Med J (Engl)* 2005;*118*(1):88.

11. Chen SL, Fang WW, Ye F, Liu YH, Qian J, Shan SJ, Zhang JJ, Chunhua RZ, Liao LM, Lin S, Sun JP. Effect on left ventricular function of intracoronary transplantation of autologous bone marrow mesenchymal stem cell in patients with acute myocardial infarction. *Am J Cardiol* 2004;*94*(1):92–95.

12. Koc ON, Day J, Nieder M, Gerson SL, Lazarus HM, Krivit W. Allogeneic mesenchymal stem cell infusion for treatment of metachromatic leukodystrophy (MLD) and Hurler syndrome (MPS-IH). *Bone Marrow Transplant* 2002;*30*(4):215–222.

13. Horwitz EM, Prockop DJ, Fitzpatrick LA, Koo WW, Gordon PL, Neel M, Sussman M, Orchard P, Marx JC, Pyeritz RE, Brenner MK. Transplantability and therapeutic effects of bone marrow-derived mesenchymal cells in children with osteogenesis imperfecta. *Nat Med* 1999;*5*(3):309–313.

14. Salasznyk RM, Westcott AM, Klees RF, Ward DF, Xiang Z, Vandenberg S, Bennett K, Plopper GE. Comparing the protein expression profiles of human mesenchymal stem cells and human osteoblasts using gene ontologies. *Stem Cells Dev* 2005;*14*(4):354–366.

15. Etheridge SL, Spencer GJ, Heath DJ, Genever PG. Expression profiling and functional analysis of wnt signaling mechanisms in mesenchymal stem cells. *Stem Cells* 2004;*22*(5):849–860.

16. Lee RH, Kim B, Choi I, Kim H, Choi HS, Suh K, Bae YC, Jung JS. Characterization and expression analysis of mesenchymal stem cells from human bone marrow and adipose tissue. *Cell Physiol Biochem* 2004;*14*(4-6):311–324.

17. Hishikawa K, Miura S, Marumo T, Yoshioka H, Mori Y, Takato T, Fujita T. Gene expression profile of human mesenchymal stem cells during osteogenesis in three-dimensional thermoreversible gelation polymer. *Biochem Biophys Res Commun* 2004;*317*(4):1103–1107.

18. Heino TJ, Hentunen TA, Vaananen HK. Conditioned medium from osteocytes stimulates the proliferation of bone marrow mesenchymal stem cells and their differentiation into osteoblasts. *Exp Cell Res* 2004;*294*(2):458–468.

19. Spees JL, Gregory CA, Singh H, Tucker HA, Peister A, Lynch PJ, Hsu SC, Smith J, Prockop DJ. Internalized antigens must be removed to prepare hypoimmunogenic mesenchymal stem cells for cell and gene therapy. *Mol Ther* 2004;*9*(5):747–756.

20. Gregory CA, Prockop DJ, Spees JL. α-Hematopoietic bone marrow stem cells: molecular control of expansion and differentiation. *Exp Cell Res* 2005;*306*(2):330–335. Epub 2005 April 15. Review.

21. Le Blanc K, Rasmusson I, Sundberg B, Gotherstrom C, Hassan M, Uzunel M, Ringden O. Treatment of severe acute graft-versus-host disease with third party haploidentical mesenchymal stem cells. *Lancet* 2004;*363*(9419):1439–1441.

22. Tlsty TD, Hein PW (2001) Know thy neighbor: stromal cells can contribute oncogenic signals. *Curr Opin Genet Dev* 2001;*11*:54–59.

23. Bissell MJ, Radisky D. Putting tumours in context. *Nat Rev Cancer* 2001;*1*:46–54.

24. Coussens LM, Werb Z. Inflammation and cancer. *Nature* **2002;**420:860–867.

25. Mueller MM, Fusenig NE. Friends or foes – bipolar effects of the tumour stroma in cancer. *Nat Rev Cancer* **2004;**4:839–849.

26. Philip M, et al. Inflammation as a tumor promoter in cancer induction. *Semin Cancer Biol* **2004;**14:433–439.

27. Bhowmick NA, et al. TGF-β signaling in fibroblasts modulates the oncogenic potential of adjacent epithelia. *Science* **2004;**303:848–851.

28. Folkman J. Fundamental concepts of the angiogenic process. *Curr Mol Med* **2003;**3:643–651.

29. Polverini PJ, Leibovich J. Induction of neovascularization in vivo and endothelial proliferation in vitro by tumor-associated macrophages. *Lab Invest* **1984;**51:635–642.

30. Mantovani A, et al. Macrophage polarization: tumor-associated macrophages as a paradigm for polarized M2 mononuclear phagocytes. *Trends Immunol* **2002;**23:549–555.

31. Silzle T, et al. The fibroblast: sentinel cell and local immune modulator in tumor tissue. *Int J Cancer* **2004;**108:173–180.

32. Dong J, et al. VEGF-null cells require PDGFRa signaling-mediated stromal fibroblast recruitment for tumorigenesis. *EMBO J* **2004;**23:2800–2810.

33. Richter G, et al. Interleukin 10 transfected into Chinese hamster ovary cells prevents tumor growth and macrophage infiltration. *Cancer Res* **1993;**53:4134–4137.

34. Kurose K, Hoshaw-Woodard S, Adeyinka A, Lemeshow S, Watson PH, Eng C. Genetic model of multi-step breast carcinogenesis involving the epithelium and stroma: clues to tumour-microenvironment interactions. *Hum Mol Genet* **2001;**10:1907–1913.

35. Kammertoens T, Schuler T, Blankenstein T. Immunotherapy: target the stroma to hit the tumor. *Trends Mol Med* **2005;**11(5):225–231. Review.

36. Kiaris H, Chatzistamou I, Trimis G, Frangou-Plemmenou M, Pafiti-Kondi A, Kalofoutis A. Evidence for nonautonomous effect of p53 tumor suppressor in carcinogenesis. *Cancer Res* **2005;**65:1627–1630.

37. Sivridis E, Giatromanolaki A, Koukourakis MI. Proliferating fibroblasts at the invading tumour edge of colorectal adenocarcinomas are associated with endogenous markers of hypoxia, acidity, and oxidative stress. *J Clin Pathol* **2005;**58(10):1033–1038.

38. Yang F, Tuxhorn JA, Ressler SJ, McAlhany SJ, Dang TD, Rowley DR. Stromal expression of connective tissue growth factor promotes angiogenesis and prostate cancer tumorigenesis. *Cancer Res* **2005;**65(19):8887–8895.

39. Di Carlo E, et al. Local release of interleukin-10 by transfected mouse adenocarcinoma cells exhibits pro- and anti-inflammatory activity and results in a delayed tumor rejection. *Eur Cytokine Netw* **1998;**9:61–68.

40. Fukino K, Shen L, Matsumoto S, Morrison CD, Mutter GL, Eng C. Combined total genome loss of heterozygosity scan of breast cancer stroma and epithelium reveals multiplicity of stromal targets. *Cancer Res* **2004;**64:7231–7236.

41. Kiaris H, Chatzistamou I, Kalofoutis C, Koutselini H, Piperi C, Kalofoutis A. Tumour-stroma interactions in carcinogenesis: basic aspects and perspectives. *Mol Cell Biochem* **2004;**261:117–122.

42. Desmouliere A, Guyot C, Gabbiani G. The stroma reaction myofibroblast: a key player in the control of tumor cell behavior. *Int J Dev Biol* **2004;**48:509–517.

43. Schmitt-Graff A, Desmouliere A, Gabbiani G. Heterogeneity of myofibroblast phenotypic features: an example of fibroblastic cell plasticity. *Virchows Arch* **1994;**425:3–24.

44. Amatangelo MD, Bassi DE, Klein-Szanto AJ, Cukierman E. Stroma-derived three-dimensional matrices are necessary and sufficient to promote desmoplastic differentiation of normal fibroblasts. *Am J Pathol* **2005;**167(2):475–488.

45. Molenaar WM, Oosterhuis JW, Oosterhuis AM, Ramaekers FC. Mesenchymal and muscle-specific intermediate filaments (vimentin and desmin) in relation to differentiation in childhood rhabdomyosarcomas. *Hum Pathol* **1985;**16:838–843.

46. Gabbiani G. Modulation of fibroblastic cytoskeletal features during wound healing and fibrosis. *Pathol Res Pract* **1994**;*190*:851–853.

47. Allinen M, Beroukhim R, Cai L, et al. Molecular characterization of the tumor microenvironment in breast cancer. *Cancer Cell* **2004**;*6*:17–32.

48. Liotta LA, Kohn EC. The microenvironment of the tumour-host interface. *Nature* **2001**;*411*:375–379.

49. Bissell MJ, Radisky D. Putting tumours in context. *Nat Rev Cancer* **2001**;*1*:46–54.

50. Bhowmick NA, Neilson EG, Moses HL. Stromal fibroblasts in cancer initiation and progression. *Nature* **2004**;*432*:332–337.

51. Mueller MM, Fusenig NE. Friends or foes – bipolar effects of the tumour stroma in cancer. *Nat Rev Cancer* **2004**;*4*:839–849.

52. Mueller MM, Fusenig NE. Tumor-stroma interactions directing phenotype and progression of epithelial skin tumor cells. *Differentiation* **2002**;*70*(9-10):486–497. Review.

53. Allers C, Sierralta WD, Neubauer S, Rivera F, Minguell JJ, Conget PA. Dynamic of distribution of human bone marrow-derived mesenchymal stem cells after transplantation into adult unconditioned mice. *Transplantation* **2004**;*78*(4):503–508.

54. Almeida-Porada G, Porada C, Zanjani ED. Plasticity of human stem cells in the fetal sheep model of human stem cell transplantation. *Int J Hematol* **2004**;*79*(1):1–6. Review.

55. Ortiz LA, Gambelli F, McBride C, Gaupp D, Baddoo M, Kaminski N, Phinney DG. Mesenchymal stem cell engraftment in lung is enhanced in response to bleomycin exposure and ameliorates its fibrotic effects. *Proc Natl Acad Sci USA* **2003**;*100*(14):8407–8411. Epub **2003** June 18.

56. Erices AA, Allers CI, Conget PA, Rojas CV, Minguell JJ. Human cord blood-derived mesenchymal stem cells home and survive in the marrow of immunodeficient mice after systemic infusion. *Cell Transplant* **2003**;*12*(6):555–556.

57. Deans RJ, Moseley AB. Mesenchymal stem cells: biology and potential clinical uses. *Exp Hematol* **2000**;*28*(8):875–884. Review.

58. Lange C, Togel F, Ittrich H, Clayton F, Nolte-Ernsting C, Zander AR, Westenfelder C. Administered mesenchymal stem cells enhance recovery from ischemia/reperfusion-induced acute renal failure in rats. *Kidney Int* **2005**;*68*(4):1613–1617.

58. Rojas M, Xu J, Woods CR, Mora AL, Spears W, Roman J, Brigham KL. Bone marrow-derived mesenchymal stem cells in repair of the injured lung. *Am J Respir Cell Mol Biol* **2005**;*33*(2):145–152. Epub **2005** May 12.

59. Phinney DG, Isakova I. Plasticity and therapeutic potential of mesenchymal stem cells in the nervous system. *Curr Pharm Des* **2005**;*11*(10):1255–1265. Review.

60. Sato Y, Araki H, Kato J, Nakamura K, Kawano Y, Kobune M, Sato T, Miyanishi K, Takayama T, Takahashi M, Takimoto R, Iyama S, Matsunaga T, Ohtani S, Matsuura A, Hamada H, Niitsu Y. Human mesenchymal stem cells xenografted directly to rat liver are differentiated into human hepatocytes without fusion. *Blood* **2005**;*106*(2):756–763. Epub **2005** April 7.

61. Natsu K, Ochi M, Mochizuki Y, Hachisuka H, Yanada S, Yasunaga Y. Allogeneic bone marrow-derived mesenchymal stromal cells promote the regeneration of injured skeletal muscle without differentiation into myofibers. *Tissue Eng* **2004**;*10*(7-8):1093–1112.

62. Silva GV, Litovsky S, Assad JA, Sousa AL, Martin BJ, Vela D, Coulter SC, Lin J, Ober J, Vaughn WK, Branco RV, Oliveira EM, He R, Geng YJ, Willerson JT, Perin EC. Mesenchymal stem cells differentiate into an endothelial phenotype, enhance vascular density, and improve heart function in a canine chronic ischemia model. *Circulation* **2005**;*111*(2):150–156. Epub **2005** January 10.

63. Kurozumi K, Nakamura K, Tamiya T, Kawano Y, Kobune M, Hirai S, Uchida H, Sasaki K, Ito Y, Kato K, Honmou O, Houkin K, Date I, Hamada H. BDNF gene-modified mesenchymal stem cells promote functional recovery and reduce infarct size in the rat middle cerebral ar-

tery occlusion model. *Mol Ther* **2004**;*9*(2):189–197. Erratum in: *Mol Ther* **2004**;*9*(5):766.

64. Dai W, Hale SL, Martin BJ, Kuang JQ, Dow JS, Wold LE, Kloner RA. Allogeneic mesenchymal stem cell transplantation in postinfarcted rat myocardium: short- and long-term effects. *Circulation* **2005**;*112*(2):214–223. Epub **2005** July 5.

65. Mansilla E, Marin GH, Sturla F, Drago HE, Gil MA, Salas E, Gardiner MC, Piccinelli G, Bossi S, Salas E, Petrelli L, Iorio G, Ramos CA, Soratti C. Human mesenchymal stem cells are tolerized by mice and improve skin and spinal cord injuries. *Transplant Proc* **2005**;*37*(1):292–295.

66. Satoh H, Kishi K, Tanaka T, Kubota Y, Nakajima T, Akasaka Y, Ishii T. Transplanted mesenchymal stem cells are effective for skin regeneration in acute cutaneous wounds. *Cell Transplant* **2004**;*13*(4):405–412.

67. Dvorak HF. Tumors: wounds that do not heal. Similarities between tumor stroma generation and wound healing. *N Engl J Med* **1986**;*315*:1650–1659.

68. Prockop DJ. Marrow stromal cells as stem cells for nonhematopoietic tissues. *Science* **1997**;*276*:71–74.

69. Bianchi G, Muraglia A, Daga A, Corte G, Cancedda R, Quarto R. Microenvironment and stem properties of bone marrow-derived mesenchymal cells. *Wound Repair Regen* **2001**;*9*:460–466.

70. Pittenger MF, Mackay AM, Beck SC, et al. Multilineage potential of adult human mesenchymal stem cells. *Science* **1999**;*284*:143–147.

71. Gronthos S, Zannettino AC, Hay SJ, et al. Molecular and cellular characterisation of highly purified stromal stem cells derived from human bone marrow. *J Cell Sci* **2003**;*116*:1827–1835.

72. Reyes M, Lund T, Lenvik T, Aguiar D, Koodie L, Verfaillie CM. Purification and ex vivo expansion of postnatal human marrow mesodermal progenitor cells. *Blood* **2001**;*98*:2615–2625.

73. Baksh D, Song L, Tuan RS. Adult mesenchymal stem cells: characterization, differentiation, and application in cell and gene therapy. *J Cell Mol Med* **2004**;*8*:301–316.

74. Devine SM, Cobbs C, Jennings M, Bartholomew A, Hoffman R. Mesenchymal stem cells distribute to a wide range of tissues following systemic infusion into nonhuman primates. *Blood* **2003**;*101*:2999–3001.

75. Fukuda K. Use of adult marrow mesenchymal stem cells for regeneration of cardiomyocytes. *Bone Marrow Transplant* **2003**;*32*(Suppl.1):S25–S27.

76. Lazarus HM, Koc ON, Devine SM, Curtin P, Maziarz RT, Holland HK, Shpall EJ, McCarthy P, Atkinson K, Cooper BW, Gerson SL, Laughlin MJ, Loberiza FR Jr, Moseley AB, Bacigalupo A. Cotransplantation of HLA-identical sibling culture-expanded mesenchymal stem cells and hematopoietic stem cells in hematologic malignancy patients. *Biol Blood Marrow Transplant* **2005**;*11*(5):389–398.

77. Koc ON, Day J, Nieder M, Gerson SL, Lazarus HM, Krivit W. Allogeneic mesenchymal stem cell infusion for treatment of metachromatic leukodystrophy (MLD) and Hurler syndrome (MPS-IH). *Bone Marrow Transplant* **2002**;*30*(4):215–222.

78. Ballas CB, Zielske SP, Gerson SL. Adult bone marrow stem cells for cell and gene therapies: implications for greater use. *J Cell Biochem Suppl* **2002**;*38*:20–28. Review.

79. McNiece I, Harrington J, Turney J, Kellner J, Shpall EJ. Ex vivo expansion of cord blood mononuclear cells on mesenchymal stem cells. *Cytotherapy* **2004**;*6*(4) 311–317.

80. Zhu W, Xu W, Jiang R, Qian H, Chen M, Hu J, Cao W, Han C, Chen Y. Mesenchymal stem cells derived from bone marrow favor tumor cell growth in vivo. *Exp Mol Pathol* **2005**; (Epub ahead of print).

81. Houghton J, Stoicov C, Nomura S, Rogers AB, Carlson J, Li H, Cai X, Fox JG, Goldenring JR, Wang TC. Gastric cancer originating from bone marrow-derived cells. *Science* **2004**;*306*(5701):1568–1571.

82. Prindull G, Zipori D. Environmental guidance of normal and tumor cell plasticity: epithelial mesenchymal transitions as a paradigm. *Blood* **2004**;*103*(8):2892–2899. Epub **2004** January 8. Review.

83. Embauer H, Minguell JJ. Selective interactions between epithelial tumour cells

and bone marrow mesenchymal stem cells. *Br J Cancer* **2000**;*82*:1290–1296.

84. De Palma M, Venneri MA, Galli R, Sergi LS, Politi LS, Sampaolesi M, Naldini L. Tie2 identifies a hematopoietic lineage of proangiogenic monocytes required for tumor vessel formation and a mesenchymal population of pericyte progenitors. *Cancer Cell* **2005**;*8*(3):211–226.

85. De Palma M, Venneri MA, Roca C, Naldini L. Targeting exogenous genes to tumor angiogenesis by transplantation of genetically modified hematopoietic stem cells. *Nat Med* **2003**;*9*(6):789–795. Epub **2003** May 12.

86. Wright N, de Lera TL, Garcia-Moruja C, Lillo R, Garcia-Sanchez F, Caruz A, Teixido J. Transforming growth factor-beta1 down-regulates expression of chemokine stromal cell-derived factor-1: functional consequences in cell migration and adhesion. *Blood* **2003**;*102*(6):1978–1984. Epub **2003** May 29.

87. Forbes SJ, Russo FP, Rey V, Burra P, Rugge M, Wright NA, Alison MR. A significant proportion of myofibroblasts are of bone marrow origin in human liver fibrosis. *Gastroenterology* **2004**;*126*(4):955–963.

88. Direkze NC, Hodivala-Dilke K, Jeffery R, Hunt T, Poulsom R, Oukrif D, Alison MR, Wright NA. Bone marrow contribution to tumor-associated myofibroblasts and fibroblasts. *Cancer Res* **2004**;*64*(23):8492–8495.

89. Ishii G, Sangai T, Ito T, Hasebe T, Endoh Y, Sasaki H, Harigaya K, Ochiai A. In vivo and in vitro characterization of human fibroblasts recruited selectively into human cancer stroma. *Int J Cancer* **2005**;*117*(2):212–220.

90. Ishii G, Sangai T, Oda T, Aoyagi Y, Hasebe T, Kanomata N, Endoh Y, Okumura C, Okuhara Y, Magae J, Emura M, Ochiya T, Ochiai A. Bone-marrow-derived myofibroblasts contribute to the cancer-induced stromal reaction. *Biochem Biophys Res Commun* **2003**;*309*(1):232–240.

91. Sangai T, Ishii G, Kodama K, Miyamoto S, Aoyagi Y, Ito T, Magae J, Sasaki H, Nagashima T, Miyazaki M, Ochiai A. Effect of differences in cancer cells and tumor growth sites on recruiting bone

marrow-derived endothelial cells and myofibroblasts in cancer-induced stroma. *Int J Cancer* **2005**;*115*(6):885–892.

92. Sugimoto T, Takiguchi Y, Kurosu K, Kasahara Y, Tanabe N, Tatsumi K, Hiroshima K, Minamihisamatsu M, Miyamoto T, Kuriyama T. Growth factor-mediated interaction between tumor cells and stromal fibroblasts in an experimental model of human small-cell lung cancer. *Oncol Rep* **2005**;*14*(4):823–830.

93. Roni V, Habeler W, Parenti A, Indraccolo S, Gola E, Tosello V, Cortivo R, Abatangelo G, Chieco-Bianchi L, Amadori A. Recruitment of human umbilical vein endothelial cells and human primary fibroblasts into experimental tumors growing in SCID mice. *Exp Cell Res* **2003**;*287*(1):28–38.

94. Jankowski K, Kucia M, Wysoczynski M, Reca R, Zhao D, Trzyna E, Trent J, Peiper S, Zembala M, Ratajczak J, Houghton P, Janowska-Wieczorek A, Ratajczak MZ. Both hepatocyte growth factor (HGF) and stromal-derived factor-1 regulate the metastatic behavior of human rhabdomyosarcoma cells, but only HGF enhances their resistance to radiochemotherapy. *Cancer Res* **2003**;*63*(22):7926–7935.

95. Yoneda T, Hiraga T. Crosstalk between cancer cells and bone microenvironment in bone metastasis. *Biochem Biophys Res Commun* **2005**;*328*(3):679–687. Review.

96. Burns MJ, Weiss W. Targeted therapy of brain tumors utilizing neural stem and progenitor cells. *Front Biosci* **2003**;*8*:e228–e234. Review.

97. Dennis JE, Cohen N, Goldberg VM, Caplan AI. Targeted delivery of progenitor cells for cartilage repair. *J Orthop Res* **2004**;*22*(4):735–741.

98. Kassem M. Mesenchymal stem cells: biological characteristics and potential clinical applications. *Cloning Stem Cells* **2004**;*6*(4):369–374. Review.

99. Schoeberlein A, Holzgreve W, Dudler L, Hahn S, Surbek DV. Tissue-specific engraftment after in utero transplantation of allogeneic mesenchymal stem cells into sheep fetuses. *Am J Obstet Gynecol* **2005**;*192*(4):1044–1052.

100. Ye J, Yao K, Kim JC. Mesenchymal stem cell transplantation in a rabbit corneal alkali burn model: engraftment and in-

volvement in wound healing. *Eye* **2005**; (Epub ahead of print).

101. Chan J, O'Donoghue K, de la Fuente J, Roberts IA, Kumar S, Morgan JE, Fisk NM. Human fetal mesenchymal stem cells as vehicles for gene delivery. *Stem Cells* **2005**;*23*(1):93–102.

102. In't Anker PS, Scherjon SA, Kleijburg-van der Keur C, de Groot-Swings GM, Claas FH, Fibbe WE, Kanhai HH. Isolation of mesenchymal stem cells of fetal or maternal origin from human placenta. *Stem Cells* **2004**;*22*(7):1338–1345.

103. Hurwitz DR, Kirchgesser M, Merrill W, Galanopoulos T, McGrath CA, Emami S, Hansen M, Cherington V, Appel JM, Bizinkauskas CB, Brackmann HH, Levine PH, Greenberger JS. Systemic delivery of human growth hormone or human factor IX in dogs by reintroduced genetically modified autologous bone marrow stromal cells. *Hum Gene Ther* **1997**;*8*(2):137–156.

104. Evans CH, Robbins PD, Ghivizzani SC, Wasko MC, Tomaino MM, Kang R, Muzzonigro TA, Vogt M, Elder EM, Whiteside TL, Watkins SC, Herndon JH. Gene transfer to human joints: progress toward a gene therapy of arthritis. *Proc Natl Acad Sci USA* **2005**;*102*(24):8698–8703. Epub **2005** June 6.

105. Pittenger MF, Mackay AM, Beck SC, Jaiswal RK, Douglas R, Mosca JD, Moorman MA, Simonetti DW, Craig S, Marshak DR. Multilineage potential of adult human mesenchymal stem cells. *Science* **1999**;*284*(5411):143–147.

106. Kurozumi K, Nakamura K, Tamiya T, Kawano Y, Ishii K, Kobune M, Hirai S, Uchida H, Sasaki K, Ito Y, Kato K, Honmou O, Houkin K, Date I, Hamada H. Mesenchymal stem cells that produce neurotrophic factors reduce ischemic damage in the rat middle cerebral artery occlusion model. *Mol Ther* **2005**;*11*(1):96–104.

107. Honma T, Honmou O, Iihoshi S, Harada K, Houkin K, Hamada H, Kocsis JD. Intravenous infusion of immortalized human mesenchymal stem cells protects against injury in a cerebral ischemia model in adult rat. *Exp Neurol* **2005** (Epub ahead of print).

108. Hamada H, Kobune M, Nakamura K, Kawano Y, Kato K, Honmou O, Houkin K, Matsunaga T, Niitsu Y. Mesenchymal stem cells (MSC) as therapeutic cytoreagents for gene therapy. *Cancer Sci* **2005**;*96*(3):149–156. Review.

109. Caplan AI, Bruder SP. Mesenchymal stem cells: building blocks for molecular medicine in the 21st century. *Trends Mol Med* **2001**;*7*:259–264.

110. Baksh D, Song L, Tuan RS. Adult mesenchymal stem cells: characterization, differentiation, and application in cell and gene therapy. *J Cell Mol Med* **2004**;*8*:301–316.

111. Feldmann RE, Jr, Bieback K, Maurer MH, Kalenka A, Burgers HF, Gross B, Hunzinger C, Kluter H, Kuschinsky W, Eichler H. Stem cell proteomes: a profile of human mesenchymal stem cells derived from umbilical cord blood. *Electrophoresis* **2005**;*26*(14):2749–2758.

112. Parham DM. Pathologic classification of rhabdomyosarcomas and correlations with molecular studies. *Mod Pathol* **2001**;*14*:506–514.

113. Iacobuzio-Donahue CA, Argani P, Hempen PM, Jones J, Kern SE. The desmoplastic response to infiltrating breast carcinoma: gene expression at the site of primary invasion and implications for comparisons between tumor types. *Cancer Res* **2002**;*62*:5351–5357.

114. Robinson SC, Coussens LM. Soluble mediators of inflammation during tumor development. *Adv Cancer Res* **2005**;*93*:159–187. Review.

115. Rowley DR. What might a stromal response mean to prostate cancer progression? *Cancer Metastasis Rev* **1998-99**;*17*(4):411–419. Review.

116. Serakinci N, Guldberg P, Burns JS, Abdallah B, Schrodder H, Jensen T, Kassem M. Adult human mesenchymal stem cell as a target for neoplastic transformation. *Oncogene* **2004**;*23*(29):5095–5098.

117. Rubio D, Garcia-Castro J, Martin MC, de la Fuente R, Cigudosa JC, Lloyd AC, Bernad A. Spontaneous human adult stem cell transformation. *Cancer Res* **2005**;*65*(8):3035–3039. Erratum in: *Cancer Res* **2005**;*65*(11):4969.

118. Burns JS, Abdallah BM, Guldberg P, Rygaard J, Schroder HD, Kassem M.

Tumorigenic heterogeneity in cancer stem cells evolved from long-term cultures of telomerase-immortalized human mesenchymal stem cells. *Cancer Res* 2005;*65*(8):3126–3135.

119. Fierro FA, Sierralta WD, Epunan MJ, Minguell JJ. Marrow-derived mesenchymal stem cells: role in epithelial tumor cell determination. *Clin Exp Metastasis* 2004;*21*(4):313–319.

120. Cunha GR, Hayward SW, Wang YZ, Ricke WA. Role of the stromal microenvironment in carcinogenesis of the prostate. *Int J Cancer* 2003;*107*(1):1–10. Review.

121. Chung LW, Baseman A, Assikis V, Zhau HE. Molecular insights into prostate cancer progression: the missing link of tumor microenvironment. *J Urol* 2005;*173*(1):10–20. Review.

122. Ohlsson LB, Varas L, Kjellman C, Edvardsen K, Lindvall M. Mesenchymal progenitor cell-mediated inhibition of tumor growth in vivo and in vitro in gelatin matrix. *Exp Mol Pathol* 2003;*75*(3):248–255.

123. Xia Z, Ye H, Choong C, Ferguson DJ, Platt N, Cui Z, Triffitt JT. Macrophagic response to human mesenchymal stem cell and poly(epsilon-caprolactone) implantation in nonobese diabetic/severe combined immunodeficient mice. *J Biomed Mater Res A* 2004;*71*(3):538–548.

124. De Kok IJ, Drapeau SJ, Young R, Cooper LF. Evaluation of mesenchymal stem cells following implantation in alveolar sockets: a canine safety study. *Int J Oral Maxillofac Implants* 2005;*20*(4):511–518.

125. Chen J, Wang C, Lu S, Wu J, Guo X, Duan C, Dong L, Song Y, Zhang J, Jing D, Wu L, Ding J, Li D. In vivo chondrogenesis of adult bone-marrow-derived autologous mesenchymal stem cells. *Cell Tissue Res* 2005;*319*(3):429–438. Epub 2005 January 26.

126. Shake JG, Gruber PJ, Baumgartner WA, Senechal G, Meyers J, Redmond JM, Pittenger MF, Martin BJ. Mesenchymal stem cell implantation in a swine myocardial infarct model: engraftment and functional effects. *Ann Thorac Surg* 2002;*73*(6):1919–1925; discussion 1926.

127. Harrington K, Alvarez-Vallina L, Crittenden M, Gough M, Chong H, Diaz RM, Vassaux G, Lemoine N, Vile R. Cells as vehicles for cancer gene therapy: the missing link between targeted vectors and systemic delivery? *Hum Gene Ther* 2002;*13*(11):1263–1280. Review.

128. Prockop DJ, Sekiya I, Colter DC. Isolation and characterization of rapidly self-renewing stem cells from cultures of human marrow stromal cells. *Cytotherapy* 2001;*3*(5):393–396.

129. Sekiya I, Larson BL, Smith JR, Pochampally R, Cui JG, Prockop DJ. Expansion of human adult stem cells from bone marrow stroma: conditions that maximize the yields of early progenitors and evaluate their quality. *Stem Cells* 2002;*20*(6):530–541.

130. Fiedler J, Etzel N, Brenner RE. To go or not to go: migration of human mesenchymal progenitor cells stimulated by isoforms of PDGF. *J Cell Biochem* 2004;*93*(5):990–998.

131. Kraitchman DL, Tatsumi M, Gilson WD, Ishimori T, Kedziorek D, Walczak P, Segars WP, Chen HH, Fritzges D, Izbudak I, Young RG, Marcelino M, Pittenger MF, Solaiyappan M, Boston RC, Tsui BM, Wahl RL, Bulte JW. Dynamic imaging of allogeneic mesenchymal stem cells trafficking to myocardial infarction. *Circulation* 2005;*112*(10):1451–1461. Epub 2005 August 29.

132. Forte G, Minieri M, Cossa P, Antenucci D, Sala M, Gnocchi V, Fiaccavento R, Carotenuto F, De Vito P, Baldini PM, Prat M, Di Nardo P. Hepatocyte growth factor effects on mesenchymal stem cells: proliferation, migration and differentiation. *Stem Cells* 2005 (in press).

133. Jiang H, Conrad C, Fueyo J, Gomez-Manzano C, Liu TJ. Oncolytic adenoviruses for malignant glioma therapy. *Front Biosci* 2003;*8*:d577–d588. Review.

134. Gomez-Manzano C, Balague C, Alemany R, Lemoine MG, Mitlianga P, Jiang H, Khan A, Alonso M, Lang FF, Conrad CA, Liu TJ, Bekele BN, Yung WK, Fueyo J. A novel E1A-E1B mutant adenovirus induces glioma regression in vivo. *Oncogene* 2004;*23*(10):1821–1828.

Part IV
Clinical Trials

Stem Cell Transplantation. Biology, Processing, and Therapy.
Edited by Anthony D. Ho, Ronald Hoffman, and Esmail D. Zanjani
Copyright © 2006 WILEY-VCH Verlag GmbH & Co. KGaA, Weinheim
ISBN: 3-527-31018-5

12

Endothelial Progenitor Cells for Cardiac Regeneration

Ulrich Fischer-Rasokat and Stefanie Dimmeler

12.1
Characterization of Endothelial Progenitor Cells

The differentiation of mesodermal cells to angioblasts and subsequent endothelial differentiation was believed to occur exclusively in embryonic development. This dogma was challenged in 1997, when Asahara and colleagues reported that purified CD34[+] hematopoietic progenitor cells from adults can differentiate *ex vivo* to an endothelial phenotype (Asahara et al., 1997). These cells were named "endothelial progenitor cells" (EPC), expressed various endothelial markers and were incorporated into neovessels at sites of ischemia. Similar findings were reported by Rafiis et al. in 1998, who showed the existence of a "circulating bone marrow-derived endothelial progenitor cell" (CEPC) in the adult (Shi et al. 1998). Again, a subset of blood-derived CD34[+] hematopoietic stem cells was shown to differentiate to the endothelial lineage and to express endothelial marker proteins such as von Willebrand factor (vWF) and incorporated Dil-acetylated LDL (DIL). Furthermore, bone marrow-transplanted, genetically tagged cells were covering implanted Dacron grafts (Shi et al., 1998). These pioneering experimental studies suggested the presence of circulating hemangioblasts in the adult. EPC or CEPC were defined as cells, which express both, hematopoietic stem cell markers such as CD34 and an endothelial marker protein such as the VEGF-receptor KDR. Since CD34 is not expressed exclusively on hematopoietic stem cells (HSC) but also on mature endothelial cells, further studies used the more immature HSC marker CD133 also known as prominin or AC133. Purified isolated CD133[+] cells also differentiated to endothelial cells *in vitro* (Gehling et al., 2000). CD133 is a highly conserved antigen with unknown biological activity, which is expressed on HSC but is absent in mature endothelial cells and monocytic cells (for review, see Handgretinger et al., 2003). Thus, CD133[+]KDR[+] cells likely reflect more immature progenitor cells. Since the biological function of CD133 is unknown, it is unclear whether CD133 only represents a surface marker or has a functional activity involved in regulation of neovascularization.

Stem Cell Transplantation. Biology, Processing, and Therapy.
Edited by Anthony D. Ho, Ronald Hoffman, and Esmail D. Zanjani
Copyright © 2006 WILEY-VCH Verlag GmbH & Co. KGaA, Weinheim
ISBN: 3-527-31018-5

EPC also can be isolated from peripheral blood mononuclear cells (PBMC) by cultivation in medium favoring endothelial differentiation. The origin of EPC is not entirely clear. In PBMC, several possible sources for endothelial cells may exist:

- the rare number of HSC;
- myeloid cells, which may differentiate to endothelial cells under the cultivation selection pressure;
- other circulating progenitor cells (e.g., "side population" cells); and
- circulating mature endothelial cells, which are shed from the vessel wall (Mutin et al., 1999) and adhere to the culture dishes.

Indeed, Hebbel and colleagues showed that morphologically and functionally distinct endothelial cell populations can be grown from PBMC (Lin et al., 2000). These authors stratified the different circulating endothelial cells according to their growth characteristics and morphological appearance as "spindle-like" cells, which exhibit a low proliferative capacity, and "outgrowing" cells. The outgrowing cells showed a high proliferative potential, were predominantly from the bone marrow donors, and therefore were considered as circulating angioblasts. The authors speculated that the spindle-like cells might likely represent circulating mature endothelial cells, which are shed from the vessel wall. However, this hypothesis is difficult to test and has not yet been proven.

One way to reduce the heterogeneity of the cultivated EPC experimentally is to pre-plate mononuclear cells, because this excludes rapidly adhering cells such as differentiated monocytic or possible mature endothelial cells (Shi et al., 1998). However, these protocols do not eliminate myeloid and nonhematopoietic progenitor cells, which may contribute to the *ex-vivo* cultivated cells. Indeed, a subset of human PBMC acts as pluripotent stem cells (Zhao et al., 2003). Moreover, there is increasing evidence that myeloid cells can also give rise to endothelial cells. Specifically, CD14$^+$/CD34$^-$ myeloid cells can co-express endothelial markers and form tube-like structures *ex vivo* (Schmeisser et al., 2001). The *ex-vivo* cultivation of purified CD14$^+$ mononuclear cells yielded cells with an endothelial characteristic, which incorporated in newly formed blood vessels *in vivo* (Urbich et al., 2003). These data would suggest that myeloid cells could differentiate (or transdifferentiate) to the endothelial lineage. Lineage tracking showed that myeloid cells are the HSC-derived intermediates, which contribute to muscle regeneration (Camargo et al., 2003), suggesting that myeloid intermediates may be part of the repair capacity after injury.

The characterization of EPC becomes particularly difficult when cells are *ex-vivo* expanded and cultured, since the culture conditions (culture supplements such as fetal calf serum and cytokines or different plastic types) rapidly change the phenotype of the cells. For example, supplementation of the medium with statins increased the number of clonally expanding endothelial cells isolated from mononuclear cells and reduced *ex-vivo* EPC senescence (Assmus et al., 2003). Moreover, continuous cultivation was shown to increase endothelial differentiation, as evidenced by elevated endothelial marker protein expression (Fujiyama et al., 2003). This may explain why published reports of EPC characterization vary so widely. Due to minor deviation in medium composition, different groups may reside in cultivating cells with different surface factor profile and functional activity, despite similar protocols having been used. Thus, the lack of endothelial marker expression reported by Rehman et al. (2003), which is in contrast to other studies (Kalka et al., 2000; Urbich et al., 2003), may be explained by these differences. Moreover, the interaction and growth factor release of cells within the heterogeneous mixture of mononuclear cells from the blood may impact the yield and the functional activity of the cultivated cells (Rookmaaker et al., 2003).

Increasing evidence suggests that nonhematopoietic stem cells also can give rise to endothelial cells. Non-bone marrow-derived cells have been shown to replace the endothelial cells in grafts (Hillebrands et al., 2002). In addition, adult bone marrow-derived stem/progenitor cells which are distinct from HSC such as the "side population" cells and multipotent adult progenitor cells, have also been shown to have the capacity to differentiate to the endothelial lineage (Jackson et al., 2001; Reyes et al., 2002). Tissue-resident stem cells have been isolated from the heart, which are capable of differentiating to the endothelial lineage (Beltrami et al., 2003). Likewise, neuroprogenitor cells can differentiate to endothelial cells (Wurmser et al., 2004). These data support the notion that it will be difficult to define the "true" endothelial progenitor cells. Better transcriptional profiling of distinct cell populations and fate mapping studies will help to identify markers, which distinguish the circulating endothelial precursor within the blood and bone marrow/non-bone marrow-derived endothelial cells.

12.2
Functions of EPC to Improve Cardiac Function

Stem or progenitor cells may have several possible ways in which they may contribute to cardiac regeneration. EPC can enhance neovascularization, thereby maintaining blood supply to the ischemic tissue. The release of growth factors by progenitor cells may additionally enhance angiogenesis and protect neighboring cardiac myocytes from apoptosis. Paracrine factors also can act as chemoattractants to stimulate the homing of endogenous circulating or tissue-residing stem/progenitor cells. One may additionally speculate that progenitor cell-mediated release of cytokines/growth factors and proteases may influence scar expansion and negative remodeling. Moreover, progenitor cells may exert immunomodulatory activities, which change scar expansion and healing. Finally, progenitor cells could contribute to cardiac regeneration in its pure sense by differentiation into cardiac myocytes, or by rescuing cardiac myocytes by fusion. The extent to which these different possibilities contribute to a functional improvement after infusion of endothelial progenitor cells at present is not yet defined.

12.2.1
Improvement of Neovascularization

Whereas the capacity of adult, isolated HSC to differentiate to cardiac myocytes in reasonable number is still the subject of debate, it has been shown by various groups that stem/progenitor cells can differentiate to endothelial cells and increase the formation of new capillaries. This was initially described for HSC by Asahara et al. (1997). Two days after creating unilateral hindlimb ischemia by excising one femoral artery, athymic mice were injected into the tail vein with Dil-labeled CD34$^+$ or CD34$^-$ cells, isolated from human peripheral blood. Histological examination after 1–6 weeks revealed numerous Dil-labeled cells in the neovascularized ischemic hindlimb. Moreover, most of the cells appeared integrated into capillaries. The amount of Dil-labeled cells was about 8-fold higher in mice injected with CD34$^+$ compared to CD34$^-$ cells. Experiments were repeated in rabbits using autologous CD34$^+$ cells, which also localized exclusively to neovascular zones of the ischemic limb.

Shortly after this, Shi et al. (1998) reported that a subset of CD34$^+$ cells, localized in the bone marrow, could be mobilized to the peripheral circulation and could colonize endothelial flow surfaces of vascular prostheses. In this study, a canine bone marrow transplantation model was used in which donor and host cells could be distinguished. After transplantation, a Dacron graft was implanted in the descending thoracic aorta, and retrieved after 12 weeks. Cells with endothelial morphology covering the graft were identified, and DNA analysis of these cells revealed only alleles from the donor. This provided evidence that a subset of bone marrow-derived cells was circulating in the peripheral blood and may have differentiated into an endothelial lineage.

First attempts to use stem/progenitor cells for therapeutic application were reported by Kalka et al. (2000). *Ex-vivo*-expanded EPC were transplanted to athymic nude mice after inducing hindlimb ischemia. The authors reported a marked improvement in blood flow recovery and capillary density in the ischemic hindlimb after 4 weeks, and the rate of limb loss was significantly reduced compared to control infusion or infusion of mature endothelial cells.

Kawamoto et al. (2001) evaluated the effect of EPC-infusion after myocardial infarction in rats. Myocardial ischemia was induced by ligating the left anterior descending coronary artery in athymic rats. A total of 10^6 EPC was injected intravenously 3 h after the induction of myocardial ischemia. Seven days later, labeled EPC were detected in the ischemic area and showed incorporation into foci of myocardial neovascularization. Echocardiography, performed 28 days after ischemia, disclosed a significant reduction in the ventricular volumes and an improved fractional shortening in the EPC group compared to controls. Immunostaining revealed that the capillary density was significantly greater in the EPC group than in the control group. Moreover, the extent of left ventricular scarring was significantly less in rats receiving EPC than in controls. Similar results were achieved when G-CSF-mobilized HSC were used (Kocher et al., 2001). Again, a rat model was used to determine the effect of cell treatment after acute myocardial infarction. Two weeks after injection of G-CSF mobilized adult human CD34$^+$ lineage cells, and in a parallel time frame with an observed neoangiogenesis, the left ventricular ejection fraction (LVEF) recovered by a mean of 22 ± 6%. This effect was well maintained up to 15 weeks, with a further improvement in LVEF to a mean of 34 ± 4% at 15 weeks after injection. In contrast, CD34$^-$ cells did not demonstrate similar effects. The therapeutic implication of stem/progenitor cells is further underscored by the finding that wild-type stem cell infusion was capable of rescuing a dysfunctional neovascularization as seen in endothelial nitric oxide synthase (eNOS)-deficient animals (Aicher et al., 2003). Likewise, young adult bone marrow-derived progenitor cells were shown to improve aging-impaired cardiac angiogenic function (Edelberg et al., 2002).

12.2.2
Paracrine Effects

As bone marrow-derived stem and progenitor cells home to sites of ischemia, this may allow the local release of factors acting in a paracrine manner on the surrounding ischemic tissue. Bone marrow-derived mononuclear cells release angiogenic growth factors such as VEGF, basic fibroblast growth factor (bFGF) and angiopoietins, thereby enhancing the local angiogenic response (Kamihata et al., 2001). Isolated human EPC also express various growth factors, which can promote cardiac myocytes survival and improve angiogenesis (personal communication). Locally released paracrine factors may additionally act via promoting arteriogenesis. Moreover, cytokines may mobilize tissue-residing stem cells to promote endogenous repair. Finally, it may be envisioned that paracrine factors modulate the immune response, thereby affecting scar expansion and healing (Musaro et al., 2004).

12.2.3
Differentiation and/or Fusion

Injection of various bone marrow-derived stem cells after myocardial infarction leads to an engraftment of these cells in the border zone of the infarcts (Jackson et al., 2001; Kocher et al., 2001). When dyes or genetic markers were used to track the cells after engraftment, the injected cells partly showed the expression of cardiac and/or endothelial marker proteins (Jackson et al., 2001; Kocher et al., 2001; Orlic et al., 2001; Toma et al., 2002). These initial results suggested that progenitor cells differentiate to the cardiac lineage and may thereby contribute to cardiac regeneration. Moreover, cocultivation of freshly isolated umbilical venous endothelial cells, EPC or isolated CD34$^+$ cells with cardiac myocytes induced differentiation of the different cells to a cardiogenic phenotype (Condorelli et al., 2001; Badorff et al., 2003; Yeh et al., 2003). However, several other studies suggested meanwhile that the expression of cardiac marker proteins is not the result of a differentiation process but rather reflects cell-to-cell fusion. Such events were initially reported for embryonic stem cells, and now have been confirmed with HSC

(Medvinsky and Smith, 2003). Interestingly, a recent study showed that murine cardiac stem cells could fuse and differentiate to a similar extent when injected after myocardial infarction (Oh et al., 2003). The overall contribution of differentiation versus fusion to cardiac regeneration by the different human cell types investigated thus far is unclear. One may speculate that different cell types may be more or less susceptible for fusion versus differentiation. Thus, mesenchymal stem cells (MSC) exerted a higher differentiation capacity than hematopoietic progenitor cells in a side-by-side comparison (Kawada et al., 2004).

12.3
Mechanisms of Homing

12.3.1
Adhesion

Despite the controversial discussion of mechanisms underlying the functional improvement observed in experimental studies, the homing of circulating cells is clearly a prerequisite for all of the putative mechanisms of action (Fig. 12.1). In a previous study assessing the *in-vivo* homing of embryonic endothelial progenitor cells derived from cord blood, the circulating cells arrested within tumor microvessels, extravasated into the interstitium, and incorporated into neovessels. This suggested that adhesion and transmigration are involved in the recruitment of endothelial progenitor cells to sites of tumor angiogenesis (Vajkoczy et al., 2003). Thus, it is conceivable that *ex-vivo*-expanded adult EPC and HSC/progenitor cells may engage similar pathways for recruitment to sites of ischemia and incorporation into newly forming vessels. The recruitment and incorporation of EPC requires a coordinated sequence of multi-step adhesive and signaling events, including chemoattraction, adhesion, transmigration and, finally, the differentiation to endothelial cells.

The initial step of homing of progenitor cells to ischemic tissue involves the adhesion of progenitor cells to endothelial cells activated by cytokines and ischemia and their transmigration through the endothelial cell monolayer (Vajkoczy et al., 2003). Integrins are known to mediate the adhesion of various cells (including HSC and leukocytes) to extracellular matrix proteins and to endothelial cells (Carlos and Harlan, 1994; Springer, 1994; Muller, 2002). Consistent with the high expression of β_2-integrins on hematopoietic stem/progenitor cells, β_2-integrins mediate adhesion and transmigration of HSC/progenitor cells (Peled et al., 1999; Kollet et al., 2001). β_2-integrins (CD18/CD11) are expressed on peripheral blood-derived EPC and are required for EPC-adhesion to endothelial cells and transendothelial migration *in vitro* (Chavakis et al., 2005). Moreover, HSC (Sca-1^+/lin$^-$) lacking β_2-integrins showed reduced homing and a lower capacity to improve neovascularization after ischemia. These studies provide the first evidence for a direct participation of β_2-integrins in neovascularization processes and particularly in stem/progenitor cell-mediated vasculogenesis (Chavakis et al., 2005).

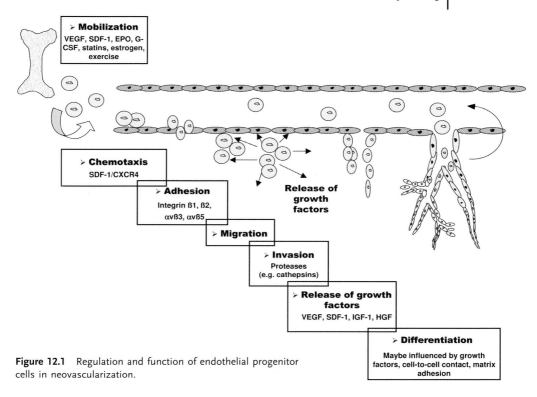

Figure 12.1 Regulation and function of endothelial progenitor cells in neovascularization.

Furthermore, *in-vitro* studies showed that MCP-1 stimulated adhesion of bone marrow-derived monocyte lineage cells to the endothelium was blocked by anti-β_1-integrin antibodies (Fujiyama et al., 2003). Interestingly, in this study, the adhesion of peripheral blood-derived CD34$^-$ CD14$^+$ monocytes to endothelial cells was not affected by MCP-1, and was not blocked by anti-β_1-integrin antibodies. Moreover, conditional deletion of the α_4-integrin selectively inhibited the homing of HSC/progenitor cells to the bone marrow but not to the spleen (Scott et al., 2003), suggesting that the homing of progenitor cells to different tissues is dependent on distinct adhesion molecules. Furthermore, the initial cell arrest of embryonic progenitor cell homing during tumor angiogenesis was suggested to be mediated by E- and P-selectin and P-selectin glycoprotein ligand-1 (Vajkoczy et al., 2003). It is important, however, to stress that this study was performed with embryonic endothelial progenitor cells. It is conceivable, that different cell types may use distinct mechanisms for homing to sites of angiogenesis.

12.3.2
Chemotaxis, Migration, and Invasion

Given the low numbers of circulating EPC, chemoattraction may be of utmost importance to allow for the recruitment of reasonable numbers of progenitor cells to the ischemic or injured tissue. Various studies have attempted to identify those factors which influence HSC engraftment to the bone marrow. These factors include chemokines such as stromal cell-derived factor-1 (SDF-1) (Lapidot, 2001; Wright et al., 2002) and lipid mediators (sphingosine-1-phosphate) (Kimura et al., 2004). Numerous data suggest that SDF-1 plays a crucial role for progenitor cell recruitment to ischemic tissue. SDF-1 expression is regulated in ischemic tissue by the hypoxia-inducible factor-1 (HIF-1) in direct proportion to reduced oxygen tension (Ceradini et al., 2004). Accordingly, SDF-1 expression is increased during the first days after induction of myocardial infarction (Askari et al., 2003). Moreover, SDF-1 stimulates the recruitment of progenitor cells to the ischemic tissue (Yamaguchi et al., 2003), while overexpression of SDF-1 augmented stem cell-homing and incorporation into ischemic tissues (Askari et al., 2003; Yamaguchi et al., 2003). HSC were shown to be exquisitely sensitive to SDF-1 and did not react to G-CSF or other chemokines (e.g. IL-8, RANTES) (Wright et al., 2002). VEGF levels are also increased during ischemia and are capable of acting as a chemoattractive factor to EPC (Kalka et al., 2000; Lee et al., 2000; Shintani et al., 2001). Interestingly, the *ex-vivo* migratory capacity of EPC or bone marrow cells towards VEGF and SDF-1 determined the functional improvement of patients after stem cell therapy (Britten et al., 2003). Beside genes, which are directly up-regulated by hypoxia, the invasion of immunocompetent cells to the ischemic tissue may further increase the levels of various chemokines within the ischemic tissue, such as monocyte chemoattractant protein-1 (MCP-1) or interleukins, which can attract circulating progenitor cells (Fujiyama et al., 2003). Whereas several studies shed some light on the mechanisms regulating attraction of EPC to ischemic tissue, less is known with respect to migration and tissue invasion. One may speculate that proteases such as cathepsins or metalloproteases may mediate the tissue invasion of EPC. Indeed, we recently demonstrated that EPC require cathepsin L to augment neovascularization after ischemia (Urbich et al., 2005).

12.4
Results from Clinical Studies

In the clinical setting, the effect of stem cell transplantation was assessed in patients with acute or chronic myocardial ischemia. Thus far, most studies used mononuclear cells isolated from bone marrow aspirates or EPC isolated from PBMC. Of note, bone marrow-derived cells consist of various different types of stem/progenitor cells including hematopoietic progenitor cells, mesenchymal progenitor cells, and endothelial progenitor cells. Intracoronary infusion as well as intramyocardial injection was used as the form of application. Overall, the in-

Table 12.1 Stem/progenitor cell therapy in patients with acute myocardial infarction.

Patients with acute myocardial infarction

Study/Reference	Cell type (No. of infused cells)	Study design	No. of patients	Follow-up	Effect
Intracoronary application					
Strauer et al. (*Circulation* 2002)	BMC (mean of 28×10^6) vs. control	open-labeled	n = 10 (BMC) n = 10 (control)	3 months	• Hypokinetic area ↓ • Contractility infarct region ↑ • End-systolic volume ↓ • Perfusion ↑
TOPCARE-AMI (*Circulation* 2002/2003) (*JACC* 2004)	BMC ($213 \pm 75 \times 10^6$) vs. EPC ($16 \pm 12 \times 10^6$)	open-labeled, randomized	n = 30 (EPC) n = 29 (BMC)	12 months	• Global & regional contractility ↑ • End-systolic volume ↓ • Viability ↑ • Coronary flow reserve ↑
BOOST (*Lancet* 2004)	BMC ($24.6 \pm 9.4 \times 10^8$) vs. control	open-labeled, randomized	n = 30 (BMC) n = 30 (control)	6 months	• LVEF ↑
MAGIC (*Lancet* 2004)	Leukocytes after G-CSF-stimulation (100×10^6) vs. G-CSF stimulation alone vs. Control	open-labeled, randomized	n = 7 (cell infusion) n = 3 (G-CSF alone) (n = 1 control)	6 months	• LVEF ↑ • Myocardial perfusion ↑ • Restenosis rate 70 %
Kuethe et al. (*Int J Cardiol* 2004)	BMC ($39 \pm 23 \times 10^6$)	open-labeled	n = 5 (BMC)	12 months	• LVEF ↔
Chen et al. (*Am J Cardiol* 2004)	BMSC (6 mL, $8{-}10 \times 10^9$ mL^{-1}) vs. Control	open-labeled, randomized	n = 34 (BMC) n = 35 (control)	3 months	• LVEF ↑ • Perfusion↑ • End-systolic/diastolic volume ↓
Fernandez-Aviles et al. (*Circ Res* 2004)	BMC ($78 \pm 41 \times 10^6$) vs. Control	open-labeled	n = 20 (BMC) n = 11 (control)	6 months	• LVEF ↑ • Regional contractility ↑ • End-systolic volume ↓
Surgical application					
Stamm et al. (*Lancet* 2003)	CD 133⁺ BMC ($1.2 \pm 0.2 \times 10^6$)	open-labeled	n = 6 (BMC)	3–10 months	• Feasible • LVEF ↑ in 4 pts. • Perfusion ↑ in 5 pts.

Table 12.2 Progenitor cell therapy in patients with chronic heart failure.

Patients with chronic ischemic heart failure

Reference	Cell type (No. of infused cells)	Study design	No. of patients	Follow-up	Effect
Intramyocardial application (NOGA-guided)					
Tse et al. (*Lancet* 2003)	BMC (cells from 40 mL BM)	open-labeled	n = 8 (BMC)	3 months	• Feasible • Wall motion & thickening ↑
Fuchs et al. (*JACC* 2003)	BMC (2.4 mL, 32.6 ± 27.5 × 10^6 mL^{-1})	open-labeled	n = 10 (BMC)	3 months	• Feasible • CCS class angina ↓
Perin et al. (*Circulation* 2003/2004)	BMC (25 ± 6.3 × 10^6) vs. control	open-labeled	n = 11 (BMC) n = 9 (control)	12 months	• Myocardial perfusion ↑ • NYHA class, CCS class angina ↓ • VO$_2$ ↑
Intracoronary application					
TOPCARE-CHF (*Circulation* 2004, suppl.)	BMC (176 ± 79 × 10^6) vs. EPC (20.7 ± 9.7 × 10^6) vs. Control	open-labeled, randomized, crossover-design	n = 40 (EPC) n = 36 (BMC) n = 19 (control)	3 months	• LVEF ↑ • regional contractility ↑ • end-systolic volume ↓ • NYHA-class ↓
Köstering et al. (*Circulation* 2004, suppl.)	BMC (mean of 9 × 10^6)	open-labeled	n = 18 (BMC)	3 months	• LVEF ↑ • myocardial glucose uptake ↑ • VO$_2$ ↑
Hambrecht et al. (*Circulation* 2004, suppl.)	EPC after G-CSF stimulation (69.6 ± 15.5 × 10^6) vs. control	double-blind, randomized	n = 26 (total)	3 months	• LVEF ↑ • coronary vascular function ↑

itial results from experimental studies summarized in Tables 12.1 and 12.2 suggest that transplantation of different types of progenitor cells is capable of improving the repair of ischemic myocardium.

12.4.1
Stem/Progenitor Cell Therapy in Patients after Acute Myocardial Infarction

Strauer and coworkers (2002) were the first to perform intracoronary infusion of bone marrow-derived mononuclear cells (BMC), which were isolated by Ficoll gradient centrifugation and cultivated overnight. In 10 patients, BMC (mean 28×10^6 cells) were infused into a coronary artery at 5 to 9 days after acute myocardial infarction. In comparison to 10 non-randomized control patients, who did not undergo cell therapy or additional catheterization, BMC infusion enhanced regional infarct region perfusion as assessed by thallium scintigraphy. Moreover, the stroke volume, end-systolic volume and contractility indices were each improved after cell therapy.

In the TOPCARE-AMI trial (Assmus et al., 2002), patients were randomized to receive either BMC or EPC. The BMC from 50 mL aspirate were isolated by Ficoll gradient centrifugation and infused immediately after isolation. The EPC were *ex-vivo* expanded out of PBMC and cultivated for 3 days. BMC or EPC were infused after a mean of 4 days post infarction. BMC and EPC each significantly improved the global LVEF, as assessed by left ventricular angiography and compared to a non-randomized control patient collective. Functional improvement was confirmed by magnetic resonance imaging in a patient subgroup (Britten et al., 2003). The effect of EPC and BMC were comparable, suggesting that both cell types have the capacity to augment cardiac regeneration.

The results of an initial randomized trial – the BOOST study – were recently published by Wollert et al. (2004) whereby 60 patients with acute myocardial infarction were randomized after successful percutaneous coronary intervention (PCI) to receive either intracoronary infusion of BMC or standard treatment (no recatheterization, no placebo). The authors reported a significant increase in LVEF after 6 months in the group of patients receiving BMC, whereas there was no improvement in heart function in patients without stem cell therapy.

In another study, Kang et al. (2004) evaluated the effect of intracoronary infusion of leukocytes isolated from peripheral blood after G-CSF stimulation in a heterogeneous collective of seven patients, at between 2 and 270 days after myocardial infarction (the MAGIC trial). Patients included in this trial received a G-CSF injection for 4 days followed by PCI plus stent implantation and cell infusion. Whereas the LVEF, myocardial perfusion and exercise capacity were significantly improved after 6 months, five out of seven patients showed an in-stent restenosis at culprit lesion at this point of time. Thus, the investigators terminated the study after assessment of the initial seven patients. This unexpected high rate of in-stent restenosis might be due to the fact that patients, after acute myocardial infarction, underwent PCI plus stent implantation only after being stimulated for several days by G-CSF. It is likely, that stent implantation at a time point of high inflammatory activity at sites of culprit lesions may enhance the development of restenosis.

In contrast to most of the above-described trials, Kuethe et al. (2004) could not detect any effect on LVEF in five patients with large anterior infarction treated with BMC at a mean of 6 days after myocardial infarction. However, the extremely small number of patients studied clearly precluded any distinct conclusion.

Chen et al. (2004) reported the first randomized trial of autologous bone marrow MSC (BMSC) in 69 patients who underwent PCI within 12 h after myocardial infarction. The patients were randomized to receive either BMSC or a saline injection into the target coronary artery. At the 3-month follow-up, the LVEF showed a significant improvement from 49 ± 9% to 67 ± 11%, whereas there was no change in the control group. However, data on safety of this study were not reported.

Most recently, Fernández-Avilés et al. (2004) also showed that intracoronary infusion of BMC at 5 to 29 days after myocardial infarction improved the heart function of patients after 6 months, whereas no changes were found in a nonrandomized control group.

Direct intramuscular injection of isolated HSC, which express AC133$^+$, was performed by Stamm et al. (2003) during bypass surgery in six patients who suffered from acute myocardial infarction between 10 days and 3 months previously. CD133-positive cells were isolated from autologous bone marrow cells by magnetic beads and injected into a distinct akinetic area corresponding to the previously infarcted area. No adverse events were reported. Four of the six patients showed an increase in LVEF associated with an improvement in the previously non- or hypoperfused infarct zone.

12.4.2
Stem/Progenitor Cell Therapy in Patients with Chronic Ischemic Heart Failure

Chronic ischemic myocardium may persist in association with variable amounts of scar tissue after myocardial infarction. In many cases, endogenous angiogenesis and myocyte regeneration seem to be too inefficient to prevent cardiac remodeling after acute ischemia, leading to progressive left ventricular dilation. Stem cell therapy may constitute a new treatment for patients with chronic ischemic heart failure. To date, the results of three clinical trials have been published, and those of another three trials were presented in 2004 at the 17th Annual Meeting of the American Heart Association.

Initially, intramuscular injection of bone marrow-derived stem/progenitor cells by catheter-based transendocardial delivery was used as the route of application. Tse et al. (2003) were the first to report intramyocardial injection of complete BMNC via the NOGA catheter. Eight patients with stable angina which was refractory to maximum medical treatment were included. No procedure-related events such as arrhythmias or elevation in troponin T levels were reported. After 3 months, the LVEF – which had been almost normal at baseline – did not improve, but a significant functional improvement could be detected in regional wall thickening during pharmacological stress using MRI imaging.

The group of Perin et al. (2003, 2004) also used the NOGA catheter to guide injection by electromechanical mapping. Patients with end-stage heart disease ineligible for percutaneous or surgical intervention were included. At 12 months' follow-up – which was available for 11 treated and nine non-treated patients – there was a significant improvement both in symptoms (NYHA class and CCS class) and in exercise capacity (assessed by metabolic equivalents) in the treatment group as compared to controls. Furthermore, SPECT analysis revealed a significant reduction in myocardial ischemia in patients after stem cell treatment.

Finally, Fuchs et al. (2003) tested the effect of total unfractionated bone marrow in 10 no-option patients with advanced coronary artery disease. Again, CCS class and stress-induced ischemia were significantly improved after a 3-month follow-up.

At present, all published studies are limited by the small patient collective and by their design as pilot safety and feasibility studies, which precludes a randomized placebo-controlled design.

The details of three clinical trials which included larger, randomized patient populations were presented at the 17th AHA meeting in November 2004. In all three trials, patients with chronic ischemic cardiomyopathy were treated by intracoronary infusion of stem/progenitor cells. Assmus et al. (2004) evaluated the effect of intracoronary infusion of BMC or EPC compared to a randomized control group in patients with severe stable ischemic chronic heart disease and a previous (minimum 3 months) myocardial infarction. Transcoronary transplantation of progenitor cells was safe and did not induce clinically significant increases in levels of either troponin T or C-reactive protein. At 3 months' follow-up, there was a significant improvement in NYHA class. The LVEF was increased significantly in the BMC group (from $41 \pm 12\%$ to $44 \pm 12\%$, p <0.001) and – to a lesser extent – in the EPC group (from $42 \pm 11\%$ to $43 \pm 11\%$, p $= 0.03$), but remained unchanged in the control group ($44 \pm 14\%$ and $43 \pm 14\%$, p $= 0.34$). The absolute increase in global LVEF was significantly greater in patients receiving BMC compared to control patients. The improvement in global LVEF could be attributed to a significantly enhanced regional contractility of the area targeted by cell infusion. Interestingly, an improved LVEF was associated with profound reductions in atrial natriuretic peptide levels.

Köstering et al. (2004) reported the details of 18 patients who were transplanted with autologous BMC. All patients had suffered a myocardial infarction at a median of 27 months previously, treated by PCI. At 3 months after cell infusion, the LVEF was increased from $52 \pm 9\%$ to $60 \pm 7\%$, myocardial glucose uptake was higher, and O_2 uptake also showed a significant increase.

Hambrecht et al. (2004) initiated an ongoing randomized, double-blind, placebo-controlled study to evaluate the impact of intracoronary EPC-infusion on vasomotion and left ventricular function in patients after recanalization of a chronically occluded coronary artery by PCI and stent implantation. The number of EPC in the peripheral blood was increased by 4 days of G-CSF stimulation. In control patients, serum was used as sham therapy. EPC application improved coronary macro- and microvascular function at a 3 months' follow-up. Furthermore, MRI revealed an increase in LVEF of 14% in patients treated with EPC, whereas nonsignificant changes were observed in control patients. The authors speculated that EPC-associated recruitment of hibernating myocardium might be the consequence of improved coronary vasodilatory capacity and neovascularization.

Despite these encouraging results, it is clear that larger randomized, double-blinded trials must be conducted in order to confirm the results of these initial clinical trials. The question of whether the infused cells in the heart act preferentially by improving tissue perfusion or also by regenerating cardiac myocytes remains to be elucidated. Clearly, stem or progenitor cells may exert synergistic effects by enhancing both neovascularization and cardiac regeneration. Moreover, the release of paracrine mediators by incorporated stem or progenitor cells may amplify the response by attracting endogenous circulating progenitor cells and/or tissue resident stem cells.

Abbreviations/Acronyms

bFGF	basic fibroblast growth factor
BMC	bone marrow-derived mononuclear cells
CCS	Canadian Cardiovascular Society
CEPC	circulating bone marrow-derived endothelial progenitor cells
DIL	Dil-acetylated LDL
eNOS	endothelial nitric oxide synthase
EPC	endothelial progenitor cells
G-CSF	granulocyte colony-stimulating-factor
HIF-1	hypoxia-inducible factor-1
HSC	hematopoietic stem cells
LVEF	left ventricular ejection fraction
MCP	monocyte chemoattractant protein
MSCs	mesenchymal stem cells
NYHA	New York Heart Association
PBMC	peripheral blood mononuclear cells
PCI	percutaneous coronary intervention
SDF-1	stromal cell-derived factor-1
SPECT	single photon emission computed tomography
VEGF	vascular endothelial growth factor
vWF	von Willebrand factor

References

Aicher, A., et al. (**2003**). Essential role of endothelial nitric oxide synthase for mobilization of stem and progenitor cells. *Nat Med* 9: 1370–1376.

Asahara, T., et al. (**1997**). Isolation of putative progenitor endothelial cells for angiogenesis. *Science* 275(5302): 964–967.

Askari, A.T., et al. (**2003**). Effect of stromal-cell-derived factor 1 on stem-cell homing and tissue regeneration in ischaemic cardiomyopathy. *Lancet* 362(9385): 697–703.

Assmus, B., et al. (**2004**). Transcoronary transplantation of progenitor cells and recovery of left ventricular function in patients with chronic ischemic heart disease: results of a randomized, controlled trial. *Circulation* 110(17): abstract No. 1142.

Assmus, B., et al. (**2002**). Transplantation of Progenitor Cells and Regeneration Enhancement in Acute Myocardial Infarction (TOPCARE-AMI). *Circulation* 106(24): 3009–3017.

Assmus, B., et al. (**2003**). HMG-CoA reductase inhibitors reduce senescence and increase proliferation of endothelial progenitor cells via regulation of cell cycle regulatory genes. *Circ Res* 92: 1049–1055.

Badorff, C., et al. (**2003**). Transdifferentiation of blood-derived human adult endothelial progenitor cells into functionally active cardiomyocytes. *Circulation* 107(7): 1024–1032.

Beltrami, A.P., et al. (**2003**). Adult cardiac stem cells are multipotent and support myocardial regeneration. *Cell* 114(6): 763–776.

Britten, M.B., et al. (**2003**). Infarct remodeling after intracoronary progenitor cell treatment in patients with acute myocardial infarction (TOPCARE-AMI): mechanistic insights from serial contrast-enhanced magnetic resonance imaging. *Circulation* 108(18): 2212–2218.

Camargo, F.D., et al. (**2003**). Single hematopoietic stem cells generate skeletal muscle through myeloid intermediates. *Nat Med* 9: 1520–1527.

Carlos, T.M. and J.M. Harlan (**1994**). Leukocyte-endothelial adhesion molecules. *Blood* 84: 2068–2101.

Ceradini, D.J., et al. (**2004**). Progenitor cell trafficking is regulated by hypoxic gradients through HIF-1 induction of SDF-1. *Nat Med* 10(8): 858–864.

Chavakis, E., et al. (**2005**). Role of β2-integrins for homing and neovascularization capacity of endothelial progenitor cells. *J Exp Med* 201(1): 63–72.

Chen, S.L., et al. (**2004**). Effect on left ventricular function of intracoronary transplantation of autologous bone marrow mesenchymal stem cell in patients with acute myocardial infarction. *Am J Cardiol* 94(1): 92–95.

Condorelli, G., et al. (**2001**). Cardiomyocytes induce endothelial cells to trans-differentiate into cardiac muscle: implications for myocardium regeneration. *Proc Natl Acad Sci USA* 98(19): 10733–10738.

Edelberg, J.M., et al. (**2002**). Young adult bone marrow-derived endothelial precursor cells restore aging-impaired cardiac angiogenic function. *Circ Res* 90(10): E89–E93.

Fernández-Avilés, F., et al. (**2004**). Experimental and clinical regenerative capability of human bone marrow cells after myocardial infarction. *Circ Res* 95(7): 742–748.

Fuchs, S., et al. (**2003**). Catheter-based autologous bone marrow myocardial injection in no-option patients with advanced coronary artery disease: a feasibility study. *J Am Coll Cardiol* 41(10): 1721–1724.

Fujiyama, S., et al. (**2003**). Bone marrow monocyte lineage cells adhere on injured endothelium in a monocyte chemoattractant protein-1-dependent manner and accelerate reendothelialization as endothelial progenitor cells. *Circ Res* 93(10): 980–989.

Gehling, U.M., et al. (**2000**). In vitro differentiation of endothelial cells from AC133-positive progenitor cells. *Blood* 95(10): 3106–3112.

Hambrecht, R., et al. (**2004**). Transplantation of blood-derived progenitor cells after recanalization of chronic coronary artery occlusions: impact on coronary vasomotion and left ventricular remodeling. *Circulation* 110(17): abstract No. 1562.

Handgretinger, R., et al. (**2003**). Biology and plasticity of CD133+ hematopoietic stem cells. *Ann N Y Acad Sci* 996: 141–151.

Hillebrands, J.L., et al. (**2002**). Bone marrow does not contribute substantially to endothelial-cell replacement in transplant arteriosclerosis. *Nat Med* 8: 194–195.

Jackson, K.A., et al. (**2001**). Regeneration of ischemic cardiac muscle and vascular endothelium by adult stem cells. *J Clin Invest* 107(11): 1395–1402.

Kalka, C., et al. (**2000**). Transplantation of ex vivo expanded endothelial progenitor cells for therapeutic neovascularization. *Proc Natl Acad Sci USA* 97(7): 3422–3427.

Kamihata, H., et al. (**2001**). Implantation of bone marrow mononuclear cells into ischemic myocardium enhances collateral perfusion and regional function via side supply of angioblasts, angiogenic ligands, and cytokines. *Circulation* 104(9): 1046–1052.

Kang, H.J., et al. (**2004**). Effects of intracoronary infusion of peripheral blood stem-cells mobilised with granulocyte-colony stimulating factor on left ventricular systolic function and restenosis after coronary stenting in myocardial infarction: the MAGIC cell randomised clinical trial. *Lancet* 363(9411): 751–756.

Kawada, H., et al. (**2004**). Nonhematopoietic mesenchymal stem cells can be mobilized and differentiate into cardiomyocytes after myocardial infarction. *Blood* 104(12): 3581–3587.

Kawamoto, A., et al. (**2001**). Therapeutic potential of ex vivo expanded endothelial progenitor cells for myocardial ischemia. *Circulation* 103(5): 634–637.

Kimura, T., et al. (**2004**). The sphingosine 1-phosphate (S1P) receptor agonist FTY720 supports CXCR4-dependent migration and bone marrow homing of human CD34+ progenitor cells. *Blood* 26: 26.

Kocher, A.A., et al. (**2001**). Neovascularization of ischemic myocardium by human bone-marrow-derived angioblasts prevents cardiomyocyte apoptosis, reduces remodeling and improves cardiac function. *Nat Med* 7(4): 430–436.

Kollet, O., et al. (**2001**). Rapid and efficient homing of human CD34(+)CD38(-/low)CXCR4(+) stem and progenitor cells to the bone marrow and spleen of NOD/SCID and NOD/SCID/B2m(null) mice. *Blood* 97: 3283–3291.

Köstering, M., et al. (**2004**). Regeneration of human infarcted heart muscle by intracoronary autologous mononuclear bone marrow cells in chronic coronary artery disease. *Circulation* 110(17): abstract No. 1141.

Kuethe, F., et al. (**2004**). Lack of regeneration of myocardium by autologous intracoronary mononuclear bone marrow cell transplantation in humans with large anterior myocardial infarctions. *Int J Cardiol* 91(1): 123–127.

Lapidot, T. (**2001**). Mechanism of human stem cell migration and repopulation of NOD/SCID and B2mnull NOD/SCID mice. The role of SDF-1/CXCR4 interactions. *Ann N Y Acad Sci* 938: 83–95.

Lee, S.H., et al. (**2000**). Early expression of angiogenesis factors in acute myocardial ischemia and infarction. *N Engl J Med* 342(9): 626–633.

Lin, Y., et al. (**2000**). Origins of circulating endothelial cells and endothelial outgrowth from blood. *J Clin Invest* 105(1): 71–77.

Medvinsky, A. and A. Smith (**2003**). Stem cells: Fusion brings down barriers. *Nature* 422(6934): 823–825.

Muller, W.A. (**2002**). Leukocyte-endothelial cell interactions in the inflammatory response. *Lab Invest* 82: 521–533.

Musaro, A., et al. (**2004**). Stem cell-mediated muscle regeneration is enhanced by local isoform of insulin-like growth factor 1. *Proc Natl Acad Sci USA* 101(5): 1206–1210.

Mutin, M., et al. (**1999**). Direct evidence of endothelial injury in acute myocardial infarction and unstable angina by demonstration of circulating endothelial cells. *Blood* 93(9): 2951–2958.

Oh, H., et al. (**2003**). Cardiac progenitor cells from adult myocardium: Homing, differentiation, and fusion after infarction. *Proc Natl Acad Sci USA* 100: 12313–12318.

Orlic, D., et al. (**2001**). Bone marrow cells regenerate infarcted myocardium. *Nature* 410(6829): 701–705.

Peled, A., et al. (**1999**). The chemokine SDF-1 stimulates integrin-mediated arrest of CD34(+) cells on vascular endothelium under shear flow. *J Clin Invest* 104: 1199–1211.

Perin, E.C., et al. (**2003**). Transendocardial, autologous bone marrow cell transplantation for severe, chronic ischemic heart failure. *Circulation* 107(18): 2294–2302.

Perin, E.C., et al. (**2004**). Improved exercise capacity and ischemia 6 and 12 months after transendocardial injection of autologous bone marrow mononuclear cells for ischemic cardiomyopathy. *Circulation* 110(11 Suppl 1): II213–II218.

Rehman, J., et al. (**2003**). Peripheral blood endothelial progenitor cells are derived from monocyte/macrophages and secrete angiogenic growth factors. *Circulation* 107(8): 1164–1169.

Reyes, M., et al. (**2002**). Origin of endothelial progenitors in human postnatal bone marrow. *J Clin Invest* 109(3): 337–346.

Rookmaaker, M.B., et al. (**2003**). Endothelial progenitor cells: mainly derived from the monocyte/macrophage-containing CD34-mononuclear cell population and only in part from the hematopoietic stem cell-containing CD34+ mononuclear cell population. *Circulation* 108: e150.

Schmeisser, A., et al. (**2001**). Monocytes coexpress endothelial and macrophagocytic lineage markers and form cord-like structures in Matrigel under angiogenic conditions. *Cardiovasc Res* 49(3): 671–680.

Scott, L.M., et al. (**2003**). Deletion of alpha4 integrins from adult hematopoietic cells reveals roles in homeostasis, regeneration, and homing. *Mol Cell Biol* 23: 9349–9360.

Shi, Q., et al. (**1998**). Evidence for circulating bone marrow-derived endothelial cells. *Blood* 92(2): 362–367.

Shintani, S., et al. (**2001**). Augmentation of postnatal neovascularization with autologous bone marrow transplantation. *Circulation* 103(6): 897–903.

Springer, T.A. (**1994**). Traffic signals for lymphocyte recirculation and leukocyte emigration: the multistep paradigm. *Cell* 76: 301–314.

Stamm, C., et al. (**2003**). Autologous bone-marrow stem-cell transplantation for myocardial regeneration. *Lancet* 361(9351): 45–46.

Strauer, B.E., et al. (**2002**). Repair of infarcted myocardium by autologous intracoronary mononuclear bone marrow cell transplantation in humans. *Circulation* 106(15): 1913–1918.

Toma, C., et al. (**2002**). Human mesenchymal stem cells differentiate to a cardiomyocyte phenotype in the adult murine heart. *Circulation* 105(1): 93–98.

Tse, H.F., et al. (**2003**). Angiogenesis in ischaemic myocardium by intramyocardial autologous bone marrow mononuclear cell implantation. *Lancet* 361(9351): 47–49.

Urbich, C., et al. (**2003**). Relevance of monocytic features for neovascularization capacity of circulating endothelial progenitor cells. *Circulation* 108: 2511–2516.

Urbich, C., et al. (**2005**). Cathepsin L is required for endothelial progenitor cell-induced neovascularization. *Nat Med* 11: 206–213.

Vajkoczy, P., et al. (**2003**). Multistep nature of microvascular recruitment of ex vivo-expanded embryonic endothelial progenitor cells during tumor angiogenesis. *J Exp Med* 197: 1755–1765.

Wollert, K.C., et al. (**2004**). Intracoronary autologous bone-marrow cell transfer after myocardial infarction: the BOOST randomised controlled clinical trial. *Lancet* 364(9429): 141–148.

Wright, D.E., et al. (**2002**). Hematopoietic stem cells are uniquely selective in their migratory response to chemokines. *J Exp Med* 195(9): 1145–1154.

Wurmser, A.E., et al. (**2004**). Cell fusion-independent differentiation of neural stem cells to the endothelial lineage. *Nature* 430(6997): 350–356.

Yamaguchi, J., et al. (**2003**). Stromal cell-derived factor-1 effects on ex vivo expanded endothelial progenitor cell recruitment for ischemic neovascularization. *Circulation* 107(9): 1322–1328.

Yeh, E.T., et al. (**2003**). Transdifferentiation of human peripheral blood CD34+-enriched cell population into cardiomyocytes, endothelial cells, and smooth muscle cells in vivo. *Circulation* 108(17): 2070–2073.

Zhao, Y., et al. (**2003**). A human peripheral blood monocyte-derived subset acts as pluripotent stem cells. *Proc Natl Acad Sci USA* 100(5): 2426–2431.

13
Stem Cells and Bypass Grafting for Myocardial and Vascular Regeneration

Christof Stamm, Dirk Strunk, and Gustav Steinhoff

13.1
Introduction

Cell therapy for myocardial regeneration is an exciting new field of medical research that has the potential to revolutionize cardiovascular medicine. Despite significant improvements in emergency treatment, myocardial infarction leads to a net loss of contractile tissue in many patients with coronary artery disease (CAD). Often, this is the beginning of a downward spiral towards congestive heart failure and life-threatening arrhythmia. Other than heart transplantation with its obvious limitations, current therapeutic means aim at preventing further episodes of myocardial ischemia and at enabling the organism to survive with a heart that is working at a fraction of its original capacity. They are far from representing a cure. In this situation, it is understandable that cardiac stem cell therapy attracts considerable attention and raises many hopes. In order to judge adequately both the potential benefits and the limitations of cardiac cell therapy, some understanding of the mechanism and the consequences of myocardial infarction and its current treatment concepts is needed. Although in Chapter 12 attention is focused on cell therapy during the acute phase of myocardial infarction, this chapter centers on cell therapy in conjunction with coronary artery bypass grafting (CABG) surgery, and therefore on patients with chronic ischemic disease. First, the basic pathophysiology of ischemic heart disease will be revised; second, the principles of CABG surgery will be summarized; third, current experimental evidence for the regeneration of infarcted myocardium will be reviewed; and fourth, a description of the preliminary clinical experience will be provided. Unfortunately, such an overview cannot be comprehensive. Cardiac cell therapy is a young and rapidly developing field in which knowledge progresses and changes with great rapidity. To date, there is no "textbook knowledge" available, and the interested researcher and clinician will need to follow the original literature in order to keep up to date with the subject.

Stem Cell Transplantation. Biology, Processing, and Therapy.
Edited by Anthony D. Ho, Ronald Hoffman, and Esmail D. Zanjani
Copyright © 2006 WILEY-VCH Verlag GmbH & Co. KGaA, Weinheim
ISBN: 3-527-31018-5

13.2
Coronary Artery Disease

Cardiovascular disease remains the number one cause of death in industrialized countries, the most important being ischemic heart disease – that is, CAD. Although the risk factors for coronary arteriosclerosis have been well established, it remains unclear why a particular individual develops CAD while another with a similar combination of risk factor does not. Other mysterious phenomena are the extent and the rate of progression of the disease. In some patients, an obstructive coronary artery lesion remains unchanged for many years, while in others it can progress to subtotal occlusion of the vessel in a matter of a few months. Similarly, many patients present with isolated proximal stenoses of one or more coronary arteries while the more distal part of the vessel appear completely normal, whereas other patients develop diffuse disease of the entire coronary vascular tree.

13.2.1
Myocardial Ischemia

Even at rest, the heart must perform a vast amount of work in order to maintain the circulation. Therefore, oxygen consumption is very high as compared with other organs, and the difference in coronary arterial and venous oxygen content (AVDO$_2$) is large, even in a resting individual. During exercise, cardiac output – and hence work load – can rapidly increase several-fold, and the myocardial demand for oxygen rises accordingly. Because AVDO$_2$ is already very high at baseline, the increased demand can only be met by an increase in coronary blood flow. This explains why the heart has such a low tolerance to coronary perfusion problems. A complex network of intrinsic and extrinsic modulators serves to adjust coronary flow when the work load changes, but this regulation is effectively shut down when a relevant obstruction to coronary flow is present. Once the narrowing of a major coronary artery exceeds 50 % of its cross-sectional area, blood flow to the downstream myocardium becomes insufficient when demand increases. A narrowing of >90 % is often associated with angina at mild exertion. Often, acute myocardial infarction occurs when the endothelial surface of an atheromatous plaque ruptures, exposing highly thrombogenic subendothelial matrix components and thus inducing platelet aggregation and thrombus formation. The resulting complete interruption of coronary blood flow leads to an immediate reduction of contractility in the downstream myocardium. Within minutes, oxygen tension falls to almost zero, and the highly energy-dependent cardiomyocytes react by switching to anaerobic energy production. This attempt at compensation is not sufficient, and the concentration of energy-rich phosphates inevitably declines, while protons, ADP, phosphate, potassium, and calcium accumulate and contribute to the complete breakdown of the fragile cellular energy balance. Unless the myocardium is reperfused within a few hours, a substantial number of cardiomyocytes will succumb to necrosis and apoptosis. The release of intracellular contents from necrotic cardiomyocytes attracts phagocytic leukocytes, leading

- *Acute myocardial infarction*
   (immediately)

- *Early after myocardial infarction*
   (< 2 weeks)

- -

- *Myocardial remodelling phase*
   (2 weeks – several months)

- *Completed remodelling, scar*
   ischemic cardiomyopathy
   (>6 months)

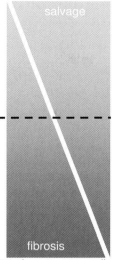

Figure 13.1 Acute myocardial infarction initiates a variety of simultaneous and sequential events that begin with myocyte necrosis and end with formation of scar tissue. Routine CABG operations are normally not performed during the early phase after myocardial infarction. If myocardial necrosis is already established and cannot be prevented by emergency CABG, the patient is usually allowed to recover for at least 2 weeks before CABG is performed (interrupted line). Although this has never been studied systematically, it is assumed that successful myocardial cell therapy becomes more difficult the longer the time since the infarction event.

to significant inflammatory infiltration of the ischemic tissue. For some time, lytic processes dominate and render the ischemic myocardium very fragile and prone to sudden rupture. Within several weeks, however, collagenous scar tissue develops and reinforces the necrotic myocardium (Fig. 13.1).

13.3
Indications for CABG Surgery

The treatment of acute myocardial infarction aims at immediate re-establishment of blood flow in the occluded coronary artery. For some time, this has mainly been attempted by intravenous administration of thrombolytic enzymes, but their efficacy is limited and there is a substantial risk of hemorrhagic complications. With the wider distribution of interventional cardiac catheterization facilities, percutaneous transluminal coronary angioplasty (PTCA) and stent placement are now the "gold standard" treatment of acute myocardial infarction. Emergency CABG surgery is limited to patients with unstable angina or beginning myocardial infarction in whom the anatomy of the lesion does not permit successful stent placement. If the 6-h window has been missed and myocardial necrosis is already irreversible, medical treatment and close observation is usually recommended. Once the patient has stabilized, CABG – if still indicated – is performed later than 2 weeks following AMI. Due to the development of novel stent models (most recently, the so-called drug-eluting stents with very good long-term patency rates), the indications for CABG surgery are constantly changing. The guidelines published jointly by the American College of Cardiology and the American Heart Association in 1999 and updated in 2004 [1] include the following indications for CABG as based on class I evidence:

- In asymptomatic patients and patients with stable angina: left main coronary artery stenosis or three-vessel disease or an anatomic equivalent, especially in patients with reduced left ventricular (LV) function. In one- or two-vessel disease, the indication is limited to certain subsets of patients. In unstable angina, the indication is limited to left main coronary artery stenosis or its anatomic equivalent, and in ongoing ischemia not responsive to maximal non-surgical therapy, while in ST-elevation infarction there is no class-I indication listed.
- In patients with poor LV function (currently the only group of candidates for cell therapy), the class-I indications include left main coronary artery stenosis or its equivalent, but also proximal left anterior descending (LAD) artery stenosis with two- or three-vessel disease.

As mentioned earlier, these indications are constantly changing and may vary significantly between individual centers.

13.3.1
Outcome of CABG Surgery

The perioperative mortality risk of CABG surgery varies greatly depending on numerous preoperative variables, most notably "emergency operation", "age >80 years", "impaired LV function", and "previous CABG surgery". Elaborate scoring systems have been developed to assess the risk for the individual patient. Generally, the operative risk for a patient aged <75 years, with normal LV function undergoing elective CABG surgery should not exceed 1–2%. Data on survival after 5 years range between 80% and 95%, with an average mortality of 10.2%, and the 10-year survival rate is said to range between 70% and 80%. When compared with the results of medical treatment, it is usually assumed that the poorer the LV function, the greater is the potential advantage of CABG surgery. Nevertheless, the long-term survival in patients with impaired LV function after CABG is still unsatisfactory. While in patients with normal LV function the 5-year mortality rate is 8.5%, it is twice as high (16.5%) in patients with abnormal LV function. In a recent multicenter study, 5-year survival after CABG was 91.2% in patients with left ventricular ejection fraction (LVEF) >50%, 85.5% with LVEF between 30–50%, and 77.7% in patients with LVEF <30% [2]. In another comparative study, 3-year survival in patients with LVEF <35% was around 70% following CABG as well as balloon intervention [3]. Without any invasive treatment, the survival of patients with myocardial infarction and LVEF <30% is less than 40% after 5 years. In the presence of concomitant disease, the survival rate drops even further. For instance, a recent large cohort study reported 8-year survival rates of 45% (after CABG), 32% (after balloon intervention), and 29% (no revascularization) in patients with additional renal disease [4]. Taken together, these data dramatically illustrate the urgent need for novel therapeutic measures, in addition to CABG surgery and/or interventional treatment, for patients with impaired LV function.

13.3.2
Technique of CABG Surgery

The basic surgical technique of CABG has not fundamentally changed for several decades. Typically, the chest is opened via a median sternotomy, the pericardium is opened, and the heart is exposed. Following cannulation of the ascending aorta and the right atrium, extracorporeal circulation is initiated. While the heart is still perfused and beating, the stenotic or occluded coronary arteries are exposed and prepared. Then, the aorta is clamped and a cardioplegic solution delivered into the coronary system; this arrests the heart and protects the myocardium from ischemic injury. The coronary artery is incised and a bypass graft (i.e., saphenous vein, internal mammary artery, or others) is anastomosed to this incision. Once all bypass-to-coronary artery anastomoses have been constructed, the aortic cross-clamp is released and the heart reperfused. Using a side-biting instrument, the anterior portion of the ascending aorta is partially clamped. Perforations are placed in the aortic wall, and the other end of the bypass graft is anastomosed to the aorta. Upon completion, blood flow through the bypass grafts is released and the patient is weaned from extracorporeal circulation. In a patient with impaired LV function, positive inotropic drugs may be necessary to augment contractility, but if this is not sufficient then intraaortic balloon counterpulsation or some other type of mechanical device can be used to support heart function until the myocardium has recovered.

Some surgeons prefer to perform CABG surgery on the beating heart, without the aid of extracorporeal circulation. This is straightforward when working on coronary arteries on the anterior surface of the heart, but can become quite problematic when the beating heart has to be lifted out of the chest in order to reach vessels on the posterior surface. This so-called "off-pump" CABG surgery clearly has its merits by avoiding the potentially harmful effects of extracorporeal circulation, but it is still used only in a small fraction of cases.

13.4
The Rationale for Cell Therapy in CABG Patients

The CABG operation in a patient with previous myocardial infarction primarily serves to bypass obstructions that are present in other coronary arteries and pose a risk for further ischemic episodes. In addition, the previously infarcted vessel, if still occluded, is revascularized when there is reason to believe that there is still vital, albeit hibernating myocardium that can be re-recruited for contractile work. SPECT or PET scans can be used to determine and quantify the presence of still vital myocardium in infarcted tissue, but in many patients CABG is indicated anyway and the infarcted vessel is bypassed if it is patent, without prior viability determination. However, a relevant amount of hibernating myocardium is present only in the minority of patients, and the improvement in LV function after bypass grafting therefore varies greatly and is generally small. In the past, the gain in LVEF by CABG surgery was assumed to be approximately 10%. With changing indications for CABG, an average improvement of 5% is nowadays more realistic. Based on our own experience with patients undergoing CABG some time after myocardial infarction, LVEF improves in very few patients in a clinically relevant fashion. Consequently, the foremost reason for myocardial cell therapy is the formation of new contractile tissue in the infarct area. The second component is the revascularization of the infarcted myocardium. In the chronic phase after myocardial infarction, perfusion of the infarct area remains impaired, even if the occluded coronary artery is bypassed. Long-standing absence of coronary blood flow and fibrotic scar formation appear to result in the obliteration of much of the myocardial microvasculature, and growth of pre-existing collateral vessels does usually not compensate for the loss of microcirculation. Therefore, cell therapy also aims at revascularization of ischemic myocardium through neoangiogenesis. Theoretically, any remaining hibernating native cardiomyocytes may hence be re-recruited, the angina may be relieved, and the blood supply to (hypothetic) contractile neo-cells may be established.

13.5
The Role of Bone Marrow Cells

13.5.1
Bone Marrow Cells and Angiogenesis

During embryonic development, the primary vascular plexus is formed by hemangioblasts – stem cells which are capable of generating both hematopoietic progeny and endothelial cells – in a process termed *vasculogenesis*. Further blood vessels are generated by both sprouting and nonsprouting angiogenesis, finally leading to the complex functional adult circulatory system [5]. Until recently, only two mechanisms of postembryonic vascular remodeling have been recognized.

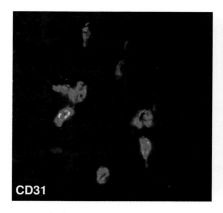

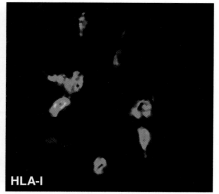

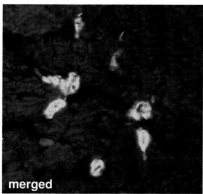

Figure 13.2 CD133⁺ bone marrow (BM) cells can differentiate along the endothelial-vasculogenic axis. The progeny of CD133⁺ human BM cells that were injected into damaged mouse myocardium are illustrated. The cells are embedded between cardiac myofibers, and clearly coexpress HLA-1 and the endothelial cell marker CD31. However, it cannot be clearly seen whether functional blood vessels have formed, illustrating the fact that expression of one phenotype-specific marker protein does not prove formation of a functioning single cell or tissue. (Illustration courtesy of M. Nan PhD, University of Rostock.)

Angiogenesis, the proliferative outgrowth of local capillaries, is one way to reinforce perfusion. Angiogenesis can occur under various conditions, including ischemia. In case of myocardial ischemia due to the occlusion of a coronary artery, pre-existing small collateral vessels also bear the capacity to enlarge in a process termed *arteriogenesis*. It has long been assumed that both mechanisms are mainly due to the local proliferation of resident cells. The advent of cellular therapy of ischemic organ damage has introduced *neoangiogenesis* (sometimes also termed vasculogenesis) due to immigrating stem cells and progenitors as a third possible mechanism operative to improved perfusion of the adult damaged heart. Accumulating evidence indicates that immigrating (stem) cells can truly differentiate along endothelial lineage, and also provide paracrine support in these three courses of action during regenerative vascular remodeling (Fig. 13.2) [6–8].

Putative progenitors for therapeutic angiogenesis have been isolated by one group from adult human peripheral blood based on their expression of CD34, a marker molecule shared by microvascular endothelial cells and hematopoietic stem cells (HSCs) [9]. The same group provided proof of concept by the transplantation of genetically marked mouse bone marrow (BM) into recipient mice that were subsequently subjected to five distinct models of vascular remodeling, including myocardial ischemia [10]. In this particular system, transgenic mice constitutively expressing beta-galactosidase under the transcriptional regulation of an endothelial cell-specific promoter were used as donors to replace the BM in the recipient animals. Definitively BM-derived endothelial (progenitor) cells were found in reproductive organ tissues as well as in healing cutaneous wounds one week after punch biopsy. Marrow-derived endothelial progenitor cells were found to incorporate into capillaries among skeletal myocytes in an additional test for peripheral post-ischemic regeneration after hindlimb ischemia, as well as into foci of neovascularization at the border of an infarct after permanent ligation of the anterior descending artery [10]. Most importantly, direct injection of the BM mononuclear cell fraction into rat models of myocardial ischemia increased the capillary density [11, 12]. An analysis of the effects of blood and BM mononuclear cell implantation into ischemic myocardium in pigs further revealed that the stem cell effects are limited not only to angiogenesis and improved collateral perfusion, but also include the supply of regulatory cytokines [13, 14]. Concerns exist, however, regarding limited efficiency owing to the minute numbers of stem cells in small sample volumes of non-enriched blood and BM that are delivered intramyocardially and the risk of foreign tissue differentiation following local stroma cell injections. Kocher et al. [15] circumvented this problem by using positively selected CD34$^+$/133$^+$ cells from human donors after stem cell mobilization with granulocyte colony-stimulating factor (G-CSF) for intravenous injection after permanent ligation of the left anterior descending coronary artery in nude rats, resulting in a five-fold increase in the number of capillaries compared to control. As a result of this stem cell-mediated angiogenesis, which was attributed to the content of marrow-derived angioblasts, the authors also found an approximately 20% increase in LVEF and cardiac index, together with a reduced severity of ventricular remodeling in human CD34-treated compared to control ischemic animals [15].

Another candidate cell population for the regeneration of ischemic cardiac muscle and vascular endothelium are CD45$^+$ hematopoietic CD34^{LOW}/c-kit$^+$, so-called side population stem cells with a specific Hoechst 33342 DNA dye efflux pattern [16, 17]. Orlic et al. [18] used an alternative method to enrich putative regenerative stem cells for local application by depleting unwanted cell lineages prior to enrichment for expression of the stem cell factor receptor c-Kit from murine BM. Thus-concentrated cells, which were considered to represent HSCs, were observed to incorporate not only into vascular structures but also dominantly led to myocardial regeneration [18]. Subsequent experiments by this group employed mobilization of stem cells by G-CSF prior to experimental myocardial infarction, which also led to a significant increase in vascular density within the scar, a reduction in mortality, and a significant reduction in infarct size [19].

Although the evidence that angiogenesis occurs in ischemic myocardium is convincing, this new therapeutic option also has a potential for serious side effects [20]. Most importantly, BM-derived endothelial cells were found as part of the tumor neovasculature in experimental colon cancer [10]. This finding might suggest a risk to trigger the growth of silent tumors by systemic use of pro-angio-genic stem cell therapy.

13.5.2
Bone Marrow Cells and Myogenesis

While the pro-angiogenic effect of marrow-derived stem cells appears to be well established, stem cell-mediated myogenesis remains a matter of debate. The traditional view implies that ischemic damage to the myocardium can only be compensated by hypertrophy, not hyperplasia, of surrounding cardiomyocytes. This dogma has recently been challenged, and intramyocardial as well as extramyocardial sources of regenerating contractile cells have been suggested [21]. Cardiomyocyte proliferation has been described, although only very infrequently [22, 23]. Furthermore, the existence of cardiomyocytes of noncardiac origin has been suggested by chimerism analyses after transplantation [24–26], though the biologic relevance of some of these data has been questioned [27, 28].

The notion that BM cells can regenerate infarcted myocardium led to great excitement. In their landmark paper, Orlic et al. [29] described that the injection of genetically labeled murine Lin^{NEG}/c-kit$^+$ stem cells (isolated from mouse BM by depletion of committed cells, and further enriched for expression of c-Kit) led to the formation of new myocardium, occupying two-thirds of the infarct region within 9 days. This report initiated a wave of enthusiasm, but also invoked critical discussion. The data were interpreted to indicate transdifferentiation of adult HSCs by crossing lineage boundaries [30]. However, the fact that cells are derived from BM does not necessarily prove that they are hematopoietic in origin, especially in the light of growing knowledge about mesenchymal, nonhematopoietic stem cells within the BM. The recognition of cell fusion as a common phenomenon in some artificial transplant models for regeneration of ischemic tissue has added to the controversy [31, 32]. From the clinicians' point of view this was no surprise, since cell fusion is an intrinsic characteristic of contractile cells. Multinucleated skeletal myotubes are a classic example of cell fusion, and cardiomyocytes have long been known to form a large syncytial union.

More serious concerns were raised by two recent publications, which could not reproduce the promising *in-vivo* transdifferentiation data. Using a modified Lin$^+$ depletion protocol for stem cell enrichment in an otherwise similar myocardial ischemia model, Balsam et al. [33] found abundant GFP$^+$ cells in the myocardium after 10 days, which nearly disappeared until day 30. The remaining donor cells lacked cardiac tissue-specific markers, and instead adopted only hematopoietic fates, as indicated by the expression of CD45. Murry et al. [34] used both cardiomyocyte-restricted and ubiquitously expressed reporter transgenes to follow murine Lin^{NEG}/c-kit$^+$ stem cells after transplantation into healthy and injured mouse hearts, and could not find evidence for relevant differentiation into cardiomyocytes. In defense of the initial report, some have argued that: (i) the cell isolation protocols were not completely identical; and (ii) both groups nevertheless observed some functional improvement in cell-treated hearts. However, it cannot been denied that the evidence for myogenesis based on hematopoietic adult stem cells myogenesis is extremely controversial [21, 35–37]. Very recently, a direct side-by-side comparison of human CD133$^+$ BM cells and human skeletal myoblasts in a myocardial ischemia model in immunoincompetent rats demonstrated similar functional improvement in both groups, although only the myoblasts reached robust engraftment. This finding underscores our limited understanding of how stem cells can elicit an improvement in heart function.

In contrast, the myogenic potential of stroma cell-derived mesenchymal stem cells (MSCs) is much better documented. Stroma cells are usually isolated based on their ability to adhere to plastic, not by selection for expression of certain surface markers. Their number in primary marrow aspirates is low, but they readily multiply for numerous cycles in culture, without apparent genotypic and phenotypic changes. Several years ago, Wakitani et al. [38] reported the *in-vitro* development of myogenic cells from rat BM MSCs exposed to the DNA-demethylating agent 5-azacytidine, and Makino et al. [39] isolated a cardiomyogenic cell line from murine BM stromal cells that were treated with 5-azacytidine and screened for spontaneous beating. Those cells which connected with adjoining cells formed myotube-like structures, and beat spontaneously and synchronously. They expressed various cardiomyocyte-specific proteins, had a cardiomyocyte-like ultrastructure, and generated several types of sinus node-like and ventricular cell-like action potentials. When isogenic BM stromal cells are implanted into rat hearts, they appear to become integrated into the cardiac myofibers, assume the histologic phenotype of cardiomyocytes, express connexins, and form gap junctions with native cardiomyocytes [40, 41]. Again, epigenetic modification with 5-azacytidine is believed to facilitate differentiation towards a cardiomyocyte phenotype *in vivo* [42]. Human MSCs derived from the BM of volunteers have also been injected in hearts of immunodeficient mice, and again it was observed that they assume cardiomyocyte morphology and express various cardiomyocyte-specific proteins [43].

Under different cultivation conditions, MSCs readily assume an osteoblast, chondrocyte, or adipocyte phenotype. In fact, preclinical research on the regeneration of skeletal components is much more advanced than that on cardiovascular applications. It is therefore no surprise that, when unmodified MSCs are implanted the heart, they may form islets resembling chondrogenic or osteogenic tissue. To date, there is very little – if any – information on stroma cell-surface markers that might be helpful in identifying subpopulations with a particular potential for myogenic differentiation. It is therefore still unclear whether unmodified stroma cells that were expanded *in vitro* following simple isolation by plastic adherence will ultimately be useful in clinical protocols, whether a certain pro-myogenic subpopulation will be identified, or whether epigenetic reprogramming prior to implantation will be necessary for functionally relevant myocardial regeneration in humans.

13.6
Combination of (Stem) Cell Treatment with CABG Surgery

CABG patients were among the first to be included in clinical trails of cell therapy for myocardial regeneration (Table 13.1). The most obvious reason is that the infarcted myocardium can be readily accessed during the operation, providing a unique opportunity to deliver cells into the center or border zone of the infarcted tissue by rather simple means.

13.6.1
Skeletal Myoblasts

The first clinical application of myocardial cell therapy in a broader sense was reported by Menasche and colleagues in 2001 [44]. A patient with previous myocardial infarction underwent CABG surgery, during which skeletal muscle progenitor cells (*skeletal myoblasts, satellite cells*) were injected into the infarct tissue. Regional contractility significantly improved, and no side effects were observed initially. Subsequently, a number of groups initiated similar clinical trials, injecting skeletal myoblasts either surgically or via transendocardial catheter delivery. However, recent reports have indicated that some of these patients develop ventricular arrhythmia, requiring anti-arrhythmic medication and/or implantable cardioverter defibrillator (ICD) implantation. Autologous skeletal myoblasts are usually isolated from a sample of skeletal muscle about 2 weeks before the planned CABG operation, and are expanded in culture. As soon as the serum concentration in the culture media is decreased *in vitro* or the myoblasts are implanted *in vivo*, they differentiate into skeletal myocytes, fuse, and form multinucleated myotubes. Several studies have demonstrated that the implantation of a large number of skeletal myoblasts in an area of infarcted myocardium improves regional contractility in rodents and large animal models. A critical issue, however, remains the integration of skeletal myoblasts or myotubes into the myocardial syncytium – that is, the expression of cardiac-specific connexins and the formation of functioning gap junctions with surrounding viable cardiomyocytes. In principle, undifferentiated myoblasts can express connexin 43 (Cx43), the predominant gap junction protein in ventricular myocardium, and at least some of the myoblasts appear to be able to form functioning cell–cell communications with cocultured cardiomyocytes *in vitro* [45]. Once they have differentiated and formed myotubes in conventional 2D culture, the Cx43 expression is markedly down-regulated, but may persist to some extent. The situation *in vivo* after implantation in infarcted myocardium is naturally more difficult to assess. Many reports have indicated that skeletal myotubes are not morphologically integrated in normal host myocardium; rather, they appear to form distinct islets in postinfarct tissue. Cx43 expression in transplanted myoblasts has been described in several animal models, but has not been detected in patients who underwent postmortem histology studies following myoblast injection. In a careful study of myoblast transplantation in mice, Rubart et al. found that the majority of the intramyocardial myoblasts/myo-

Table 13.1 Clinical cellular therapy for myocardial tissue repair.[a]

Disease	Cell type/source	Application	Patients (n)	Effects	Readout	Additional readout	Reference
cIHF post AMI	CD133+ BM-MNC (85–195 mL)	i.my./CABG	12	EF ↑; perfusion defect ↓ LVEDV unchanged	Echo, SPECT	24-h ECGFCM	49
cIHF	BM-MNC (50 mL) vs. CTRL; prospective, nonrandomized	i.my./NOGA (15 × 0.2 mL)	30	EF ↑; LVESV ↓ perfusion defect ↓ NYHA/CCSAS class ↓	LVA SPECT	Clinical/laboratory evaluation; treadmill; 2-D Doppler; 24-h ECG	52
cIHF	BMC (20 mL) Pilot study	i.my./NOGA (12 × 0.2 mL)	27	Stress perfusion defect ↓ CCSAS class ↓ (Exercise capacity ↑)	D-SPECT Treadmill	ECG Clinical/laboratory evaluation ELISA (VEGF; MCP1)	53, 54
cIHF	BM-MNC (40 mL)	i.my./NOGA	8	Wall motion and thickening ↑ perfusion defect ↓; no arrhythmia EF unchanged; NYHA class ↓	MRI (7/8) SPECT; 24-h ECG	ECG; medication use Clinical/laboratory evaluation CFU-GM; FCM	55
cIHF post AMI EF <35%	Cultured skeletal muscle cells (myoblasts ~8.7 × 10^8)	i.my./CABG	10	EF ↑; wall thickness ↓; NYHA class ↓; 4 delayed tachycardia → ICD 1 unrelated death	TTE 24-h ECG	Clinical/laboratory evaluation	44, 56
cIHF post AMI EF <35%	Cultured skeletal muscle cells (myoblasts ~1.8 × 10^8)	i.my./CABG	12	Perfusion defect ↓; no arrhythmia EF ↑; wall thickness ↑	^{18}F-FDG PET TTE	ECG; Clinical/laboratory evaluation ^{13}N-PET; FCM	57

[a] Published studies with more than five patients all showing feasibility.
Abbreviations: Aph. = Apheresis; Appl. = application mode; BMC = filtered heparinized whole bone marrow; DSE = dobutamine stress echocardiography; iAP = intractable angina pectoris; LVA = left/right ventricular angiography; TTE = transthoracic echocardiography. For other abbreviations, see below.

tubes are functionally isolated from the surrounding myocardium, and suggested that the remaining cells connect with host cardiomyocytes as a result of cell fusion [46]. These observations might not only explain the conflicting results of other studies, but also provide an explanation for a significant clinical problem. The duration of calcium transients recorded from intramyocardial skeletal myoblasts were heterogeneous compared with those in neighboring host cardiomyocytes, which may interfere with the propagation of excitation across the ventricular myocardium and place the heart at risk of ventricular arrhythmia.

13.6.2
Bone Marrow Mononuclear Cells

Probably the simplest approach to myocardial cell therapy in the clinical setting is the transfer of BM mononuclear cells into the myocardium. The proponents of this approach argue that by using unmodified marrow or unselected mononuclear cells, the "ideal" cell for myocardial regeneration, which has not yet been identified, is not lost during the preparation process. Conversely, opponents argue that the vast majority of the BM mononuclear cells are blood cells of all lineages and their immediate progenitors, while only few cells formally meet the stem cell criteria. Whether the local concentration of relevant stem- or progenitor cells will surpass the hypothetical threshold for induction of regeneration processes remains unclear. Indubitably, marrow mononuclear cells can be easily collected and prepared during a standard CABG operation, which is an obvious and important logistic advantage.

The first such report came from Yamaguchi University, Japan. Hamano and colleagues described five patients who underwent CABG with simultaneous BM collection from the iliac crest [47]. The mononuclear cell fraction was prepared using a commercially available apheresis system, and between five and 22 injections of 5×10^7 to 1×10^8 cells were made into the ischemic myocardium that was not directly revascularized by bypass grafting. In three of the five patients, improved perfusion of the cell-treated tissue was noted postoperatively. No complications such as arrhythmia or local calcification were noted, but no statement was made with respect to LV function. In a similar trial, Galinanes and colleagues from Leicester University, UK, collected BM by sternal bone aspirate at the time of CABG surgery [48]. This was diluted with autologous serum and injected into LV scar tissue. Postoperatively, regional contractility in the LV wall segments that did or did not receive BM cells was assessed by dobutamine stress echocardiography. Subsequently, only the segmental wall motion score of those areas injected with BM and receiving a bypass graft in combination improved upon dobutamine stress. It is most likely that many more patients have been subjected to similar treatment protocols elsewhere, but very little or no information as to the functional outcome is available. Most importantly, no controlled trial has so far demonstrated the superiority of CABG and mononuclear cell injection over CABG alone.

13.6.3
Bone Marrow Stem Cells

Our own group has focused on the intramyocardial injection of purified hematopoietic BM stem cells since 2001 [49]. We chose not to simply inject an unmodified mononuclear BM cell suspension, because the large number of leukocytes and their progenitors may primarily induce local inflammation, rendering the actual stem cell effects insignificant. Instead, we prepare a purified stem cell suspension using clinically approved methods. Two monoclonal antibodies are currently available for the clinical selection of BM stem cells, anti-CD34 and anti-CD133. Approximately 60–70 % of the $CD34^+$ BM cells coexpress the CD133 antigen, and 70–80 % of the $CD133^+$ cells are also $CD34^+$. The $CD133^+$ BM cell population contains a small proportion of clonogenic cells, which have a very high potential to induce neoangiogenesis (Figs. 13.2 and 13.3) [50]. Furthermore, there is accumulating evidence that the $CD133^+/CD34^-$ subpopulation includes multipotent stem cells with a significant potential for differentiation into mesenchymal and other nonhematopoietic lineages (Fig. 13.4).

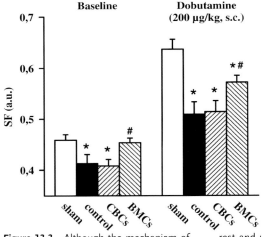

Figure 13.3 Although the mechanism of action has not been completely clarified, the functional benefit of bone marrow (BM) cell delivery to damaged myocardium does exist. Echocardiographic measurements of LV contractility were performed in immunodeficient mice: the "sham" group underwent a sham operation alone and represents a normal control. Mice in the "control" group had myocardial necrosis induced by local cryoinjury. Left ventricular contractility is impaired, both at rest and under dobutamine stress. In the "CBCs" group mononuclear cells obtained from human umbilical cord blood were injected directly into the injured myocardium, but contractility appeared unchanged. In the "BMCs" group, however, human BM mononuclear cells were injected, resulting in improved LV contractility both at rest and under dobutamine stimulation. (Illustration courtesy of Y. Ladilov PhD, University of Rostock.)

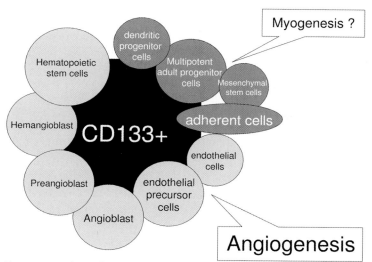

Figure 13.4 The surface antigen CD133 has been identified not only on hematopoietic cells but also on stem- and progenitor cell types involved in angiogenic processes. One point of discussion is the presence of CD133 on certain subpopulations of plastic adherent bone marrow cells, which might include mesenchymal stem cells with the potential for cross-lineage differentiation into myocyte-type contractile cells.

Between 2001 and 2003, we conducted a formal Phase I safety and feasibility trial in 15 patients, including a dose-escalation protocol (Fig. 13.5). Since 2003, an open-label controlled Phase II trial has been undertaken that will eventually include 100 patients. Fifty patients will undergo CABG plus intramyocardial stem cell delivery, and 50 patients with comparable characteristics will have CABG alone. The inclusion criteria were defined as follows:

- Documented transmural myocardial infarction more than 10 days and less than 3 months prior to admission for surgery.
- Presence of a localized area of akinetic LV wall without paradoxical systolic movement that corresponded with the infarct localization.
- The infarct area should not be amenable to surgical or interventional revascularization.
- Elective CABG indicated to bypass stenoses or occlusions of coronary arteries other than the infarct vessel.
- Absence of severe concomitant disease (i.e., terminal renal failure, malignoma, debilitating neurological disease).

Patients who underwent emergency surgery for unstable angina, reoperations, concomitant valve procedures, or who had a history of significant ventricular arrhythmia were excluded. In our experience, it proved rather time-consuming to recruit patients who met the inclusion criteria (approximately 10 patients per

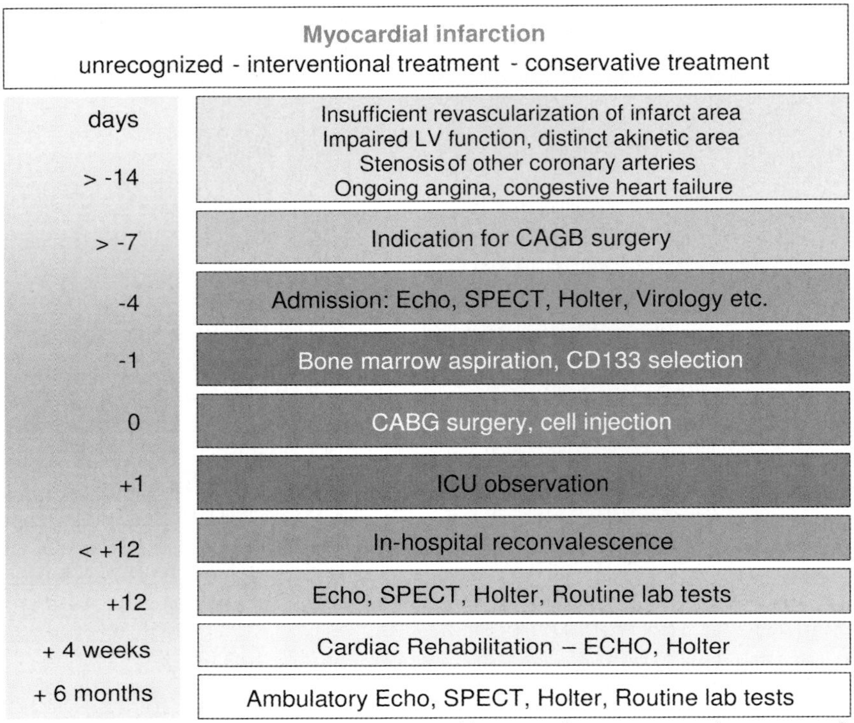

	Myocardial infarction unrecognized - interventional treatment - conservative treatment
days	Insufficient revascularization of infarct area
	Impaired LV function, distinct akinetic area
> -14	Stenosis of other coronary arteries
	Ongoing angina, congestive heart failure
> -7	Indication for CAGB surgery
-4	Admission: Echo, SPECT, Holter, Virology etc.
-1	Bone marrow aspiration, CD133 selection
0	CABG surgery, cell injection
+1	ICU observation
< +12	In-hospital reconvalescence
+12	Echo, SPECT, Holter, Routine lab tests
+ 4 weeks	Cardiac Rehabilitation – ECHO, Holter
+ 6 months	Ambulatory Echo, SPECT, Holter, Routine lab tests

Figure 13.5 Clinical protocol of adult stem cell injection for myocardial regeneration in CABG patients (University of Rostock). Usually, a patient had suffered from myocardial infarction that was unrecognized or treated too late, so that substantial myocardial necrosis occurred. Subsequently, the indication for surgery is made to bypass stenoses of other coronary arteries (or the infarct vessel, provided that viable myocardium is present downstream of the occlusion). Patients are admitted several days before surgery, preoperative tests are performed, and bone marrow is aspirated 1 day before surgery. Postoperatively, patients remain in hospital for a minimum 12-day recovery before being transferred to a cardiac rehabilitation program.

year), probably because the modern rapid catheter interventions in acute myocardial infarction prevent the development of completely akinetic LV wall areas in many patients [51].

13.6.3.1 *Cell Preparation*

One day before surgery, BM aspiration is performed under local anesthesia, with 100–200 mL of BM being aspirated into heparin-filled 20-mL syringes. The BM transplant is then prepared in the hematology laboratory under hygienic conditions that meet the criteria of EuGMP Level B; open containers with cells or other material are handled under a laminar-flow hood in accordance with EuGMP Level A (Fig. 13.6). Before and after every preparation step, cell samples

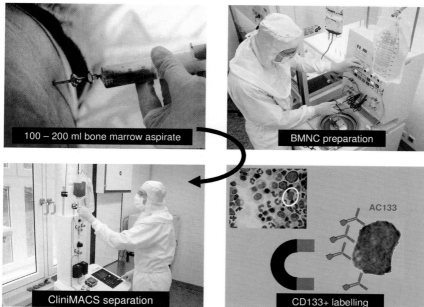

Figure 13.6 To collect adult stem cells, 100–200 mL bone marrow is aspirated from the hip bone under local anesthesia. The mononuclear cells (BMNC) are prepared by Ficoll centrifugation, followed by several washing steps in a closed system. The BMNC are incubated with antibody against CD133 that is conjugated with ferrite particles embedded in a dextran sacculus. Labeled cells are separated with the Miltenyi CliniMacs™ system. Typically, between 5×10^6 and 1×10^7 CD133$^+$ cells are obtained, with both purity and viability exceeding 90%.

are drawn for determination of stem cell number, viability, and sterility. The mononuclear cell fraction is isolated by Ficoll density centrifugation, after which the cells are transferred into the single-use cell processing system of a Cobe 2991 cell processor. Monoclonal CD133 antibody conjugated to superparamagnetic ferrite crystals within a dextran-sacculus is injected into the cell-processing bag and the suspension is incubated for 30 min. The cell-processing bag is removed from the processing system, and the cells are resuspended in a transfer bag. The transfer bag is connected to the CliniMACS Magnetic Cell Separation device. Inside the CliniMACS system, the cells run through an iron matrix-filled column, which is placed inside a strong permanent magnet. Hence, cells bound to the ferrite crystal-conjugated CD133 antibody are retained within the column, while unlabeled cells pass through and are collected in a waste bag. After removal of the magnetic field, the CD133$^+$ cells are washed out of the column and the procedure is repeated twice, yielding a purified CD133$^+$ cell suspension. The cells are resuspended and aliquoted into 2-mL vials, resulting in final dosages of 1.0×10^6 to 5×10^6 target cells per vial, as required by the study protocol. The stem cells are filled into pre-sterilized 2-mL plastic tubes, packed into a sterile container, and stored at 4 C overnight before transplantation.

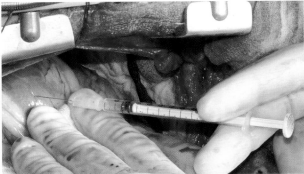

Figure 13.7 During CABG surgery, bone marrow cells are injected into the border zone between infarcted and normal myocardium while the heart is arrested with high-potassium cardioplegic solution under extracorporeal circulation. Typically, 10 injections each of 0.2 mL cell suspension are made; the injection site is occluded with a swab for several seconds to minimize cell suspension reflux.

13.6.3.2 **Surgery**

CABG is performed as described above. Once the bypass-to-coronary artery anastomoses have been completed, the infarct area is visualized. Using a standard 1-mL syringe fitted with a 20-gauge needle, 10 injections of 0.2 mL cell suspension each are placed circumferentially along the infarct border zone (Fig. 13.7). Care is taken to inject obliquely and to distribute the cell suspension intramurally during each injection. Each injection site is occluded using a swab for approximately 10 s in order to minimize reflux of cell suspension. Following the last injection, the aortic clamp is released, the myocardium is reperfused, the aortic anastomoses are constructed, and the operation is completed as usual.

13.6.3.3 **Preliminary Results**

In the Phase I trial, no complications occurred that could be attributed to the cell injection. Occasionally, patients developed temporary supraventricular arrhythmia, respiratory tract infection, or transient neuropsychologic disturbances, but these are complications that inevitably occur with a certain frequency during the postoperative course of CABG patients. With respect to cardiac function, we observed a significant improvement of LVEF, associated with a decrease in LV end-diastolic volume (Fig. 13.8). Furthermore, there was a striking improvement of perfusion in the infarcted and cell-treated myocardial segments, which in some patients was even more pronounced at late follow-up than early postoperatively (Figs. 13.9 and 13.10). Naturally, it is not possible to distinguish clearly between the effects of the CABG operation and the stem cell injection. We refrained from reporting changes in regional contractility, because the effects of bypass grafting and cell treatment will undoubtedly overlap in the infarct border zone of many patients, even if the infarct vessel is not revascularized. To determine accurately

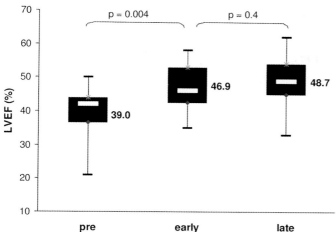

Figure 13.8 Changes in left ventricular contractility (as LVEF) in patients (n = 15) who underwent CABG and CD133⁺ cell injection in a Phase I safety and feasibility trial (University of Rostock). The mean pre-injection LVEF (39 %) increased to approx. 47 % after 2 weeks ("early"), and to approx. 49 % after 6 months ("late").

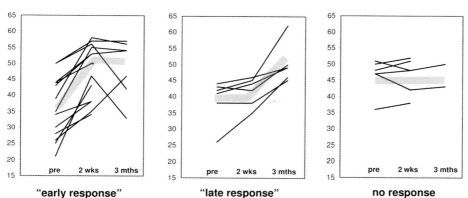

Figure 13.9 Among patients undergoing CABG and intramyocardial stem cell injection (University of Rostock), many had an increase in LVEF early after surgery, followed by a plateau phase ("early response"). In other patients the improvement in LV function appeared to occur rather late after the procedure ("late response"). The latter patients may benefit most from the cell injection, whereas early responders may owe any improvement mainly to the bypass graft (this notion is unsubstantiated). In some patients there was no relevant improvement in LV function ("no response").

the efficacy of stem cell treatment in conjunction with CABG, controlled trials are necessary. At present, the data of our ongoing Phase II trial indicate that patients who undergo CABG and CD133⁺ cell injection benefit more from the operation than those patients who have CABG alone, though as yet no definitive statement can be made.

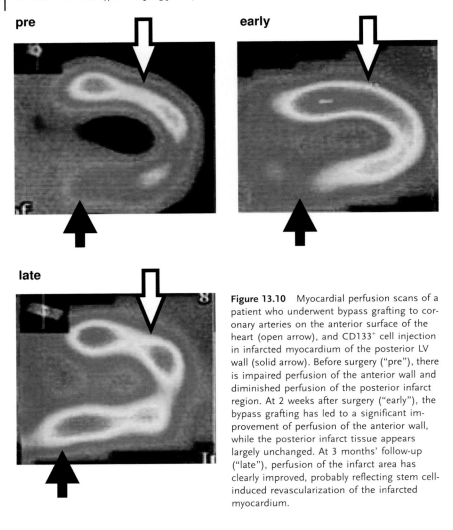

pre   **early**

late

Figure 13.10 Myocardial perfusion scans of a patient who underwent bypass grafting to coronary arteries on the anterior surface of the heart (open arrow), and CD133+ cell injection in infarcted myocardium of the posterior LV wall (solid arrow). Before surgery ("pre"), there is impaired perfusion of the anterior wall and diminished perfusion of the posterior infarct region. At 2 weeks after surgery ("early"), the bypass grafting has led to a significant improvement of perfusion of the anterior wall, while the posterior infarct tissue appears largely unchanged. At 3 months' follow-up ("late"), perfusion of the infarct area has clearly improved, probably reflecting stem cell-induced revascularization of the infarcted myocardium.

13.7
Outlook

Based on the limited existing experience, it should be justified to conclude that transplantation of autologous BM cells in the infarct border zone can be safely performed in patients with ischemic heart disease. Whether neoangiogenesis, neomyogenesis, or both, occur in the human situation remains unclear at this point, and carefully designed controlled studies are needed to determine further the efficacy of clinical cell transplantation for ischemic heart disease. The recent controversy regarding the myogenic potential of HSCs in mice has fueled a heated debate about the scientific and ethical justification of clinical pilot studies.

Even the most progressive clinical stem cell user will admit that efficacy has not yet been clearly demonstrated in the clinical situation. It may well be that relevant myocardial regeneration cannot be induced by using adult stem/progenitor cells that have not been expanded and modified *ex vivo*. However, given the vast amount of data demonstrating functional benefits in large animal models, it is essential that clinical pilot trials are performed. It is nothing unusual in medical history that a novel treatment concept is pushed into the clinic, only to find that it does not keep its promise, and must be returned to the laboratory for further fine-tuning. Human heart transplantation was started before the discovery of the HLA system and before the development of powerful drugs for immunosuppression. In fact, the negative experience with donor organ rejection formed the basis for much of the present research investigations. Of course, the situation has changed fundamentally ever since, and there is no longer any place for heroic surgical procedures. Therefore, each clinical trial must be designed and conducted with the greatest care so as to minimize the risk for the individual patient. Cell therapy for myocardial regeneration holds great promise, and the expectations of our patients are high. We must not disappoint them.

Abbreviations/Acronyms

AMI	acute myocardial infarction
Aph.	apheresis
$AVDO_2$	arterial and venous oxygen content
CABG	coronary artery bypass grafting
CAD	coronary artery disease
cIHF	chronic ischemic heart failure
Cx43	connexin 43
DSE	dobutamine stress echocardiography
EF	ejection fraction
G-CSF	granulocyte colony-stimulating factor
GFP	green fluorescent protein
HSC	hematopoietic stem cell
iAP	intractable angina pectoris
ICD	implantable cardioverter defibrillator
LAD	left anterior descending (artery)
LV	left ventricular
LVEDV	left ventricular end-diastolic volume
LVESV	left ventricular end-systolic volume
LVEF	left ventricular ejection fraction
MSC	mesenchymal stem cell
NOGA	percutaneous intracardiac electromechanical mapping and catheeter-based transendocardial cell injection
PET	positron emission tomography

PTCA   percutaneous transluminal coronary angioplasty
SPECT   single photon emission computed tomography

References

1. Eagle KA, Guyton RA, Davidoff R, et al. ACC/AHA **2004** guideline update for coronary artery bypass graft surgery: summary article. *J Am Coll Cardiol* **2004**;*44*:1146–1154.

2. Appoo J, Norris C, Merali S, et al. Long-term outcome of isolated coronary artery bypass surgery in patients with severe left ventricular dysfunction. *Circulation* **2004**;*110*:II13–II17.

3. Sedlis SP, Ramanathan KB, Morrison DA, et al. Outcome of percutaneous coronary intervention versus coronary bypass grafting for patients with low left ventricular election fractions, unstable angina pectoris, and risk factors for adverse outcomes with bypass (the AWESOME randomized trial and registry). *Am J Cardiol* **2004**;*94*:118–120.

4. Hemmelgarn BR, Southern D, Culleton BF, et al. Survival after coronary revascularization among patients with kidney disease. *Circulation* **2004**;*110*:1890–1895.

5. Risau W. Mechanisms of angiogenesis. *Nature* **1997**;*386*:671–674.

6. Kinnaird T, Stabile E, Burnett MS, et al. Bone marrow-derived cells for enhancing collateral development – Mechanisms, animal data, and initial clinical experiences. *Circ Res* **2004**;*95*:354–363.

7. Rafii S, Lyden D. Therapeutic stem and progenitor cell transplantation for organ vascularization and regeneration. *Nat Med* **2003**;*9*:702–712.

8. Urbich C, Dimmeler S. Endothelial progenitor cells: characterization and role in vascular biology. *Circ Res* **2004**;*95*:343–353.

9. Asahara T, Murohara T, Sullivan A, et al. Isolation of putative progenitor endothelial cells for angiogenesis. *Science* **1997**;*275*:964–967.

10. Asahara T, Masuda H, Takahashi T, et al. Bone marrow origin of endothelial progenitor cells responsible for postnatal vasculogenesis in physiological and pathological neovascularization. *Circ Res* **1999**;*85*:221–228.

11. Tomita S, Li RK, Weisel RD, et al. Autologous transplantation of bone marrow cells improves damaged heart function. *Circulation* **1999**;*100*:II247–II256.

12. Kobayashi T, Hamano K, Li TS, et al. Enhancement of angiogenesis by the implantation of self bone marrow cells in a rat ischemic heart model. *J Surg Res* **2000**;*89*:189–195.

13. Kamihata H, Matsubara H, Nishiue T, et al. Implantation of bone marrow mononuclear cells into ischemic myocardium enhances collateral perfusion and regional function via side supply of angioblasts, angiogenic ligands, and cytokines. *Circulation* **2001**;*104*:1046–1052.

14. Kamihata H, Matsubara H, Nishiue T, et al. Improvement of collateral perfusion and regional function by implantation of peripheral blood mononuclear cells into ischemic hibernating myocardium. *Arterioscler Thromb Vasc Biol* **2002**;*22*:1804–1810.

15. Kocher AA, Schuster MD, Szabolcs MJ, et al. Neovascularization of ischemic myocardium by human bone-marrow-derived angioblasts prevents cardiomyocyte apoptosis, reduces remodeling and improves cardiac function. *Nat Med* **2001**;*7*:430–436.

16. Goodell MA, Rosenzweig M, Kim H, et al. Dye efflux studies suggest that hematopoietic stem cells expressing low or undetectable levels of CD34 antigen exist in multiple species. *Nat Med* **1997**;*3*:1337–1345.

17. Jackson KA, Majka SM, Wang H, et al. Regeneration of ischemic cardiac muscle and vascular endothelium by adult stem cells. *J Clin Invest* **2001**;*107*:1395–1402.

18. Orlic D, Kajstura J, Chimenti S, et al. Transplanted adult bone marrow cells repair myocardial infarcts in mice. *Ann N Y Acad Sci* **2001**;*938*:221–229.

19. Orlic D, Kajstura J, Chimenti S, et al. Mobilized bone marrow cells repair the infarcted heart, improving function and survival. *Proc Natl Acad Sci USA* **2001**;*98*:10344–10349.

20. Epstein SE, Kornowski R, Fuchs S, et al. Angiogenesis therapy – Amidst the hype, the neglected potential for serious side effects. *Circulation* **2001**;*104*:115–119.

21. Mathur A, Martin JF. Stem cells and repair of the heart. *Lancet* **2004**;*364*:183–192.

22. Kajstura J, Leri A, Finato N, et al. Myocyte proliferation in end-stage cardiac failure in humans. *Proc Natl Acad Sci USA* **1998**;*95*:8801–8805.

23. Beltrami AP, Urbanek K, Kajstura J, et al. Evidence that human cardiac myocytes divide after myocardial infarction. *N Engl J Med* **2001**;*344*:1750–1757.

24. Quaini F, Urbanek K, Beltrami AP, et al. Chimerism of the transplanted heart. *N Engl J Med* **2002**;*346*:5–15.

25. Laflamme MA, Myerson D, Saffitz JE, et al. Evidence for cardiomyocyte repopulation by extracardiac progenitors in transplanted human hearts. *Circ Res* **2002**;*90*:634–640.

26. Muller P, Pfeiffer P, Koglin J, et al. Cardiomyocytes of noncardiac origin in myocardial biopsies of human transplanted hearts. *Circulation* **2002**;*106*:31–35.

27. Spangrude GJ, Torok-Storb B, Little MT. Chimerism of the transplanted heart. *N Engl J Med* **2002**;*346*:1410–1412.

28. Bianchi DW, Johnson KL, Salem D. Chimerism of the transplanted heart. *N Engl J Med* **2002**;*346*:1410–1412.

29. Orlic D, Kajstura J, Chimenti S, et al. Bone marrow cells regenerate infarcted myocardium. *Nature* **2001**;*410*:701–705.

30. Korbling M, Estrov Z. Adult stem cells for tissue repair – A new therapeutic concept? *N Engl J Med* **2003**;*349*:570–582.

31. Goodell MA. Stem-cell "plasticity": befuddled by the muddle. *Curr Opin Hematol* **2003**;*10*:208–213.

32. Camargo FD, Chambers SM, Goodell MA. Stem cell plasticity: from transdifferentiation to macrophage fusion. *Cell Prolif* **2004**;*37*:55–65.

33. Balsam LB, Wagers AJ, Christensen JL, et al. Haematopoietic stem cells adopt mature haematopoietic fates in ischaemic myocardium. *Nature* **2004**;*428*:668–673.

34. Murry CE, Soonpaa MH, Reinecke H, et al. Haematopoietic stem cells do not transdifferentiate into cardiac myocytes in myocardial infarcts. *Nature* **2004**;*428*:664–668.

35. Chien KR. Stem cells: lost in translation. *Nature* **2004**;*428*:607–608.

36. Couzin J, Vogel G. Cell therapy – Renovating the heart. *Science* **2004**;*304*:192–194.

37. Honold J, Assmus B, Lehman R, et al. Stem cell therapy of cardiac disease: an update. *Nephrol Dialysis Transplant* **2004**;*19*:1673–1677.

38. Wakitani S, Saito T, Caplan AI. Myogenic cells derived from rat bone marrow mesenchymal stem cells exposed to 5-azacytidine. *Muscle Nerve* **1995**;*18*:1417–1426.

39. Makino S, Fukuda K, Miyoshi S, et al. Cardiomyocytes can be generated from marrow stromal cells in vitro. *J Clin Invest* **1999**;*103*:697–705.

40. Chedrawy EG, Wang JS, Nguyen DM, et al. Incorporation and integration of implanted myogenic and stem cells into native myocardial fibers: anatomic basis for functional improvements. *J Thorac Cardiovasc Surg* **2002**;*124*:584–590.

41. Wang JS, Shum-Tim D, Galipeau J, et al. Marrow stromal cells for cellular cardiomyoplasty: feasibility and potential clinical advantages. *J Thorac Cardiovasc Surg* **2000**;*120*:999–1005.

42. Bittira B, Kuang JQ, Al Khaldi A, et al. In vitro preprogramming of marrow stromal cells for myocardial regeneration. *Ann Thorac Surg* **2002**;*74*:1154–1159.

43. Toma C, Pittenger MF, Cahill KS, et al. Human mesenchymal stem cells differentiate to a cardiomyocyte phenotype in the adult murine heart. *Circulation* **2002**;*105*:93–98.

44. Menasche P, Hagege AA, Scorsin M, et al. Myoblast transplantation for heart failure. *Lancet* **2001**;*357*:279–280.

45. Reinecke H, MacDonald GH, Hauschka SD, et al. Electromechanical coupling between skeletal and cardiac muscle. Implications for infarct repair. *J Cell Biol* **2000**;*149*:731–740.

46. Rubart M, Soonpaa MH, Nakajima H, et al. Spontaneous and evoked intracellular calcium transients in donor-derived myocytes following intracardiac myoblast transplantation. *J Clin Invest* **2004**;*114*:775–783.

47. Hamano K, Nishida M, Hirata K, et al. Local implantation of autologous bone marrow cells for therapeutic angiogenesis in patients with ischemic heart disease: clinical trial and preliminary results. *Jpn Circ J* **2001**;*65*:845–847.

48. Galinanes M, Loubani M, Davies J, et al. Autotransplantation of unmanipulated bone marrow into scarred myocardium is safe and enhances cardiac function in humans. *Cell Transplant* **2004**;*13*:7–13.

49. Stamm C, Westphal B, Kleine HD, et al. Autologous bone-marrow stem-cell transplantation for myocardial regeneration. *Lancet* **2003**;*361*:45–46.

50. Peichev M, Naiyer AJ, Pereira D, et al. Expression of VEGFR-2 and AC133 by circulating human CD34(+) cells identifies a population of functional endothelial precursors. *Blood* **2000**;*95*:952–958.

51. Stamm C, Kleine HD, Westphal B, et al. CABG and bone marrow stem cell transplantation after myocardial infarction. *Thorac Cardiovasc Surg* **2004**;*52*:152–158.

52. Perin EC, Dohmann HF, Borojevic R, et al. Transendocardial, autologous bone marrow cell transplantation for severe, chronic ischemic heart failure. *Circulation* **2003**;*107*:2294–2302.

53. Fuchs S, Satler LF, Kornowski R, et al. Catheter-based autologous bone marrow myocardial injection in no-option patients with advanced coronary artery disease – A feasibility study. *J Am Coll Cardiol* **2003**;*41*:1721–1724.

54. Fuchs S, Kornowski R, Weisz G, et al. Transendocardial autologous bone marrow cell transplantation in patients with advanced ischemic heart disease: Final results from a multi-center feasibility study. *J Am Coll Cardiol* **2004**;*43*:99A.

55. Tse HF, Kwong YL, Chan JKF, et al. Angiogenesis in ischaemic myocardium by intramyocardial autologous bone marrow mononuclear cell implantation. *Lancet* **2003**;*361*:47–49.

56. Menasche P, Hagege AA, Vilquin JT, et al. Autologous skeletal myoblast transplantation for severe postinfarction left ventricular dysfunction. *J Am Coll Cardiol* **2003**;*41*:1078–1083.

57. Herreros J, Prosper F, Perez A, et al. Autologous intramyocardial injection of cultured skeletal muscle-derived stem cells in patients with non-acute myocardial infarction. *Eur Heart J* **2003**;*24*:2012–2020.

14

Adoptive Immunotherapy: Guidelines and Clinical Practice

Hans-Jochem Kolb, Christoph Schmid, Iris Bigalke, Raymund Buhmann,
Belinda Simoes, Ting Yang, Johanna Tischer, Michael Stanglmaier, Horst Lindhofer,
Christine Falk, and Georg Ledderose

Abstract

Allogeneic stem cell transplantation may induce lasting remissions in leukemia
and hematological malignancies that otherwise are refractory to chemotherapy.
An important role in the control of the disease is played by immune competent
cells of the graft. This role has been most convincingly demonstrated by the effect
of donor lymphocyte transfusions for the treatment of recurrent chronic myelogenous leukemia. In other forms of leukemia and hematological malignancies
this role has been less clearly defined. In this chapter, strategies are discussed
to improve the effect of donor lymphocyte transfusions. Methods to improve
responses have involved the use of cytokines, natural killer cells, and antibodies.

14.1
Introduction

The effect of allogeneic lymphocytes on leukemia was originally demonstrated in
mice by Barnes and Loutit in 1957 [1], and later introduced into clinical practice
by Mathé [2], who coined the term "adoptive immunotherapy". However, patients
cured of leukemia died from graft-versus-host disease (GVHD) or sequela thereof.
Subsequently, Weiden et al. reported the beneficial effect of GVHD on the control
of leukemia [3]. The role of donor lymphocytes was demonstrated by the depletion
of T cells from the graft [4]; this depletion increased the relapse rate of chronic
myelogenous leukemia (CML) as compared to transplantation of unmodified
marrow with and without the development of GVHD.

Stem Cell Transplantation. Biology, Processing, and Therapy.
Edited by Anthony D. Ho, Ronald Hoffman, and Esmail D. Zanjani
Copyright © 2006 WILEY-VCH Verlag GmbH & Co. KGaA, Weinheim
ISBN: 3-527-31018-5

14.2
Animal Experiments

The principles of adoptive immunotherapy were defined in experimental transplantation in DLA-identical littermate dogs [5]. Depletion of T cells from the marrow graft using antithymocyte globulin prior to transplantation was effective for the prevention of GVHD in littermate dogs [6]. Stable chimerism and tolerance could be induced in most DLA-haploidentical and in DLA-identical littermates. Transfusion of donor lymphocytes was studied at various times after marrow transplantation [5]. Intense GVHD was produced when transfusion was carried out within the first days and weeks, but did not occur when transfusion was delayed for 2 months and longer. In DLA-identical littermates, mixed chimerism was persistent for many years, if animals were conditioned with 10 Gy total body irradiation and transplanted with a limited number of marrow cells treated with antithymocyte globulin to deplete T cells. Transfusion of donor lymphocytes converted mixed into complete chimerism, indicating that residual hematopoiesis of the host could be eliminated without producing GVHD. In the same experiments, immunity against tetanus toxin could be transferred from the donor to the host, and immunity against diphtheria toxin could be improved.

14.3
The First Clinical Results in CML

Encouraged by the finding in dogs that residual hematopoiesis of the host could be eliminated by the transfusion of donor lymphocytes, three patients with recurrent CML following allogeneic stem cell transplantation were treated with transfusion of donor lymphocytes. Each patient responded with a complete cytogenetic remission [7], and two patients developed GVHD but recovered when treated with immunosuppressive therapy. A fourth patient developed severe pancytopenia that required the transfusion of donor marrow. This patient's hematopoiesis was subsequently reconstituted and the disease went into molecular remission.

14.4
The EBMT Study

The results of the above-mentioned clinical study were confirmed by more than 50 centers in Europe [8] and North America [9]. The best results were obtained in patients with recurrent CML in chronic phase (hematological relapse) or in cytogenetic relapse. Less effective were donor lymphocyte transfusions (DLT) in patients with blast crisis, acute myeloid leukemia (AML) and myeloma, while poor results were seen in patients with acute lymphoid leukemia (ALL) and lymphoma (Table 14.1; Figs. 14.1 and 14.2). Control of leukemia was not achieved in patients

Table 14.1 Graft-versus-leukemia effect of donor lymphocyte transfusions. EBMT data.

Diagnosis	Patients studied (n)	Patients evaluable (n)[a]	Patients in complete remission (n)
CML			
cytogenetic relapse	57	50	40 (80)
hematologic relapse	124	114	88 (77)
transformed phase	42	36	13 (36)
Polycythemia vera/MPS	2	1	1
AML/M DS	97	58	15 (26)
ALL	55	20	3 (15)
MMY	25	17	5 (29)

Values in parentheses are percentages.
[a] Survival >30 days.

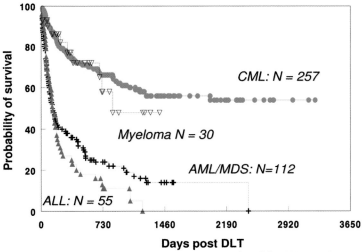

Figure 14.1 Donor lymphocyte transfusions for the treatment of leukemia and myeloma after allogeneic stem cell transplantation. Results of the EBMT study: patients with CML have a chance of long-term survival in molecular remission.

given prophylactic immunosuppressive treatment and in patients without evidence of persistent chimerism.

Myelosuppression was observed in 10 to 20 % of patients, and was more frequent among those patients with hematological relapse than in those with only

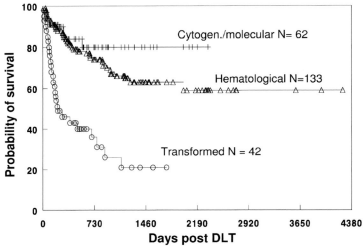

Figure 14.2 Donor lymphocyte transfusions for treatment of recurrent CML according to the stage at relapse. Cytogenetic and molecular relapses have the best survival chance after DLT, hematological relapse in chronic phase an intermediate chance, and transformed phase a poor chance of long-term survival, though some patients in the transformed phase have benefited.

cytogenetic relapse. The use of G-CSF mobilized blood cells instead of lymphocytes from the marrow donor could not prevent myelosuppression [10].

GVHD occurred in about 60 % of patients, and required immunosuppressive treatment in about 40 %. In the absence of leukemia, deaths were due to GVHD and infections. GVHD could be prevented in the majority of patients using donor lymphocytes depleted of CD8-positive cells [11, 12], or by escalating the cell doses, starting with small doses below 10^7 CD3-positive cells kg^{-1} body weight [13]. In an EBMT survey, GVHD could be significantly reduced by starting doses of below 2×10^7 CD3 cells kg^{-1} body weight [14]. Treatment with small doses of donor lymphocytes at the time of cytogenetic or molecular relapse is presently the preferred treatment.

14.5
The Graft-versus-Leukemia Effect

Evidence available to date indicates that there is a strong graft-versus-leukemia (GVL) effect exerted by T cells contained in the donor lymphocyte transfusion (Table 14.2). CD8-positive T cells may not be necessary as they can be depleted without totally ablating the GVL effect. Indeed, CD4-positive T cells may recruit CD8 T cells *in vivo*. The target antigens are minor histocompatibility antigens presented by HLA-antigens of class II and class I. A GVL effect may result from the graft-versus-host (GVH) reaction directed towards minor histocompat-

Table 14.2 Graft-versus-leukemia effect.

Effectors	Targets
CD 4 T cells	HLA-class II restricted
CD 8 T cells	HLA-class I restricted
	Leukemia-specific antigens
	Minor histocompatibility antigens
	Only on hematopoietic cells
	On all cells
NK cells	Alloreactive group
Dendritic cells, macrophages, cytokines	

ibility antigens present on hematopoietic cells, as minor histocompatibility antigens present on all cells would induce a GVH reaction also against skin, liver, and gut. Natural killer (NK) cells and NK-T cells may also exert a strong GVL effect, if they are not inhibited by their own HLA-associated cross-reactive group. Finally, the reaction of both T cells and NK cells is modified by dendritic cells macrophages and cytokines providing stimulatory signals.

14.6
Cytokines

The better result in leukemia of myeloid origin was the basis of the hypothesis that myeloid dendritic cells (DC1) of leukemic origin were important for the GVL effect to occur (Fig. 14.3). This hypothesis was supported by the demonstration of dendritic cells of leukemic origin by the present authors [15] (Fig. 14.4) and others [16, 17] in both CML and AML [18]. However, fresh leukemic cells of patients with CML and AML are poor stimulators in mixed lymphocyte reactions [19, 20]. After culture in the presence of GM-CSF and interleukin (IL)-4 or interferon-α (IFN-α) cells with properties of dendritic cells develop, and cytotoxic T cells against autologous leukemia cells may be produced [18, 20].

These principles were applied to the treatment of patients not responding to DLT alone. Patients with relapse of CML were treated with GM-CSF, IFN-α and DLT simultaneously (Table 14.3). In six out of nine patients with CML relapse in an advanced stage a complete cytogenetic remission was induced, while no remission was obtained in two patients and a partial remission in one patient (unpublished results). Two of three patients given chemotherapy prior to DLT developed severe GVHD.

In relapsed AML the combination of mobilized donor cells and GM-CSF improved the response rate from 25 % [21] to 67 % [22]. Among 24 patients with

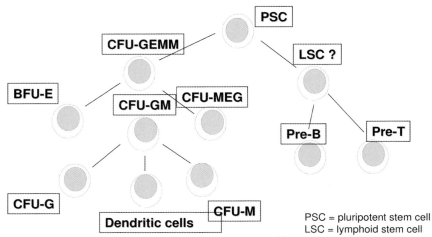

Figure 14.3 Differentiation of leukemic cells to dendritic cells *in vivo* may represent a factor for good response to DLT. Myeloid leukemia may respond better due to differentiation to dendritic cells I.

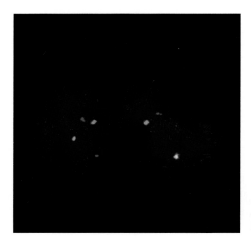

Figure 14.4 Demonstration of dendritic cells of leukemia origin by FISH analysis of Ph-1-positive cells: co-localization of bcr and abl is indicated by the arrow.

AML in relapse, 11 responded to low-dose cytarabine (Ara-C) and 10 to DLT; four of these patients are still in remission after 2, 4, and 8 years, and one patient died after more than 7 years of refractory relapse in the CNS in the presence of a systemic remission. Response to the treatment with low-dose Ara-C and relapse later than 180 days were favorable prognostic factors for long-term remission. Two of the patients in continuous complete remission had received repeated DLT from their donors.

Table 14.3 GM-CSF and interferon-α for adoptive immunotherapy of chronic myelogenous leukemia (CML).30-year-old
male, CML 3/91; no response to IFN-α.

Date	Treatment/outcome
9.05.96	BMT from MUD foll. TBI, ATG, CY
1.04.97	Cytogenetic relapse: t(2;8), t(2;21), t(9;22) in all metaphases no response to I FN-a
20.12.97	DLT I + IFN-α: no cytogenetic nor molecular response
28.10.98	DLT II + IFN-α: no cytogenetic nor molecular response
27.11.99	DLT III + IFN-α + GM-CSF
10.02.00	Partial cytogenetic response (73 / 127)
5.06.00	Major cytogenetic response (36 / 228)
8.12.00	PCR negative
6.02.01	PCR negative
17.10.02	PCR negative
10.02.03	PCR negative
15.11.04	PCR negative

ATG: antithymocyte globulin; BMT: bone marrow transplantation;
CML: chronic myelogenous leukemia; CY: cyclophosphamide;
DLT: donor lymphocyte transfusion;
GM-CSF: granulocyte-monocyte colony-stimulating factor;
IFN-α: interferon-α; MUD: matched unrelated donor;
PCR: polymerase chain reaction; TBI: total body irradiation.

14.7
Bispecific Antibodies

Another possible means of activating GVL responses is to use bispecific antibodies [23]. The antibodies used were hybrids of mouse and rat monoclonal antibodies directed against CD3 and CD20 and containing a biologically active Fc-portion. This unique isotype combination [24] may be termed "trifunctional", because they bind T cells via the CD3 antigen, B lymphoma and leukemia cells via the second arm, and macrophages and dendritic cells via the Fc-portion. This class of antibodies promotes phagocytosis [25] which may lead to the processing of lymphoma proteins and the presentation of relevant peptides (Fig. 14.5). Basically, the mechanism has been demonstrated *in vitro* using donor lymphocytes and cells of patients with chronic lymphocytic leukemia (CLL) [26]. Cytotoxic T cells could be produced against allogeneic CLL cells *in vitro*. Clinical responses were observed with reduction or clearance of leukemia cells in CLL [26]; however, longlasting cellular immunity against CLL cells could not be demon-

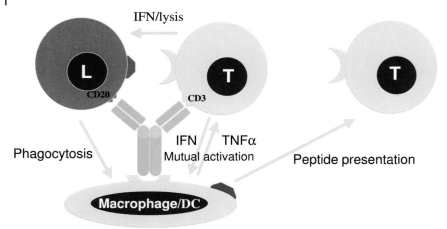

Figure 14.5 Bispecific antibodies combining CD20 and CD3 and an active Fc part stimulate T cells in the neighborhood of CD20-positive lymphoma cells, leading to the death of lymphoma cells. Activated dendritic cells process antigens of the lymphoma cells and stimulate unbound T cells.

strated following the treatment of patients with bispecific antibodies. Further attempts are in progress to overcome resistance to cellular immunity in CLL.

14.8
NK and NK-T Cells and HLA-Haploidentical Transplantation

Natural killer (NK) cells and natural killer T cells produce a GVL effect using a different pathway from T cells. Ligands for killer immunoglobulin-like receptors are expressed in association with HLA-antigens, the expression of autologous HLA-antigens inhibits NK cells [27]. The selection of HLA-mismatched donors and HLA-matched unrelated donors may allow for the activation of NK cells, if donor and patient do not share the same cross-reactive group of inhibitory receptors. HLA-haploidentical transplants have been performed using mega doses of CD34-selected progenitor cells [28, 29], and the anti-leukemia effect was exerted by the intensive conditioning treatment and development of NK cells from the progenitor cells [30]. In these patients severe immune deficiency may persist for prolonged periods of time.

We have chosen to use unmodified bone marrow on day 0 and to add mobilized blood cells that have been depleted of CD6-positive cells on day 6 for transplantation. CD6-depleted mobilized blood cells contain CD34-positive progenitor cells, NK cells and a minority of CD8-positive cells that are immunosuppressive and anti-leukemic. This regimen has been applied successfully to the treatment of patients with acute leukemia and lymphoma in advanced stages, as well as to one patient with severe aplastic anemia refractory to immunosuppressive therapy.

The intensity of conditioning could be reduced by decreasing the dose of total body irradiation from 12 Gy to 4 Gy, and in the majority of patients the immuno-suppressive treatment after grafting could be discontinued after 1 or 2 months. The mechanism of anti-leukemia activity of CD6-negative, CD8-positive cells is presently the subject of extensive analysis.

14.9
Outlook of Adoptive Immunotherapy in Chimerism

The state of chimerism allows adoptive immunotherapy with donor cells that are tolerated by the patient's immune system and are not themselves tolerant to the leukemia. T cells, NK cells and even B cells [31] may be involved in the GVL reaction. The prevention of GVHD and simultaneous immune escape from leukemia remain the goals of further research. Moreover, treatments utilizing the activation of cytokines, bispecific antibodies and immunomodulatory cells with anti-leukemic activity offer great promise and require further investigation.

Abbreviations/Acronyms

ALL	acute lymphoid leukemia
AML	acute myeloid leukemia
Ara-C	cytarabine
ATG	antithymocyte globulin
BMT	bone marrow transplantation
CLL	chronic lymphocytic leukemia
CML	chronic myelogenous leukemia
CY	cyclophosphamide
DLT	donor lymphocyte transfusion
EBMT	European Group for Blood and Marrow Transplantation
G-CSF	granulocyte colony-stimulating factor
GM-CSF	granulocyte macrophage colony-stimulating factor
GVHD	graft-versus-host disease
GVL	graft-versus-leukemia
IFN-α	interferon-α
IL	interleukin
MUD	matched unrelated donor
TBI	total body irradiation

References

1. Barnes, D.H.W., Loutit, J.F., *Br. J. Haematol.* **1957**, *3*, 241–252.
2. Mathé, G., Amiel, J.L., Schwarzenberg, L., Cattan, A., Schneider, M., *Cancer Res.* **1965**, *25*, 1525–1530.
3. Weiden, P.L., Sullivan, K.M., Flournoy, N., Storb, R., Thomas, E.D., and The Seattle Marrow Transplant Team, *N Engl J Med* **1981**, *304*, 1529–1531.
4. Horowitz, M.M., Gale, R.P., Sondel, P.M., Goldman, J.M., Kersey, J., Kolb, H.J., Rimm, A.A., Ringden, O., Rozman, C., Speck, B., et al., *Blood* **1990**, *75*, 555–562.
5. Kolb, H.J., Günther, W., Schumm, M., Holler, E., Wilmanns, W., Thierfelder, S., *Transplantation* **1997**, *63*, 430–436.
6. Kolb, H.J., Rieder, I., Rodt, H., Netzel, B., Grosse Wilde, H., Scholz, S., Schaffer, E., Kolb, H., Thierfelder, S., *Transplantation* **1979**, *27*, 242–245.
7. Kolb, H.J., Mittermueller, J., Clemm, C., Ledderose, G., Brehm, G., Heim, M., Wilmanns, W., *Blood* **1990**, *76*, 2462–2465.
8. Kolb, H.J., Schattenberg, A., Goldman, J.M., Hertenstein, B., Jacobsen, N., Arcese, W., Ljungman, P., Ferrant, A., Verdonck, L., Niederwieser, D., van Rhee, F., Mittermüller, J., De Witte, T., Holler, E., Ansari, H., *Blood* **1995**, *86*, 2041–2050.
9. Collins, R.H., Shpilberg, O., Drobyski, W.R., Porter, D.L., Giralt, S., Champlin, R., Goodman, S.A., Wolff, S.N., Hu, W., Verfaillie, C., List, A., Dalton, W., Ognoskie, N., Chetrit, A., Antin, J.H., Nemunaitis, J., *J Clin Oncol* **1997**, *15*, 433–444.
10. Flowers, M.E.D., Sullivan, K.M., Martin, P., Buckner, C.D., Beach, K., Higano, T., Radich, J.P., Clift, R.A., Bensinger, W., Deeg, H.J., Chauncey, T.R., Rowley, S., Storb, R., Appelbaum, F.R., *Blood* **1995**, *86*(Suppl.1), 564a.
11. Giralt, S., Hester, J., Huh, Y., Hirsch-Ginsberg, C., Rondon, G., Seong, D., Lee, M., Gajewski, J., Van Besien, K., Khouri, I., Mehra, R., Przepiorka, D., Körbling, M., Talpaz, M., Kantarjian, H., Fischer, H., Deisseroth, A., Champlin, R., *Blood* **1995**, *86*, 4337–4343.
12. Alyea, E.P., Soiffer, R.J., Canning, C., Neuberg, D., Schlossman, R., Pickett, C., Collins, H., Wang, Y., Anderson, K.C., Ritz, J., *Blood* **1998**, *91*(10), 3671–3680.
13. Mackinnon, S., Papadopoulos, E.B., Carabasi, M.H., Reich, L., Collins, N.H., Boulad, F., Castro-Malaspina, H., Childs, B.H., Gillio, A.P., Kernan, N.A., Small, T.M., Young, J.W., O'Reilly, R.J., *Blood* **1995**, *86*, 1261–1268.
14. Guglielmi, C., Arcese, W., Dazzi, F., Brand, R., Bunjes, D., Verdonck, L.F., Schattenberg, A., Kolb, H.J., Ljungman, P., Devergie, A., Bacigalupo, A., Gomez, M., Michallet, M., Elmaagacli, A., Gratwohl, A., Apperley, J., Niederwieser, D., *Blood* **2002**, *100*(2), 397–405.
15. Chen, X., Regn, S., Kolb, H.J., Roskrow, M., *Blood* **1999**, *94*(10 Suppl.1), 529a.
16. Eibl, B., Ebner, S., Duba, Ch., Böck, G., Romani, N., Gächter, A., Nachbaur, D., Schuler, G., Niederwieser, D., *Bone Marrow Transplant* **1997**, *19*(Suppl.1), S33.
17. Smit, W.M., Rijnbeck, M., van Bergen, C.A.M., de Paus, R.A., Willemze, R., Falkenburg, J.H.F., *Br J Haematol* **1996**, *93*(Suppl.2), 313-abstract 1186.
18. Woiciechowsky, A., Regn, S., Kolb, H.J., Roskrow, M., *Leukemia* **2001**, *15*, 246–255.
19. Schneider, E.M., Chen, Z.Z., Ellwart, J., Wilmanns, W., Kolb, H.J., *Bone Marrow Transplant* **1996**, *17*(Suppl.1), S69, abstract 319.
20. Choudhury, B.A., Liang, J., Thomas, E.K., Flores-Romo, L., Xie, Q.S., Agusala, K., Sutaria, S., Sinha, I., Champlin, R.E., Claxton, D., *Blood* **1999**, *93*(3), 780–786.
21. Kolb, H.J., Schattenberg, A., Goldman, J.M., Hertenstein, B., Jacobsen, N., Arcese, W., Ljungman, P., Ferrant, A., Verdonck, L., Niederwieser, D., *Blood* **1995**, *86*(5), 2041–2050.
22. Schmid, C., Schleuning, M., Aschan, J., Ringden, O., Hahn, J., Holler, E., Hegenbart, U., Niederwieser, D., Dugas, M., Ledderose, G., Kolb, H.J., *Leukemia* **2004**, *18*(8), 1430–1433.
23. Ruf, P., Lindhofer, H., *Blood* **2001**, *98*, 2526–2534.
24. Lindhofer, H., Mocikat, R., Steipe, B., Thierfelder, S., *J Immunol* **1995**, *155*, 219–225.
25. Zeidler, R., Mysliwietz, J., Csanady, M., Walz, A., Ziegler, I., Schmitt, B., Wollenberg, B., Lindhofer, H., *Br J Cancer* **2000**, *83*(2), 261–266.

26. Pinto-Simoes, B., Stanglmair, M., Faltin, M., Schmid, C., Bergmann, M., Humann, M., Wegner, H., Ledderose, G., Lindhofer, H., Kolb, H.J., *Onkologie* **2003**, *26*(S5), 19.

27. Karre, K., Ljunggren, H.G., Piontek, G., Kiessling, R., *Nature* **1986**, *319*(6055), 675–678.

28. Aversa, F., Tabilio, A., Terenzi, A., Velardi, A., Falzetti, F., Giannoni, C., Iacucci, R., Zei, T., Martelli, M.P., Gambelunghe, C., Rosetti, M., Caputo, P., Latini, P., Aristei, C., Raymondi, C., Reisner, Y., Martelli, M.F., *Blood* **1994**, *84*, 3948–3955.

29. Handgretinger, R., Klingebiel, T., Lang, P., Schumm, M., Neu, S., Geiselhart, A., Bader, P., Schlegel, P.G., Greil, J., Stachel, D., Herzog, R.J., Niethammer, D., *Bone Marrow Transplant* **2001**, *27*(8), 777–783.

30. Ruggeri, L., Capanni, M., Casucci, M., Volpi, I., Tosti, A., Perruccio, K., Urbani, E., Negrin, R.S., Martelli, M.F., Velardi, A., *Blood* **1999**, *94*(1), 333–339.

31. Bellucci, R., Wu, C.J., Chiaretti, S., Weller, E., Davies, F.E., Alyea, E.P., Dranoff, G., Anderson, K.C., Munshi, N.C., Ritz, J., *Blood* **2004**, *103*(2), 656–663.

15

Immune Escape and Suppression by Human Mesenchymal Stem Cells

Katarina Le Blanc and Olle Ringdén

Abstract

Mesenchymal stem cells (MSC) can be obtained from bone marrow, fat and several fetal tissues. MSC can be expanded *in vitro* to millions of cells. MSC have the capacity to differentiate to mesenchymal tissues such as bone, cartilage, and fat. MSC escape the immune system *in vitro,* they do not stimulate alloreactivity, and they escape lysis by cytotoxic T cells and NK cells. Therefore, MSC may be transplantable between major histocompatibility complex (MHC) mismatched individuals. They may be used for therapeutic purposes. Furthermore, MSC have immunomodulatory effects and inhibit T-cell proliferation in mixed lymphocyte cultures, prolong skin allograft survival, and may decrease graft-versus-host disease (GVHD) when cotransplanted with hematopoietic stem cells. Immunosuppression by MSC is MHC-independent. MSC induce their immunosuppressive effect via a soluble factor. Some candidates and various mechanisms have been proposed, though contradictory data exist. The immunosuppressive properties of first-trimester fetal MSC are less pronounced, but inducible with interferon-γ. Recently, MSC were found to reverse grade IV acute GVHD of gut and liver, but tolerance was not induced. To conclude, MSC have the potential to repair damaged tissues and to manipulate the immune response.

15.1
Introduction

Friedenstein and coworkers were the first to identify an adherent, fibroblast-like population in bone marrow that, *in vivo*, could regenerate rudiments of normal bone including cartilage, bone, adipose and stroma [1–4]. These so-called mesenchymal stem cells (MSCs) have also been isolated from adipose tissue as well as from fetal liver, blood, bone marrow, lung, and cord blood [5–8]. MSCs, which are capable of differentiating *in vitro* and *in vivo* to several mesenchymal tissues, including bone, cartilage, tendon, muscle, adipose and bone marrow stro-

Stem Cell Transplantation. Biology, Processing, and Therapy.
Edited by Anthony D. Ho, Ronald Hoffman, and Esmail D. Zanjani
Copyright © 2006 WILEY-VCH Verlag GmbH & Co. KGaA, Weinheim
ISBN: 3-527-31018-5

ma [9–11], have been characterized phenotypically as nonhematopoietic cells. By flow cytometry, MSCs stain negative for CD34 or CD45 but express a specific pattern of adhesion molecules (CD105/CD166/CD44/CD29/SH3/SH4) [12]. Because MSCs have the capacity to differentiate into multiple mesenchymal tissues and to support hematopoiesis *in vitro*, researchers have been encouraged to investigate their use in cellular therapy. Experimental studies conducted in animals have shown that MSCs not only engraft in multiple tissues after intravenous infusion but also demonstrate site-specific differentiation [13–15]. MSCs also secrete a number of cytokines and promote the expansion and differentiation of hematopoietic stem cells (HSCs) both *in vitro* [16, 17] and *in vivo* [6, 18]. Isolated from other bone marrow cells by their adherence to plastic, MSCs, although rare in the body, can be expanded *in vitro* to millions of cells [19]. MSCs possess immunomodulatory properties and inhibit T-cell proliferation *in vitro* [19–24]. In a baboon model, the infusion of *ex-vivo*-expanded matched donor or third-party MSCs delayed the time to rejection of histoincompatible skin grafts [21]. In humans, the immunosuppressive effect of MSCs *in vitro* and in preclinical models, suggests that these cells may be used for the prevention and treatment of graft-versus-host disease (GVHD) in allogeneic stem cell transplantation (SCT), in organ transplantation to prevent rejection, and also in autoimmune disorders.

15.2
MSCs Escape the Immune System

MSCs share cell surface markers with the thymic epithelium, and adhesion molecules expressed by MSCs are essential for their interaction with T cells. VCAM-1, ICAM-2 and LFA-3 are present on unstimulated MSCs [10, 19, 23, 26]. Although adult MSCs express intermediate levels of human leukocyte antigen (HLA) major histocompatibility complex (MHC) class I molecules, their expression on fetal MSC is lower [10,19,23,26–29]. HLA class II is detectable by Western blot on whole-cell lysates of unstimulated MSCs, which suggests that they contain intracellular deposits of HLA class II alloantigens [27]. Cell surface expression can be induced by treatment of the cells with interferon-γ (IFN-γ) for 1–2 days. In contrast to postnatal MSCs, the fetal liver-derived cells have no class II either intracellularly or on the cell surface [28]. Intracellular synthesis can be detected by Western blotting on whole-cell lysates after 2 days of exposure to IFN-γ, but after 7 days of culture in the presence of IFN-γ class II alloantigens are detectable on the cell surface. MSCs fail to induce proliferation of allogeneic lymphocytes in coculture experiments [20,22–30]. When MSCs are exposed to IFN-γ, more than 90% of them express HLA class II. Despite IFN-γ-induced expression of class II alloantigens, MSCs do not stimulate alloreactivity [23, 27, 28]. Neither do MSCs express Fas ligand or costimulatory molecules such as B7-1, B7-2, CD40 or CD40L [23, 29]. When costimulation is inadequate, T-cell proliferation can be induced by the addition of exogenous costimulation. Thus, epithelial cells and fibroblasts acquire the ability to present alloantigen in the presence of a costimulatory antibody

[31, 32]. No T-cell proliferation was observed when MSCs were cultured with lymphocytes in the presence of a CD28-stimulating antibody [23]. Infusion or implantation of allogeneic MHC-mismatched MSCs into baboons has been well tolerated [15, 33, 34]. MSCs are not targets of cytotoxic lymphocytes [35], and even lymphocytes allowed to proliferate against those derived from a specific donor, will lyse lymphocytes from that individual in a chromium-release assay, but not MSCs derived from the same donor. Even killer-cell inhibitory receptor- (KIR-ligand) mismatched NK cells are unable to lyse MSCs, suggesting that MSCs escape recognition by alloreactive NK cells. Natural killer (NK) cell-mediated lysis of ^{51}Cr-labeled K562 cells, a CML cell-line which is sensitive to NK cells due to a lack of MHC class I molecules, is not affected by MSCs [35].

15.3
Immunosuppression by MSCs

MSCs inhibit T-cell alloreactivity induced in mixed lymphocyte cultures (MLC) or by mitogens, such as phytohemagglutinin (PHA) and concanavalin A (Con A) [20–25,27,29,35–39]. MSCs suppress both naïve and memory T cells. MSC immunosuppression is not HLA- restricted; rather, immunosuppression occurs both when the MSC are autologous with the stimulatory or the responder lymphocytes or derived from a third party [21, 22, 24]. Therefore, MSCs used in allogeneic SCT may not need to be derived from the HSC donor. MSC-induced T-cell suppression is dose-dependent [22–25]. MSCs, representing proportionally 10–40 % of lymphocytes, are inhibitory, while low doses (0.1–1 %) may sometimes enhance lymphocyte proliferation in MLC (Fig. 15.1).

The T cells do not become apoptotic or anergic, and can be restimulated if the MSCs are removed [21, 25]. Cytotoxic T-lymphocyte (CTL)-mediated lysis is inhibited by MSCs in a time- and dose-dependent manner [35]. When MSCs (ratio 1:10) are present in the MLC from the beginning of the 6-day culture, MSCs completely prevent the lysis of target blasts in a ^{51}Cr-release assay (Fig. 15.2). MSCs only slightly reduce cytotoxicity if added on day 3. Lower numbers of MSCs (ratio 1:1000) have no effect on the CTL, and no inhibition is seen if the MSCs are added during the cytotoxic phase.

MSCs decrease the expression of CD4+ activation markers, CD25, CD38 and CD69 on PHA-simulated lymphocytes (Table 15.1) [37]. Furthermore, MSC increase the numbers of regulatory T cells [38]. MSC increase the transcription and translation of interleukin (IL)-2 and soluble IL-2 receptor in MLCs, while the levels decrease if MSC are present among PHA-stimulated lymphocytes. Thus, the suppression may be mediated by different mechanisms, depending on the T-cell stimulus [39]. MSCs inhibit lymphocyte proliferation induced by Con A, although if IL-2 is added to Con A-stimulated lymphocytes the inhibition is partly abrogated [21]. Furthermore, cocultures of MSC with purified activated dendritic cells lead to a decreased tumor necrosis factor-α (TNF-α) secretion and an increased IL-10 secretion [38]. The coculture of MSCs with effector T

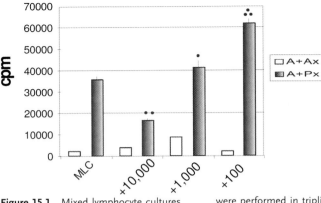

Figure 15.1 Mixed lymphocyte cultures (MLC) where (A + Ax) shows background values using nonirradiated and irradiated lymphocytes from the same individual (Ax) and stimulation with a pool of lymphocytes from five donors (Px). Responder cells (1×10^5) were cultured with 1×10^5 stimulator cells (x) cultured in 0.2 mL medium. Mesenchymal stem cells (MSC) were added in various amounts (10 000, 1000, and 100). Cultures were performed in triplicate, one experiment out of more than 60. High concentrations of MSCs inhibited the MLC. In some experiments, 1000 or 100 MSCs had a stimulatory effect on the MLC. Vertical bars indicate SD. Statistically significant differences (Student's t-test)P ●, p <0.05; ●●, p <0.01; ●●●, p <0.001. Reproduced with permission from [22].

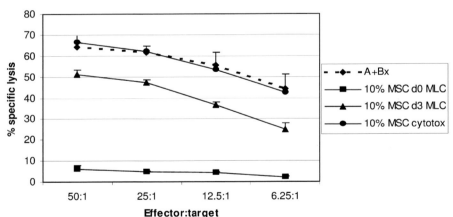

Figure 15.2 Cell-mediated lympholysis (CML) showing specific lysis with mesenchymal stem cells (MSC) added at various times and in different amounts. CML between effector (A) and target cells (B) in different ratios. The figure shows specific lysis of target peripheral blood lymphocytes (PBL) by cytotoxic T lymphocytes with or without 10% MSCs. When no MSCs were added, lysis exceeded 60% in the 50:1 ratio (mean ± SD of triplicates; —). Addition of MSCs at the start of the mixed lymphocyte culture (■) inhibited lysis. MSCs added on day 3 (▲) reduced lysis by 15%, but addition to the cytotoxic assay had no effect (●). Reproduced with permission from [35].

Table 15.1 Percentage of positive lymphocyte subsets (determined by fluorescence-activated cell sorting) after stimulation with phytohemagglutinin (PHA) in the absence or presence of 10% allogeneic mesenchymal stem cells (MSC) (five experiments). Reproduced with permission from [37].

Cell subset studied	Percentage of lymphocytes[a]		p-value
	PBL	PBL + MSC	
CD3$^+$	76 ± 8	58 ± 11	<0.05
CD3$^+$, CD4$^+$	42 ± 6	29 ± 8	<0.01
CD3$^+$, CD8$^+$	39 ± 12	23 ± 8	n.s.
CD14$^+$	8 ± 4	10 ± 5	<0.05
CD3$^+$, CD25$^+$	70 ± 19	26 ± 24	<0.01
CD3$^+$, CD38$^+$	45 ± 23	18 ± 24	<0.01
CD3$^+$, CD69$^+$	37 ± 13	30 ± 13	n.s.

[a]Values are mean ± SD.
n.s.: not significant; PBL, peripheral blood lymphocytes.

cells or purified NK cells lead to a decreased IFN-γ secretion or an increased IL-4 secretion, respectively. Depending upon the kinetics, MSCs can either enhance or depress IL-10 levels in MLCs [39]. MSCs which are differentiated to osteocytes, adipocytes and chondrocytes continue to suppress MLC [27] (Fig. 15.3). Osteocytes and adipocytes show a more pronounced inhibition compared to undifferentiated MSCs, while chondrocytes have less inhibitory capacity. This finding may be clinically important, and suggests that allogeneic MSCs may, for example, be used to repair bone defects. The suppression by both and differentiated MSCs is enhanced if the cells undifferentiated are exposed to IFN-γ.

The immunomodulatory properties of first-trimester fetal liver-derived MSCs differ from those of adult MSCs [28–30]. Fetal MSCs suppress mitogenic responses, but not alloreactivity in MLC. However, when precultured with IFNγ for full HLA class II expression, fetal MSCs inhibit lymphocyte proliferation at a magnitude similar to that seen with adult MSCs. Thus, in spite of the up-regulation of class II alloantigens, IFN-γ appears to enhance the antiproliferative effect that fetal MSCs exert on lymphocyte proliferation.

Several mechanisms have been proposed for the way in which MSCs suppress T-cell activation. MSCs are suppressive even if the T cells are separated from MSCs in a transwell system [20, 23, 35]. Thus, suppression is probably mediated by a soluble factor(s) produced by human MSCs, but this is unlikely to be a factor constitutively secreted by MSCs, because cell-free MSC culture supernatants fail to suppress alloreactivity, in contrast to supernatants from MSC-lymphocyte co-cultures which are immunosuppressive [25, 37, 40]. The soluble factors mediating the suppressive effect were suggested to be hepatocyte growth factor (HGF) and transforming growth factor (TGF-β) [20]. The addition of anti-HGF and anti-

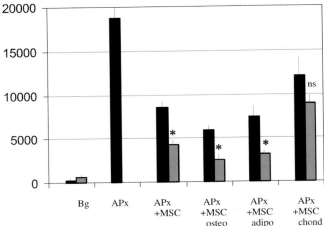

Figure 15.3 MSC inhibition of lymphocyte proliferation is increased by differentiation or exposure to IFN-γ. Undifferentiated MSCs (10⁴) or MSCs grown in induction media for 1 week (black bars), differentiated to osteoblasts, adipocytes or chondrocytes, and then exposed to IFN-γ for 48 h (gray bars), were added to the MLC (100 000 responding and stimulating PBL, respectively). Only autologous cells were used. Lymphocyte proliferation was significantly suppressed when undifferentiated MSCs, or MSCs cultured in osteogenic, adipogenic or chrondrogenic media were added to the MLC (*p <0.05; Student's *t*-test for paired samples). Exposure of different cells to IFN-γ enhance the suppression significantly when compared to differentiated cells not exposed to INF-γ (p <0.05). Bg: background values; ns: not significant. Reproduced with permission from [27].

TGF-β restored T-cell proliferation in the presence of MSCs. Contradictory data also exist, however [37]. Indeed, it was recently suggested that MSC-produced prostaglandin E2 is the factor which induces T-cell suppression [38]. The results of another study suggested that indoleamine 2,3-dioxygenase (IDO)-mediated tryptophan depletion by MSC can act as a T-cell inhibitory effector mechanism [41]. IDO, which is induced by IFN-γ, catalyzes the conversion from tryptophan to kynurenine and inhibits T-cell responses [42]. However, in contrast one report showed that neither MSC production of IL-10, TGFβ-1, prostaglandin E2, nor tryptophan depletion in the culture medium was responsible for the immunosuppressive effect by MSCs [23]. In addition, MSCs produce bone morphogenetic protein 2 (BMP 2), which may mediate immunosuppression via the generation of CD8+ regulatory cells [40]. These controversial data may be due to the use of MSCs generated by different techniques, the use of different fetal calf sera, the stimuli used, or the culture conditions, doses and kinetics, as well as to the different lymphocyte populations tested. This may in turn affect cytokine and chemokine secretion, with seemingly contradictory results. Some investigators may wish to select a few experiments with uniform outcome to simplify heterogeneous results. Others include all data showing more variation and a more complex picture [22, 25]. Apparent species-specific differences exist, particularly between murine and human MSCs, which adds to the confusion. In mice – in contrast to humans – MSCs require cell-to-cell contact in order to induce immunosuppression. To date, all available data suggest that several mechanisms are involved in the MSC-mediated immunosuppression.

15.4
MSC in the Clinic

A Phase I trial was performed in 15 patients to determine the feasibility of collection, expansion and intravenous infusion of human MSCs in the autologous setting [43]. Five patients in each group were administered 1, 10 or 50 × 10^6 MSCs, respectively, with no adverse reactions being observed in any patient.

Following intravenous infusion into the human body, MSCs circulate for a short period of time. Koç et al. detected circulating clonogenic MSCs in some (but not all) patients within the first hours of infusion, but not at later time points [44]. Human MSCs were shown to engraft in multiple tissues and to demonstrate site-specific differentiation after intrauterine transplantation into sheep [14, 45]. An *in-vivo* engraftment potential in bone has also been shown after intravenous MSC infusion in children with osteogenesis imperfecta [46] and a patient with severe aplastic anemia [47]. Engraftment, although in low numbers, was demonstrated using gene-marked MSCs [46].

Following allogeneic SCT, the immune and hematopoietic system is in most cases entirely donor in origin [48–50]. After high-dose chemoirradiation therapy prior to autologous or allogeneic SCT, the marrow stroma is damaged and slow to reconstitute [51–56]. Because damage to the stroma may affect hematopoietic engraftment after SCT, it is possible that reconstitution of stromal cells by the infusion of MSCs may enhance hematopoiesis after transplantation.

In a first study to explore if MSCs enhance engraftment in autologous SCT in humans, MSCs were isolated and expanded in breast cancer patients receiving peripheral blood stem cell infusions [44]. Twenty-eight patients were infused with 1–2 × 10^6 MSCs given intravenously, after which no toxicity was seen and there was rapid hematopoietic recovery. Clearly, the results of this safety study have paved the way for prospective randomized trials.

MSCs express high levels not only of arylsulfatase A, the deficiency of which is the cause of metachromatic leukodystrophy (MLD), but also α-L-iduronidase [57], the deficiency of which is the cause of Hurler's disease. Donor-derived MSCs were expanded and given intravenously to patients with MLD and Hurler's disease who previously had undergone allogeneic SCT [58]. Following MSC infusion, there was clear evidence of improvement of nerve conduction velocity in four of five patients with MLD.

A 20-year-old woman with acute myeloid leukemia received peripheral blood stem cells (PBSCs) combined with MSCs from her HLA haploidentical father [59]. The patient engrafted rapidly, had no acute or chronic GVHD, and was well at 31 months after allogeneic SCT, though it must be remembered that this was a single case with a remarkable outcome. Using haploidentical grafts which are not T-cell-depleted, the risk of GVHD is overwhelming and the risk of rejection is also significant.

In a multicenter clinical trial, HSCs and MSCs derived from HLA-identical sibling donors were infused in order to promote hematopoietic engraftment and limit GVHD [60]. There was no toxicity noted after MSC infusion, and grade II–IV acute GVHD was 15 % in the MSC group, compared to 40 % in a matched control group (p = 0.01). Survival at 6 months in the patients receiving MSCs was 88 %, compared to 68 % in controls. The results of this preliminary study suggest that the infusion of MSCs together with allogeneic SCT may enhance engraftment, decrease GVHD, and improve survival. Currently, prospective randomized trials with MSCs are needed in allogeneic SCT, and we have initiated two such studies in Europe – one in HLA-identical sibling transplants and one using unrelated donors.

Recently, we reported a male patient with grade IV acute GVHD of the gut and liver who had undergone allogeneic SCT with a female HLA-A, -B, -DR-compatible unrelated donor [61]. The patient was unresponsive to all types of immunosuppression, including prednisolone 2 mg kg^{-1} daily, repeated infusions of intravenous methylprednisolone, extracorporeal treatment with psoralen and ultraviolet-A light one to four times weekly for 6 weeks, infliximab and daclizumab for 4 weeks, mucophenolate mofetil, and methotrexate. He was treated for repeated bacterial, viral, and invasive fungal infections. After infusion of 2×10^6 MSCs kg^{-1} from his HLA haploidentical mother, he had a miraculous response of GVHD with a decline in bilirubin and normalization of stool. After infusion of MSCs, a DNA analysis of the patient's bone marrow showed the presence of minimal residual disease (MRD) of his acute lymphoid leukemia (ALL) [62]. Cyclosporine treatment was discontinued to allow for a maximum graft-versus-leukemia effect. When immunosuppression was discontinued, the patient again developed severe acute GVHD. He received a repeat infusion of MSCs from his mother (1×10^6 cells kg^{-1}). After 1 week, his stools were normal, he started to eat again, and his bilirubin subsequently normalized. At one year after transplant, the patient was living at home and was well, with no MRD in the blood or bone marrow. Two additional patients with grades II–IV acute GVHD of the gut responded to MSC therapy from their respective HLA-identical allogeneic SCT donors at our unit. It has also been suggested that MSCs might be used to treat rejections of organ allografts. Indeed, a rat cardiac allograft study showed that MSCs home to the site of allograft rejection [63]. Furthermore, MSCs may have an implication in autoimmune inflammatory bowel disease, because of their immunomodulatory effect and the possibility of healing damaged gut epithelium [61].

Abbreviations/Acronyms

ALL	acute lymphoid leukemia
BMP 2	bone morphogenetic protein 2
CML	cell-mediated lympholysis
Con A	concanavalin A
CTL	cytotoxic T-lymphocyte
GVHD	graft-versus-host disease
HGF	hepatocyte growth factor
HLA	human leukocyte antigen
HSC	hematopoietic stem cell
ICAM-2	intracellular adhesion molecule-2
IDO	indoleamine 2,3-dioxygenase
KIR	killer-cell inhibitory receptor
LFA	leukocyte functional antigen
MHC	major histocompatibility complex
MLC	mixed lymphocyte cultures
MRD	minimal residual disease
MSC	mesenchymal stem cell
NK	natural killer
PBL	peripheral blood lymphocytes
PBSC	peripheral blood stem cell
PHA	phytohemagglutinin
SCT	stem cell transplantation
TGF-β	transforming growth factor-beta
VCAM	vascular cell adhesion molecule

Acknowledgments

The authors thank Inger Hammarberg for typing the manuscript. These studies were supported by grants from the Swedish Cancer Society (0070-B03-17XBC, 4562-B03-03XAC), The Children's Cancer Foundation (PROJ03/039, PROJ03/007), The Swedish Research Council (K2003-32X-05971-23A, K2003-32XD-14716-01A), The Cancer Society in Stockholm, The Tobias Foundation, and the Karolinska Institutet.

References

1. Friedenstein AJ, Petrakova KV, Kurolesova AI, et al. Heterotypic transplants of bone marrow: analysis of precursor cells for osteogenic and hematopoietic tissues. *Transplantation* **1968**; *6*: 230–247.

2. Friedenstein AJ. Precursor cells of mechanocytes. *Int Rev Cytol* **1976**; *47*: 327–345.

3. Friedenstein AJ, Chailakhyan RK, Gerasimov UV. Bone marrow osteogenic stem cells: in vitro cultivation and transplantation into diffusion chambers. *Cell Tissue Kinet* **1987**; *20*: 263–272.

4. Owen ME, Friedenstein AJ. Stromal stem cell: marrow-derived osteogenic precursors. *SIBA Foundation Symposium* **1988**; *136*: 42–60.

5. Campagnoli C, Roberts IA, Kumar S, Bennett PR, Bellantuono I, Fisk NM. Identification of mesenchymal stem/progenitor cells in human first-trimester fetal blood, liver, and bone marrow. *Blood* **2001**; *98*: 2396–2402.

6. Noort WA, Kruysselbrink AB, in't Anker PS, Kruger M, van Bezooijen RL, de Paus RA, Heemskerk MH, Lowik CV, Falkenburg JH, Willemze R, Fibbe WE. Mesenchymal stem cells promote engraftment of human umbilical cord blood-derived CD34+ cells in NOD/SCID mice. *Exp Hematol* **2002**; *30*: 870–878.

7. Erices A, Conget P, Minguell JJ. Mesenchymal progenitor cells in human umbilical cord blood. *Br J Haematol* **2000**; *109*: 235–242.

8. De Ugarte D, Morizono K, Elbarbary A, Alfonso Z, Zuk P, Zhu M, Dragoo J, Ashjian P, Thomas B, Benhaim P, Chen I, Fraser J, Hedrik M. Comparison of multilineage cells from human adipose tissue and bone marrow. *Cells Tissues Organs* **2003**; *174*: 101–109.

9. Haynesworth SE, Goshima J, Goldberg VM, Caplan AI. Characterization of cells with osteogenic potential from human marrow. *Bone* **1992**; *13*: 81–88.

10. Pittenger MF, Mackay AM, Beck SC, et al. Multilineage potential of human mesenchymal stem cells. *Science* **1999**; *284*: 143–147.

11. Prockop DJ. Marrow stromal cells as stem cells for non-hematopoietic tissues. *Science* **1997**; *276*: 71–74.

12. Haynesworth SE, Baber MA, Caplan AI. Cell surface antigens on human marrow-derived mesenchymal cells are detected by monoclonal antibodies. *Bone* **1992**; *13*: 69–80.

13. Pereira RF, Halford KW, O'Hara MD, et al. Cultured adherent cells from marrow can serve as long-lasting precursors for bone, cartilage, and lung in irradiated mice. *Proc Natl Acad Sci USA* **1995**; *92*: 4857–4861.

14. Liechty KW, MacKenzie TC, Shaaban AF, Radu A, Moseley AM, Deans R, Marshak DR, Flake AW. Human mesenchymal stem cells engraft and demonstrate site-specific differentiation after in utero transplantation in sheep. *Nat Med* **2000**; *6*(11): 1282–1286.

15. Devine SM, Cobbs C, Jennings M, Bartholomew A, Hoffman R. Mesenchymal stem cells distribute to a wide range of tissues following systemic infusion into non-human primates. *Blood* **2003**; *101*: 2999–3001.

16. Campagnoli C, Roberts IA, Kumar S, et al. Identification of mesenchymal stem/progenitor cells in human first-trimester fetal blood, liver, and bone marrow. *Blood* **2001**; *98*: 2396–2402.

17. Majumdar MK, Thiede MA, Mosca JD, et al. Phenotypic and functional comparison of cultures of marrow-derived mesenchymal stem cells (MSC) and stromal cells. *J Cell Physiol* **1998**; *176*: 186–192.

18. Angeloupolou M, Novelli E, Grove J, et al. Cotransplantation of human mesenchymal stem cells enhances human myelopoiesis and megakaryocytopoiesis in NOD/SCID mice. *Exp Hematol* **2003**; *31*: 413–420.

19. Deans RJ, Moseley A-M. Mesenchymal stem cells: Biology and potential clinical uses. *Exp Hematol* **2000**; *28*: 875–884.

20. Di Nicola M, Carlostella C, Magni M, Milanesi M, Longoni PD, Matteucci P, et al. Human bone marrow stromal cells suppress T-lymphocyte proliferation induced by cellular or nonspecific mito-genic stimuli. *Blood* **2002**; *99*: 3838–3843.

21. Bartholomew A, Sturgeon C, Siatskas M, et al. Mesenchymal stem cells suppress lymphocyte proliferation in vitro and prolong skin graft survival in vivo. *Exp Hematol* **2002**; *30*: 42–48.

22. Le Blanc K, Tammik C, Sundberg B, Haynesworth S, Ringdén O. Mesenchymal stem cells inhibit and stimulate mixed lymphocyte cultures and mitogenic responses independently of the major histocompatibility system. *Scand J Immunol* **2003**; *57*: 11–20.

23. Tse WT, Pendleton JD, Beyer WM, Egalka MC, Guinan EC. Suppression of allogeneic T-cell proliferation by human marrow stromal cells: Implications in transplantation. *Transplantation* **2003**; *75*: 389–397.

24. Krampera M, Glennie S, Dyson J, Scott D, Laylor R, Simpson E, et al. Bone marrow mesenchymal stem cells inhibit the response of naïve and memory antigen-specific T cells to their cognate peptide. *Blood* **2003**; *101*: 3722–3729.

25. Maitra B, Szekely E, Gjini K, Laughlin MJ, Dennis J, Haynesworth SE, Koç ON. Human mesenchymal stem cells support unrelated donor hematopoietic stem cells and suppress T-cell activation. *Bone Marrow Transplant* **2004**; *33*: 597–604.

26. Majumdar M, Keane-Moore M, Buyaner D *et al*. Characterization and functionality of cell surface molecules on human mesenchymal stem cells. *J Biomed Sci* **2003**; *10*: 228–241.

27. Le Blanc K, Tammik C, Götherström C, Zetterberg E, Ringdén O. HLA-expression and immunologic properties of differentiated and undifferentiated mesenchymal stem cells. *Exp Hematol* **2003**; *31*: 890–896.

28. Götherström C, Ringdén O, Tammik C, Zetterberg E, Westgren M, Le Blanc K. Immunological properties of human fetal mesenchymal stem cells. *Am J Obstet Gynecol* **2004**; *190*: 239–245.

29. McIntosh K, Bartholomew A. Stromal cell modulation of the immune system. *Graft* **2000**; *3*: 324–328.

30. Götherström C, Ringdén O, Westgren M, Tammik C, Le Blanc K. Immunomodulatory effects of human foetal liver-derived mesenchymal stem cells. *Bone Marrow Transplant* **2003**; *32*: 265–272.

31. Wilson JL, Proud G, Forsythe J, et al. Renal allograft rejection. Tubular epithelial cells present alloantigen in the presence of costimulatory CD28 antibody. *Transplantation* **1995**; *59*: 91–97.

32. Laning J, Deluca JE, Isaacs C, et al. In vitro analysis of CD40-CD154 and CD28-CD80/86 interactions in the primary T-cell response to allogeneic "nonprofessional" antigen presenting cells. *Transplantation* **2001**; *71*: 1467–1474.

33. Bartholomew A, Patil S, Mackay A, et al. Baboon mesenchymal stem cells can be genetically modified to secrete human erythropoietin *in vivo*. *Human Gene Ther* **2001**; *12*: 1527–1591.

34. Devine SM, Bartholomew AM, Mahmud N, et al. Mesenchymal stem cells are capable of homing to the bone marrow of non-human primates following systemic infusion. *Exp Hematol* **2001**; *29*: 244–255.

35. Rasmusson I, Ringdén O, Sundberg B, Le Blanc K. Mesenchymal stem cells inhibit the formation of cytotoxic T lymphocytes, but not activated cytotoxic T lymphocytes or natural killer cells. *Transplantation* **2003**; *76*: 1208–1213.

36. Potian J, Aviv H, Ponzio N, Harrison J, Rameshwar P. Veto-like activity of mesenchymal stem cells: Functional discrimination between cellular responses to allo-antigens and recoll antigens. *J Immunol* **2003**; *171*: 3426–3434.

37. Le Blanc K, Rasmusson I, Götherström C, Seidel C, Sundberg B, Rosendahl K, Tammik C, Ringdén O. Mesenchymal stem cells inhibit the expression of IL-2 receptor (CD25) and CD38 on phytohemagglutinin activated lymphocytes. *Scand J Immunol* **2004**; *60*: 307–315.

38. Aggarwal S, Pittenger F. Mechanism of human mesenchymal stem cell modulation of allogeneic T cell response. 10th Annual Meeting of ISCT, Dublin, May 7–10, **2004**, p.82, abstract 201.

39. Rasmusson I, Ringdén O, Sundberg B, Le Blanc K. Mesenchymal stem cells inhibit lymphocyte proliferation by mitogens and allogens by different mechanisms. Exp Cell Res 2005; 305; 33–41.

40. Djouad F, Plence P, Bony C, et al. Immunosuppressive effect of mesenchymal stem cells favors tumor growth in allogeneic animals. *Blood* **2003**; *102*: 3837–3844.

41. Meisel R, Zibert A, Laryea M, Göbel U, Däubener W, Dilloo D. Human bone marrow stromal cells inhibit allogeneic T-cell responses by indoleamine 2,3-dioxygenase mediated tryptophan degradation. *Blood* **2004**; *103*: 4619–4621.

42. Munn DH, Zhou M, Attwood JT, Bondarev I, Conway SJ, Marshall B, Brown C, Mellor AL. Prevention of allogeneic fetal rejection by tryptophan catabolism. *Science* **1998**; *281*: 1191–1193.

43. Lazarus HM, Haynesworth SE, Gerson SL, Rosenthal NS, Caplan AI. *Ex vivo* expansion and subsequent infusion of human bone marrow derived stromal progenitor cells (mesenchymal progenitor cells): implications for therapeutic use. *Bone Marrow Transplant* **1995**; *16*: 557–564.

44. Koç ON, Gerson SL, Cooper BW, Dyhouse SM, Haynesworth SE, Caplan AI, Lazarus HM. Rapid hematopoietic recovery after co-infusion of autologous-blood stem cells and culture-expanded marrow mesenchymal stem cells in advanced breast cancer patients receiving high-dose chemotherapy. *Clin Oncol* **2000**; *18*: 307.

45. Airey J, Almeida-Porada G, Colletti E, Porada C, Chamberlain J, Movsesian M, Sutko J, Zanjani E. Human mesenchymal stem cells from form Purkinje fibers in fetal sheep heart. *Circulation* **2004**; *109*: 1401–1407.

46. Horwitz EM, Gordon PL, Koo WK, Marx JC, Neel MD, McNall RY, Muul L, Hofmann T. Isolated allogeneic bone marrow-derived mesenchymal cells engraft and stimulate growth in children with osteogenesis imperfecta: Implications for cell therapy of bone. *Proc Natl Acad Sci USA* **2002**; *99*: 8932–8937.

47. Fouillard L, Bensidhoum M, Bories D, et al. Engraftment of allogeneic mesenchymal stem cells in the bone marrow of a patient with severe idiopathic aplastic anemia improves stroma. *Leukemia* **2003**; *17*: 474–476.

48. Durnam DM, Anders KR, Fisher L, O'Quigley J, Bryant EM, Thomas ED. Analysis of the origin of marrow cells in bone marrow transplant recipients using a Y-chromosome-specific in situ hybridization assay. *Blood* **1989**; *74*: 2220.

49. Mattsson J, Uzunel M, Remberger M, Ringdén O. T-cell mixed chimerism is significantly correlated to a decreased risk of acute graft-versus-host disease after allogeneic stem cell transplantation. *Transplantation* **2001**; *71*: 433–439.

50. Stute N, Fehse B, Schroder J, Arps S, Adamietz P, Held KR, Zander AR. Human mesenchymal stem cells are not of donor origin in patients with severe aplastic anemia who underwent sex-mismatched allogeneic bone marrow transplant. *J Hematother Stem Cell Res* **2002**; *11*: 977–984.

51. Chamberlain W, Barone J, Kedo A, Fried W. Lack of recovery of murine hematopoietic stromal cells after irradiation-induced damage. *Blood* **1974**; *44*: 385–392.

52. Fried W, Chamberlain W, Kedo A, Barone J. Effect of radiation on hematopoietic stroma. *Exp Hematol* **1976**; *4*: 310–314.

53. O'Flaherty E, Sparrow R, Szer J. Bone marrow stromal function from patients after bone marrow transplantation. *Bone Marrow Transplant* **1995**; *15*: 207–212.

54. Carlostella C, Tabilio A, Regazzi E, et al. Effect of chemotherapy for acute myelogenous leukaemia on hematopoietic and fibroblast marrow progenitors. *Bone Marrow Transplant* **1997**; *20*: 465–471.

55. Galotto M, Berisso G, Delfino L, et al. Stromal damage as consequence of high-dose chemo/radiation therapy in bone marrow transplant recipients. *Exp Hematol* **1999**; *27*: 1460–1466.

56. Awaya N, Rupert K, Bryant E, Torok-Storb B. Failure of adult marrow-derived stem cells to generate marrow stroma after successful hematopoietic stem cell transplantation. *Exp Hematol* **2002**; *30*: 937–942.

57. Koç O, Peters C, Raghavan S, DeGasperi R, Kolodny E, Ben Yoseph Y, et al. Bone marrow derived mesenchymal stem cells of patients with lysosomal and peroxisomal storage diseases remain host type following allogeneic bone marrow transplantation. *Exp Hematol* **1999**; *27*: 1675–1681.

58. Koç ON, Day J, Nieder M, Gerson SL, Lazarus HM, Krivit W. Allogeneic mesenchymal stem cell infusion for treatment of metachromatic leukodystrophy (MLD) and Hurler syndrome (MPS-IH). *Bone Marrow Transplant* **2002**; *30*: 215–222.

59. Lee ST, Jang JH, Cheong J-W, Kim JS, Maemg H-Y, Hahn JS, Ko YW, Min YH. Treatment of high-risk acute myelogenous leukaemia by myeloablative chemoradiotherapy followed by co-infusion of T cell-depleted haematopoietic stem cells and culture-expanded marrow mesenchymal stem cells from a related donor with one fully mismatched human leucocyte antigen haplotype. *Br J Haematol* **2002**; *118*: 1128–1131.

60. Frassoni F, Labopin M, Bacigalupo A, et al. Expanded mesenchymal stem cells (MSC), co-infused with HLA identical hematopoietic stem cell transplants, reduce acute and chronic graft-versus-host disease: a matched pair analysis. *Bone Marrow Transplantation* **2002**; *29* (Suppl.2): S2.

61. Le Blanc K, Rasmusson I, Sundberg B, Götherström C, Hassan M, Uzunel M, Ringdén O. Treatment of severe acute graft-versus-host disease with third party haploidentical mesenchymal stem cells. *Lancet* **2004**; *363*: 1439–1441.

62. Uzunel M, Mattsson J, Jaksch M, Remberger M, Ringdén O. The significance of graft-versus-host disease and pre-transplant minimal residual disease status to outcome after allogeneic stem cell transplantation in patients with acute lymphoblastic leukemia. *Blood* **2001**; *98*: 1982–1984.

63. Wu GD, Nolta JA, Yin J-S, Marr ML, Yu H, Starnes VA, Cramer DV. Migration of mesenchymal stem cells to heart allografts during chronic rejection. *Transplantation* **2003**; *75*: 679–685.

16

Stem Cell Transplantation: The Basis for Successful Cellular Immunotherapy

Lessons Learned from Indolent B-Cell Lymphoma

Peter Dreger, Matthias Ritgen, and Anthony D. Ho

16.1
Introduction

Although allogeneic stem cell transplantation (allo-SCT) has been in clinical use for more than 30 years, it has been discovered only recently that in several situations its success essentially relies on immunotherapeutic mechanisms. Although originally developed to circumvent hematopoietic toxicity of high-dose radiochemotherapy, in many instances allo-SCT serves as a tool for establishing lympho-hematopoietic chimerism as the prerequisite for effective induction of graft-versus-leukemia (GVL) activity. The aim of this chapter is to illustrate the clinical potential of allo-SCT-based immunotherapy by the example of indolent B-cell lymphoma. The major focus will be on chronic lymphocytic leukemia (CLL) as an entity which is studied best in this context, though extrapolations will be performed to follicular lymphoma and Waldenström's disease.

16.2
Allogeneic Stem Cell Transplantation in CLL

Although CLL is one of the most common hematological malignancies and not curable with conventional treatment, until recently it has been a relatively infrequent indication for allo-SCT. This had to do with the fact that patients with far advanced disease – for whom allo-SCT was deemed particularly helpful – often were ineligible for classical allo-SCT with myeloablative conditioning due to high age or CLL-related morbidity. On the other hand, valid prognostic parameters for the early segregation of aggressive versus indolent CLL to allow "preemptive" allo-SCT early during the course of the disease phases were unavailable.

Therefore, until the late 1990s only a few young patients with advanced or refractory CLL underwent allo-SCT, mostly outside of clinical protocols. All of these

Stem Cell Transplantation. Biology, Processing, and Therapy.
Edited by Anthony D. Ho, Ronald Hoffman, and Esmail D. Zanjani
Copyright © 2006 WILEY-VCH Verlag GmbH & Co. KGaA, Weinheim
ISBN: 3-527-31018-5

Table 16.1 Selected studies on allogeneic SCT with myelo-ablative conditioning in chronic lymphocytic leukemia (CLL).

Parameter	Trial			
	IBMTR [1, 6]	Int. Project [3]	NMDP [2]	Barcelona [35]
Design	Registry analysis	Registry analysis	Registry analysis	Single center
No. of patients	54	46	38	23
Alternative donors (%)[a]	0	0	100	0
TRM (%)	44 (10 y)	31 (3 m)	38 (5 y)	17 (5 y)
Survival (%)	41 (10 y)	56 (5 y)	33 (5 y)	67 (5 y)
Relapse rate (%)	15 (10 y)	23 (5 y)	32 (5 y)	12 (5 y)
Late relapses (>2 years)	3	2	1	3
Follow-up (months)[b]	120 (60–192)	35 (11–176)	72 (36–108)	62 (17–215)

[a]Matched unrelated donors or mismatched family donors.
[b]Values are mean (range).
TRM, Treatment-related mortality.

transplantations were performed with standard myeloablative conditioning – generally total body irradiation with high-dose cyclophosphamide (TBI/CY) – to deliver maximum cytotoxic activity for achieving direct complete tumor cell elimination. Subsequent transfusion of the graft was merely regarded as source of hematopoietic reconstitution.

Altogether, the results of allo-SCT with classical conditioning in CLL were disappointing, largely because of an excessive transplant-related mortality which in some series was up to 50 % [1–5]. The reasons for this extraordinary high complication rate are not completely clear, but the relatively high patient age, unfavorable patient selection, extensive pretreatment, and reduced performance status due to advanced disease together with the genuine toxicity of standard allo-SCT all may have contributed to the high mortality observed.

In spite of these unsatisfying outcome data, these early series already suggested that allo-SCT could provide long-term disease control, even in refractory CLL (Table 16.1). A first registry analysis reported on 54 patients with advanced CLL who were allografted from HLA-identical sibling donors after classical myeloablative conditioning [1]. Although 45 % of the patients had refractory disease at the time of SCT, only two relapses occurred later than 2 years post transplant. Due to high treatment-related mortality (TRM), however, overall survival after 3 years was not higher than 46 %. A recent update after a median follow-up of 10 years showed only three additional CLL-related deaths (two in patients who had received a T-cell-depleted graft), which translated into a 10-year leukemia-free and overall survival of 37 % and 41 %, respectively [6].

16.3
Graft-versus-Leukemia Effect in CLL

In addition to the observation that late relapses rarely occur after allo-SCT, there is ample circumstantial evidence that GVL activities are effective in CLL. This includes findings that long-term molecular responses can be obtained with allo-SCT but not with auto-SCT [7–10], a reduced relapse risk in the presence of chronic graft-versus-host disease (GVHD) [11], an increased relapse risk associated with the use of T-cell-depleted allografts [12], and anecdotal reports on the efficacy of donor lymphocyte infusion (DLI) [10, 12, 13].

A direct proof of the existence and immunotherapeutic capacity of the GVL principle in CLL, however, requires evidence of tumor response closely correlated to immune interventions or GVHD in the absence of effective cytostatic treatment. To this end, a recent study analyzed the kinetics of minimal residual disease (MRD) by quantitative RQ-PCR in nine patients with poor-risk CLL who were allografted after nonmyeloablative conditioning with fludarabine and cyclophosphamide. The results showed that profound and sustained complete molecular responses occurred only after establishing chronic GVHD or DLI, whereas the influence of the conditioning regimen on the tumor cell load was very limited. The dynamic pattern of this process and its close correlation with immune modulating maneuvers or chronic GVHD strongly suggested that GVL activity was responsible for tumor disappearance. Of note, all patients studied had an unmutated VH gene status, indicating that genetically unfavorable unmutated CLL is sensitive to GVL effects [14]. The fact that the GVL-mediated anti-leukemic activity observed was clearly superior to that of the intensified fludarabine-cyclophosphamide regimen used for conditioning, as well as to that of the fully myeloablative treatment delivered to controls in the context of autologous SCT, provides a rationale basis for studying allo-SCT with reduced-intensity conditioning in CLL.

16.4
Allo-SCT with Reduced-Intensity Conditioning in CLL

The finding that GVL activity, but not the conditioning regimen's cytotoxicity, is the crucial factor for disease eradication in at least some allo-SCT indications, such as chronic myeloid leukemia (CML), has led to the development of novel conditioning strategies for allo-SCT during recent years; this is termed reduced-intensity conditioning transplantation (RICT). The hallmark of RICT is that conditioning is tailored in a way that it has immunosuppressive potential sufficient to allow full lymphohematopoietic engraftment, but does not necessarily exert relevant cytotoxic effects on the tumor itself. Accordingly, acute toxic complications can be kept very low in comparison to standard myeloablative conditioning. This makes the procedure suitable for older and comorbid persons who are ineligible for standard allo-SCT. Given the documented GVL sensitivity of CLL, RICT

Table 16.2 Selected studies on allogeneic SCT with nonmyeloablative conditioning in chronic lymphocytic leukemia (CLL).

	Study					
	Seattle [17]		**GCTSG [10]**		**GCLLSG [18]**	
	ID	**UD**	**ID**	**UD**	**ID**	**UD**
Study design	Multicenter		Multicenter		Multicenter	
Conditioning	TBI 2 Gy, fludarabine 90 mg m^2		Busulfan 8 mg kg^{-1}, fludarabine 180 mg m^{-2}		Cyclophosphamide 2.5 g m^2, fludarabine 150 mg m^2	
No. of patients	44	20	13	17	19	18
TRM (%)	22	20 (2 y)	15 (4 y)		0	19 (2 y)
Early death (<100 days) (%)	8	18	6		0	6
Survival (%)	56	74 (2 y)	69 (4 y)[a]		92	61 (2 y)[b]
Relapse rate (%)	34	5 (2 y)	30 (4 y)		17	30 (2 y)
Late relapses (>2 years)	0	0	4		2	0
Follow-up (months)[c]	24 (3–63)		50 (29–65)		15 (2–64)	

[a]No significant difference, ID versus UD.
[b]p <0.05.
[c]Values are mean (range).
ID, matched sibling donor; TBI, total body irradiation;
TRM, treatment-related mortality; UD, matched unrelated donor.

is increasingly being studied as potentially curative treatment for patients with poor-risk CLL.

A retrospective analysis which compared RICT with myeloablative transplantation for the treatment of advanced CLL indeed demonstrated a significant advantage of RICT in terms of TRM [15]. Since relapse risk was increased after RICT, however, overall survival was not different between the two types of conditioning. Although this observation might be due to risk profile imbalances between the two cohorts, it questions the hypothesis that GVL effects might be able to eradicate CLL independent of the conditioning regimen's intensity, thereby underlining the necessity of prospective clinical trials on RICT in CLL.

To date, information on four prospective CLL-specific RICT series is available (Table 16.2). Despite the fact that none of these considered defined risk situations by intent-to-treat, the majority of patients in all three trials not using T-cell depletion were characterized by clinical or biological factors which are well known to be associated with a very unfavorable prognosis, such as refractoriness to purine analogues or poor-risk genetics. Nevertheless, the incidence of relapse appeared to be remarkably low after RICT in the absence of T-cell depletion. Particular encouraging was the paucity of late recurrences, paralleling the situation after standard myeloablative conditioning and strongly suggesting GVL-mediated disease control. In contrast, a clear-cut reduction of the relapse incidence with time could not be observed in a prospective analysis of 41 patients with CLL who had undergone allo-SCT with the Mackinnon regimen (fludarabine-melphalan-alemtuzumab conditioning) which includes *in-vivo* T-cell depletion [16]. However, the median follow-up of these patients was only 15 months. Only the other hand, TRM in the prospective trials was intriguingly modest, resulting in 2-year overall survivals of 56 to 80 %; these appear to compare favorably with those reported for standard allo-SCT in CLL [10, 17, 18].

In conclusion, there is no doubt that the crucial therapeutic principle of allo-SCT is GVL activity. All published information suggests that reduction of the conditioning regimen intensity can help to reduce TRM after SCT in CLL without affecting GVL effectiveness. However, there might be clinical settings, such as aggressive disease with 17p13 deletion, where conditioning intensity could play a role. Thus, according to the individual situation, the optimum choice of conditioning regimen may vary. Whereas in the presence of comorbidity and sensitive disease reduced-intensity regimens appear to be more appropriate, high-intensity regimens might be preferable in patients with robust performance status but poorly controlled disease [19]. Prospective clinical trials will help to guide precisely the choice of conditioning intensity.

16.5
RICT from Unrelated Donors

Basically, allo-SCT can be performed using an HLA-identical related donor (ID-SCT) or an HLA-matched unrelated donor (UD-SCT). Due to the larger immunogenetic difference, GVHD-mediated toxicity might be more pronounced after UD-SCT. For the same reason, UD-SCT promises to provide more effective disease control by GVL. To date, data on UD-SCT in CLL are available from five larger trials. Exclusively UD-SCT with standard myeloablative conditioning was studied by investigators from the National Marrow Donor Programs (NMDP) [2]. As in comparable series using matched related donors, the results were characterized by high TRM but low relapse rate, with only one disease recurrence later than 2 years post transplant (see Table 16.1). All four prospective RICT trials described earlier allowed inclusion of unrelated donors. Data are available for two T-cell-replete studies, with no major differences to sibling donors in terms of feasibility and toxicity (see Table 16.2). The largest case numbers were reported by the Seattle group; interestingly, in this study the overall survival tended to be better after UD-SCT than after ID-SCT, due to better disease control [17]. Altogether, the therapeutic potential of UD-SCT seems to be at least equivalent to that of ID-SCT. In the future, adherence to recommendations based on recent analyses of the impact of HLA-class-II high-resolution and HLA-C matching might help to further reduce TRM after UD-SCT [20].

Accordingly, the recent *Guidelines on the diagnosis and treatment of chronic lymphocytic leukemia* of the British Society for Haematology, as well as the *Consensus on CLL Transplants* produced by the European Blood and Marrow Transplantation Group (EBMT) do not differentiate their recommendations for allo-SCT in CLL by donor source [21, 22]. Anyhow, if used with modern conditioning regimens, UD-SCT provides overall survival rates in patients with fludarabine-resistant CLL which are clearly superior to any other antibody- or purine analogue-based salvage therapy.

16.6
T-Cell Depletion

Investigators from the Dana Farber Cancer Institute were the first to study allo-SCT for advanced CLL in a larger series of patients. Bone marrow grafts from HLA-identical sibling donors underwent *ex-vivo* T-cell depletion (TCD) with an anti-CD6 monoclonal antibody and complement and were subsequently transplanted to recipients who had been conditioned with myeloablative TBI and high-dose cyclophosphamide. Although early analyses suggested that sustained MRD-negative remissions can be achieved in a large proportion of patients [23], a recent follow-up of this study revealed disappointing results, with 17 of 25 patients (68 %) relapsing. In contrast to the lack of late relapses observed in the published CLL series without TCD, these recurrences were not restricted to the first two post-transplant years, but showed a continuous pattern resembling that found after auto-SCT for CLL. Accordingly, progression-free survival 6 years after allo-SCT was only 24 % and, thus, significantly worse than that of 137 patients who had been reconstituted with purged autografts due to lack of a compatible sibling donor [12].

During recent years, the focus has moved from *ex-vivo* TCD to *in-vivo* TCD with alemtuzumab, which is infused to the recipient (usually from day −8 through day −4) along with the conditioning regimen. This approach is currently widely used by investigators from the United Kingdom in the context of fludarabine/melphalan nonmyeloablative conditioning. As described earlier, the data are not mature enough to evaluate whether long-term disease control is possible by this type of treatment [16]. The 15 % incidence of graft failure was higher than observed with alemtuzumab/ fludarabine/melphalan conditioning in other lymphoproliferative diseases, and is ascribed to CLL-specific interaction with alemtuzumab pharmacokinetics and CLL-related imbalances of the recipient immune system. However, relevant engraftment problems are not reported for RICT strategies without TCD. In the GCLLSG CLL3X trial, only two graft failures (5 %) were reported among 41 patients not receiving *in-vivo* TCD, whereas three cases (27 %) of non-engraftment occurred in 11 patients who were treated with alemtuzumab as part of the conditioning regimen (p = 0.057, Fisher's Exact test).

Altogether, a beneficial effect of TCD on the outcome of allo-SCT in CLL could not be documented to date, thereby underlining the importance of immune mechanisms for the success of allo-SCT in this disease.

16.7
Allo-SCT in Follicular Lymphoma

Similar to CLL, follicular lymphoma (FL) has been an infrequent indication for allo-SCT, due largely to the high median age of onset and the effectiveness of conventional treatment which, nevertheless, has no curative potential. Another similarity to CLL is that myeloablative conditioning followed by allo-SCT was associated with unacceptably high rates of TRM [24, 25]. A study from the International Bone Marrow Transplant Registry (IBMTR) compared outcomes in 176 patients with FL who had undergone allo-SCT from an HLA-identical sibling after myeloablative conditioning, with 728 patients who had received auto-SCT. Better disease control after allo-SCT was largely offset by its high TRM, resulting in a similar 5-year overall survival. Nevertheless, the lack of late recurrences in the allogeneic group (only 2 % beyond 1 year) and an apparent plateau of the survival curve suggest a potentially curative effect of allo-SCT [26]. Evidence of GVL sensitivity of FL is further substantiated by anecdotal reports of response to DLI in patients having relapsed after allo-SCT [27, 28], and the observation of continuing relapses after the Mackinnon RICT regimen with *in-vivo* TCD [28]. On the other hand, a low incidence of FL recurrence plateauing at 10 % could be achieved with an *in-vivo* TCD approach using the more intensive BEAM regimen (carmustin, etoposide, ara-C, melphalan) instead of fludarabine/melphalan, indicating that lack of GVL activity might be overcome by conditioning intensification [29]. This conclusion must be considered as preliminary as the follow-up time in this study was only 1.4 years, but is in line with the observation that the relapse rate after myeloablative conditioning followed by syngeneic SCT is not inferior to that of T-cell-depleted or unmanipulated allo-SCT [30].

Nevertheless, both studies suggest that RICT might be able to strongly reduce the toxicity of allo-SCT in FL, as the 3-year TRM was 11 % and 16 %, respectively, in these two trials. In the BEAM trial, TRM was excessively high in patients who had previously undergone auto-SCT [29]. This is of concern, since the majority of patients who become candidates for allo-SCT will have had auto-SCT as part of first-line or salvage therapy. Thus, a requirement for an appropriate allo-SCT regimen for FL is that it is feasible in patients who have failed auto-SCT.

To this end, we have studied T-replete RICT in patients with indolent B-cell lymphoma for treatment of relapse after auto-SCT. In order to further elucidate the therapeutic potential of (myeloablative) cytotoxic treatment versus that of GVL-mediated immunotherapy, we compared disease control achieved by RIC allo-SCT and myeloablative auto-SCT, respectively, on an intra-individual basis in 20 patients with FL (n = 10) or CLL (n = 10) who had been allografted for relapse after auto-SCT. High-dose preparations for auto-SCT consisted of standard

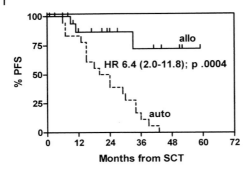

Figure 16.1 Intra-individual comparison of progression-free survival (PFS) after auto-SCT and allo-SCT, respectively, in patients with indolent lymphoma (follicular lymphoma, n = 10; chronic lymphocytic leukemia, n = 10) undergoing reduced-intensity conditioning allo-SCT for relapse after auto-SCT.

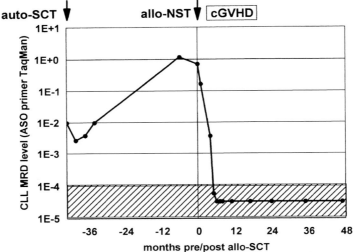

Figure 16.2 Minimal residual disease (MRD) kinetics in a patient with chronic lymphocytic leukemia after auto-SCT and allo-SCT.

TBI/ cyclophosphamide or BEAM, whereas fludarabine/cyclophosphamide or fludarabine/low-dose TBI were used for conditioning for allo-SCT. The median progression-free survival (PFS) after auto-SCT was 24 (range: 6 to 47) months, and time from auto-SCT to allo-SCT was 34 (range: 10 to 117) months. The median number of salvage regimens after auto-SCT was 1 (range: 1 to 8). With a median follow-up of 22 (range: 2 to 59) months, there were no toxic deaths and only three relapses after allo-SCT (two with CLL, one with FL), resulting in a significantly inferior PFS of auto-SCT in comparison to RICT (Hazard ratio 6.4 (range: 2-11.8); p = 0.0004; Fig. 16.1). Five patients had samples available for quantitative comparison of MRD levels by real-time quantitative IgH PCR using allele-specific primers (ASO RQ-PCR). At 12 months after auto-SCT, MRD was present in all five patients at a median level of 9E-04 (range: 2E-04 to 1.5E-02) which increased to 1.1E-01 (range: 1.3E-02 to 1.4E-00) immediately prior to allo-SCT. One year post RICT, however, MRD was undetectable (<5E-05) in four of these five cases (Fig. 16. 2). The remaining patient had a MRD level of 3E-03 with early signs of clinical relapse.

Although these results indicate unequivocally that GVL, and not conditioning intensity, is the crucial factor determining the success of allo-SCT in FL, the potential role of RICT is less clear than in CLL as a lack of GVL activity might be at least partly compensated by preparation regimen intensity. Moreover, the low rates of TRM associated with RICT must be re-evaluated in the light of the observation that also the toxicity of myeloablative allo-SCT has significantly decreased over time [26]. Thus, similar to CLL, in FL the optimum choice of transplant strategy may depend on the clinical situation with preference for RICT regimen in comorbid patients or those with sensitive relapse after auto-SCT, and preference for BEAM-alemtuzumab or fully myeloablative conditioning with or without TCD in patients with good performance status but more aggressive disease. Again, further prospective clinical trials are mandatory to guide precisely the choice of conditioning in FL.

16.8
Allo-SCT in Waldenström's Disease

Waldenström's macroglobulinemia (WM) is a rare B-cell neoplasia characterized by uncontrolled proliferation of IgM-producing clonal plasmacytic lymphocytes. According to the World Health Organization classification, it must be regarded as monoclonal IgM-producing lymphoplasmacytic lymphoma. Clinical features include hyperviscosity-related symptoms, organomegaly, cytopenia, and peripheral neuropathy. Though generally indolent, the course of the disease can be aggressive with a median overall survival of less than 5 years [31]. As conventional therapy based on alkylating agents, purine analogues or antibodies can achieve only temporary remissions, immunotherapy by allo-SCT might be an option similar to related disorders, such as CLL or multiple myeloma.

Although the series were small, myeloablative auto-SCT in WM has been studied since the mid-1990s, suggesting that this approach is effective, but not curative [32, 33]. In contrast, only six cases of allo-SCT for WM had been reported until 2003 [33], when first conclusions on the possible role of allo-SCT in the treatment of this disease became available from a registry analysis by the Société Francaise de Greffe de Moelle (SFGM) [34]. This group presented data from 10 patients who had undergone allo-SCT for advanced WM, in nine cases after myeloablative conditioning. The outcome was excellent for those patients who were chemotherapy-sensitive at SCT, with a 4-year PFS rate of more than 80%, whilst refractory patients had a poor outcome. TRM was reported to be 40%, but it must be taken into account that three of four patients dying from treatment-related causes had disease progression prior to death. By showing a lack of late relapses and response to DLI, this study was the first to provide evidence for the existence of GVL effects in WM.

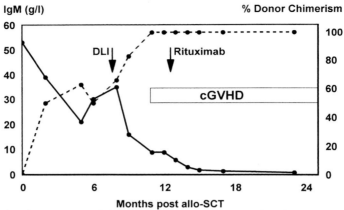

Figure 16.3 IgM and chimerism kinetics in a patient with Waldenström's macroglobulinemia after allo-SCT and donor lymphocyte infusion (DLI).

To further substantiate this hypothesis, we sought to investigate the role of RICT for the treatment of advanced WM. Four symptomatic patients of median age 53 (range: 37 to 54) years who had failed conventional treatment were eligible. Previous therapy comprised four (range: 2 to 5) regimens. Three patients were chemosensitive and one patient was refractory at the time of SCT. Conditioning consisted of daily fludarabine (30 mg m^{-2}) and cyclophosphamide (500 mg m^{-2}) over 5 days. GVHD prophylaxis was performed with cyclosporine A and short-course methotrexate or mycophenolat. Peripheral blood stem cells (PBSCs) were obtained from HLA-identical sibling or unrelated donors. Initially, prompt hematopoietic recovery occurred in all patients. However, the refractory patient remained so immediately after SCT and developed WM-related secondary pancytopenia. Subsequent infusion of donor lymphocytes induced chronic GVHD, followed by conversion to full donor chimerism and complete remission of the disease (Fig. 16.3). Currently, this patient is alive and disease-free with slowly recovering hematopoiesis at 42 months post transplant. Two patients achieved ongoing complete remission in the context of chronic GVHD, and the remaining patient is too early. Apart from GVHD, non-hematopoietic toxicity was low. The median follow-up period is presently 31 (range: 4 to 42) months. Thus, RICT can mediate potent immunotherapeutic activities by GVL effects in patients with poor-risk WM, and is currently under prospective investigation by the German CLL Study Group.

16.9
Conclusions and Perspectives

Taken together, there are several lines of evidence that GVL activity conferred by allo-SCT is a powerful treatment modality in all three indolent B-cell neoplasms discussed in this chapter. Allo-SCT can be regarded as trigger of a permanent immunotherapeutic process in the patient, which is capable of providing long-term disease control and possibly cure in these otherwise incurable diseases. To date, it is not clear whether this is based merely on a general allo-effect or if tumor-specific components are involved. Whereas in FL intensive conditioning might support immune-mediated anti-lymphoma activity, for CLL and WM if remains to be shown if increased dose intensity of the conditioning regimen can add anything but toxicity to the anti-tumor effects exerted by the GVL mechanism itself. Future studies should aim at clarification of these issues and help to define individual transplant-based immunotherapeutic strategies for defined clinical situations where allo-SCT promises to have a better benefit–risk ratio than alternative treatments.

Acknowledgments

These studies were supported in part by grants DJCLS-R18, Deutsche José Carreras Leukämie-Stiftung.

Abbreviations/Acronyms

allo-SCT	allogeneic stem cell transplantation
BEAM	carmustin, etoposide, ara-C, melphalan
CLL	chronic lymphocytic leukemia
CML	chronic myeloid leukemia
CY	cyclophosphamide
DLI	donor lymphocyte infusion
FL	follicular lymphoma
GVL	graft-versus-leukemia
MRD	minimal residual disease
PBSC	peripheral blood stem cell
PFS	progression-free survival
RICT	reduced-intensity conditioning transplantation
TBI	total body irradiation
TCD	T-cell depletion
TRM	treatment-related mortality
WM	Waldenström's macroglobulinemia

References

1. Michallet M, Archimbaud E, Rowlings PA, et al. HLA-identical sibling bone marrow transplants for chronic lymphocytic leukemia. *Ann Intern Med* 1996;*124*:311–315.

2. Pavletic SZ, Khouri IF, Haagenson M, et al. Unrelated donor marrow transplantation for B-cell chronic lymphocytic leukemia after using myeloablative conditioning: results from the center for international blood and marrow transplant research. *J Clin Oncol* 2005;*23*:5788–5794.

3. Esteve J, Montserrat E, Dreger P, et al. Stem cell transplantation (SCT) for chronic lymphocytic leukemia (CLL): Outcome and prognostic factors after autologous and allogeneic transplants [abstract]. *Blood* 2001;*98* (Suppl.1): 482a.

4. Pavletic ZS, Arrowsmith ER, Bierman PJ, et al. Outcome of allogeneic stem cell transplantation for B cell chronic lymphocytic leukemia. *Bone Marrow Transplant* 2000;*25*:717–722.

5. Dreger P, Montserrat E. Autologous and allogeneic stem cell transplantation for chronic lymphocytic leukemia. *Leukemia* 2002;*16*:985–992.

6. Michallet M, Michallet AS, Le QH, et al. Conventional HLA-identical sibling bone marrow transplantation is able to cure chronic lymphocytic leukemia [abstract]. *Blood* 2003;*102*:474a.

7. Mattsson J, Uzunel M, Remberger M, et al. Minimal residual disease is common after allogeneic stem cell transplantation in patients with B cell chronic lymphocytic leukemia and may be controlled by graft-versus-host disease. *Leukemia* 2000;*14*:247–254.

8. Esteve J, Villamor N, Colomer D, et al. Stem cell transplantation for chronic lymphocytic leukemia: different outcome after autologous and allogeneic transplantation and correlation with minimal residual disease status. *Leukemia* 2001;*15*:445–451.

9. Esteve J, Villamor N, Colomer D, Montserrat E. Different clinical value of minimal residual disease after autologous and allogeneic stem cell transplantation for chronic lymphocytic leukemia. *Blood* 2002;*99*:1873–1874.

10. Schetelig J, Thiede C, Bornhauser M, et al. Evidence of a graft-versus-leukemia effect in chronic lymphocytic leukemia after reduced-intensity conditioning and allogeneic stem-cell transplantation: the Cooperative German Transplant Study Group. *J Clin Oncol* 2003;*21*:2747–2753.

11. Dreger P, Brand R, Hansz J, et al. Low treatment-related mortality but retained graft-versus-leukemia activity after allogeneic stem cell transplantation for chronic lymphocytic leukemia using reduced-intensity conditioning. *Leukemia* 2003;*17*:841–848.

12. Gribben JG, Zahrieh D, Stephans K, et al. Autologous and allogeneic stem cell transplantation for poor risk chronic lymphocytic leukemia. *Blood* 2005 (E-pub August 30).

13. Rondon G, Giralt S, Huh Y, et al. Graft-versus-leukemia effect after allogeneic bone marrow transplantation for chronic lymphocytic leukemia. *Bone Marrow Transplant* 1996;*18*:669–672.

14. Dreger P, Ritgen M, Böttcher S, Schmitz N, Kneba M. The prognostic impact of minimal residual disease assessment after stem cell transplantation for chronic lymphocytic leukemia: Is achievement of molecular remission worthwhile? *Leukemia* 2005;*19*:1135–1138.

15. Dreger P, Brand R, Milligan D, et al. Reduced-intensity conditioning lowers treatment-related mortality of allogeneic stem cell transplantation for chronic lymphocytic leukemia: a population-matched analysis. *Leukemia* 2005;*19*:1029–1033.

16. Delgado J, Thomson K, Russell N, et al. Results of alemtuzumab-based reduced-intensity allogeneic transplantation for chronic lymphocytic leukemia: a British Society of Blood and Marrow Transplantation study. *Blood* 2005 (E-pub October 20).

17. Sorror ML, Maris MB, Sandmaier BM, et al. Hematopoietic cell transplantation after nonmyeloablative conditioning for advanced chronic lymphocytic leukemia. *J Clin Oncol* 2005;*23*:3819–3829.

18. Dreger P, Böttcher S, Stilgenbauer S, et al. Long-term disease control of poor-risk chronic lymphocytic leukemia by allogeneic stem cell transplantation with nonmyeloablative conditioning: Interim results of a prospective study [abstract]. *Ann Oncol* 2005;*16* (Suppl.5):v43–v44.

19. Montserrat E, Moreno C, Esteve J, Urbano-Ispizua A, Gine E, Bosch F. How we treat chronic lymphocytic leukemia. *Blood* 2005 (E-pub October 4).

20. Flomenberg N, Baxter-Lowe LA, Confer D, et al. Impact of HLA class I and class II high-resolution matching on outcomes of unrelated donor bone marrow transplantation: HLA-C mismatching is associated with a strong adverse effect on transplantation outcome. *Blood* 2004;*104*:1923–1930.

21. Oscier D, Fegan C, Hillmen P, et al. Guidelines on the diagnosis and management of chronic lymphocytic leukaemia. *Br J Haematol* 2004;*125*:294–317.

22. Dreger P, Corradini P, Kimby E, et al. Consensus on indications for allogeneic stem cell transplantation in chronic lymphocytic leukemia: A proposal by the EBMT subcommittee [abstract]. *Leukemia Lymphoma* 2005;*46* (Suppl.1):S97.

23. Rabinowe SN, Soiffer RJ, Gribben JG, et al. Autologous and allogeneic bone marrow transplantation for poor prognosis patients with B-cell chronic lymphocytic leukemia. *Blood* 1993;*82*:1366–1376.

24. van Besien K, Sobocinski KA, Rowlings PA, et al. Allogeneic bone marrow transplantation for low-grade lymphoma. *Blood* 1998;*92*:1832–1836.

25. Hosing C, Saliba RM, McLaughlin P, et al. Long-term results favor allogeneic over autologous hematopoietic stem cell transplantation in patients with refractory or recurrent indolent non-Hodgkin's lymphoma. *Ann Oncol* 2003;*14*:737–744.

26. van Besien K, Loberiza FR, Jr., Bajorunaite R, et al. Comparison of autologous and allogeneic hematopoietic stem cell transplantation for follicular lymphoma. *Blood* 2003;*102*:3521–3529.

27. Mandigers CM, Raemaekers JM, Schattenberg AV, et al. Allogeneic bone marrow transplantation with T-cell-depleted marrow grafts for patients with poor-risk relapsed low-grade non-Hodgkin's lymphoma. *Br J Haematol* 1998;*100*:198–206.

28. Morris E, Thomson K, Craddock C, et al. Outcome following alemtuzumab (CAMPATH-1H)-containing reduced intensity allogeneic transplant regimen for relapsed and refractory non-Hodgkin's lymphoma. *Blood* 2004;*104*:3865–3871.

29. Faulkner RD, Craddock C, Byrne JL, et al. BEAM-alemtuzumab reduced-intensity allogeneic stem cell transplantation for lymphoproliferative diseases: GVHD, toxicity, and survival in 65 patients. *Blood* 2004;*103*:428–434.

30. Bierman PJ, Sweetenham JW, Loberiza FR, Jr., et al. Syngeneic hematopoietic stem-cell transplantation for non-Hodgkin's lymphoma: a comparison with allogeneic and autologous transplantation – The Lymphoma Working Committee of the International *Bone Marrow Transplant* Registry and the European Group for *Blood* and Marrow Transplantation. *J Clin Oncol* 2003;*21*:3744–3753.

31. Dimopoulos MA, Hamilos G, Zervas K, et al. Survival and prognostic factors after initiation of treatment in Waldenstrom's macroglobulinemia. *Ann Oncol* 2003;*14*:1299–1305.

32. Dreger P, Glass B, Kuse R, et al. Myeloablative radiochemotherapy followed by reinfusion of purged autologous stem cells for Waldenström's macroglobulinaemia. *Br J Haematol* 1999;*106*:115–118.

33. Anagnostopoulos A, Aleman A, Giralt S. Autologous and allogeneic stem cell transplantation in Waldenstrom's macroglobulinemia: review of the literature and future directions. *Semin Oncol* 2003;*30*:286–290.

34. Tournilhac O, Leblond V, Tabrizi R, et al. Transplantation in Waldenstrom's macroglobulinemia – the French experience. *Semin Oncol* 2003;*30*:291–296.

35. Moreno C, Villamor N, Colomer D, et al. Allogeneic stem-cell transplantation may overcome the adverse prognosis of unmutated VH gene in patients with chronic lymphocytic leukemia. *J Clin Oncol* 2005;*23*:3433–3438.

Index

Stem Cell Transplantation. Biology, Processing, and Therapy.
Edited by Anthony D. Ho, Ronald Hoffman, and Esmail D. Zanjani
Copyright © 2006 WILEY-VCH Verlag GmbH & Co. KGaA, Weinheim
ISBN: 3-527-31018-5